Contents

Contents

vi

SOIL EROSION AND CONSERVATION

SOIL EROSION AND
CONSERVATION

THIRD EDITION

R. P. C. Morgan

National Soil Resources Institute,
Cranfield University

Blackwell
Publishing

BLACKWELL PUBLISHING
350 Main Street, Malden, MA 02148-5020, USA
108 Cowley Road, Oxford OX4 1JF, UK
550 Swanston Street, Carlton, Victoria 3053, Australia

First published 1986 Longman Group Limited
Second edition 1995
Third edition published 2005 by Blackwell Publishing Ltd

Library of Congress Cataloging-in-Publication Data

Morgan, R. P. C. (Royston Philip Charles), 1942–
    Soil erosion and conservation / R. P. C. Morgan. – 3rd ed.
        p. cm.
    Includes bibliographical references and index.
    ISBN 1-4051-1781-8 (pbk. : alk. paper)
    1. Soil erosion.   2. Soil conservation.   I. Title.
    S623.M68   2005
    631.4′5 – dc22

                                                2004009787

A catalogue record for this title is available from the British Library.

Set in 9 on 11½ pt Minion
by SNP Best-set Typesetter Ltd., Hong Kong
Printed and bound in the United Kingdom
by MPG Books Ltd, Bodmin, Cornwall

The publisher's policy is to use permanent paper from mills that operate a sustainable forestry policy, and which has been manufactured from pulp processed using acid-free and elementary chlorine-free practices. Furthermore, the publisher ensures that the text paper and cover board used have met acceptable environmental accreditation standards.

For further information on
Blackwell Publishing, visit our website:
www.blackwellpublishing.com

Foreword

Soil erosion has been an environmental concern in such countries as China and those bordering the Mediterranean Sea for millennia. In the United States the major impetus to scientific research on soil erosion and conservation came from Hugh Hammond Bennett, who led the soil conservation movement in the 1920s and 1930s. In Western Europe there was growing realization from the 1970s that soil erosion could have a major effect on soils, even on lowland arable areas. More recently the topic has come on to the political agenda with the Commission of the European Communities developing a thematic strategy for soil protection. Integral to this is the recognition that soils perform a range of key functions, including the production of food, the storage of organic matter, water and nutrients, the provision of a habitat for a huge variety of organisms and preserving a record of past human activity. Any degradation in the quality of the soil resource through erosion can have an impact on the ability of soils to perform this range of functions. In the twentieth century and earlier, the main practical concern about soil erosion was with reference to impacts on food production. This is still the case in many parts of the world, but now the more frequent concerns relate to the reduction in soil carbon, the movement of nitrogen and the removal of phosphorus in soluble and particulate forms. There are also concerns about the effects of erosion on landscape quality as well as on cultural records. As an example, many archaeologists are concerned about the effects of soil surface lowering through erosion and the consequential impacts of deeper ploughing on archaeological features. The publication of this third edition of Roy Morgan's book *Soil Erosion and Conservation* is thus very timely and reflects the wider concerns regarding the issue. The book is also permeated with Roy Morgan's own extensive and international experience in soil erosion research.

A key theme of this book is that a soil conservation strategy must evolve from detailed knowledge and understanding of actual erosional processes. Thus Chapters 2 to 6 deal with the processes of soil erosion, the assessment of erosion risk at different scales and the monitoring and modelling of erosion. The treatment of modelling in Chapter 6 is particularly comprehensive through discussion of empirical and physically based models, sensitivity analysis and model validation. The inclusion of many worked examples is of great assistance to the reader. The remaining chapters focus on conservation strategies with emphases on crop and vegetation management, soil management and mechanical methods of erosion control. A section dealing with tillage erosion reflects recent research that indicates the potential magnitude of this process. In Chapter 11 Roy Morgan argues that the successful implementation of soil conservation measures is only possible through a combination of scientific, socio-economic and political considerations, exemplified by the highly successful and integrated approach of the Australian Land Care Programme.

He argues in the concluding chapter that the weakest part of soil conservation programmes has been the lack of effective legal frameworks. This will only be remedied if there is wider appreciation of soil erosion processes and the need for management. This book makes a major contribution to achieving that objective.

Donald A. Davidson

Preface

Soil erosion is a hazard traditionally associated with agriculture in tropical and semi-arid areas and is important for its long-term effects on soil productivity and sustainable agriculture. It is, however, a problem of wider significance occurring additionally on land devoted to forestry, transport and recreation. Erosion also leads to environmental damage through sedimentation, pollution and increased flooding. The costs associated with the movement and deposition of sediment in the landscape frequently outweigh those arising from the long-term loss of soil in eroding fields. Major problems can result from quite moderate and frequent erosion events in both temperate and tropical climates. Erosion control is a necessity in almost every country of the world under virtually every type of land use. Further, eroded soils may lose 75–80 per cent of their carbon content, with consequent emission of carbon to the atmosphere. Erosion control has the potential to sequester carbon as well as restoring degraded soils and improving water quality.

Since the second edition of *Soil Erosion and Conservation* was published in 1995, soil erosion has assumed even greater importance because of the higher priority now being given to the environmental issues associated with sediment. This revised edition recognizes more strongly that erosion is not just an agricultural problem and that loss of soil from construction sites, road banks and pipeline corridors can also result in unwanted and costly downstream damage, as well as hindering attempts at land restoration. Nevertheless, rather than these issues and environmental protection being discussed in detail, the decision has been taken to maintain the philosophy of the original book, namely to provide a text that covers soil conservation from a substantive treatment of erosion. A thorough understanding of the processes of erosion and their controlling factors is a prerequisite for designing erosion control measures on a sound scientific basis wherever they are needed. The aim of producing a text with a global perspective on research and practice is also retained.

The text follows the structure of previous editions but substantial changes have been made to some chapters and minor revisions to the others. Following major advances in research over the past ten years, new material is included on the importance of tillage in moving soil over the landscape, the use of terrain analysis in erosion risk assessment, the use of tracers in erosion measurement, the validation of erosion models and problems of uncertainty in their output, defining soil loss tolerance by performance-related criteria, traditional soil conservation measures, incentives for soil conservation and community approaches to land care. The sections on gully erosion, the mechanics of wind erosion, the dynamic nature of soil erodibility, the effects of vegetation on wind erosion and mass movement, economic evaluation of erosion control, the use of geotextiles and the use of legislative instruments in promoting soil conservation have been substantially rewritten. Updates have been made throughout the text. In line with the comments of reviewers of the previous edition, the chapter on measurement is now placed before the chapter on modelling. In the revised text, this is certainly a more logical order. In addition, selected topics have been removed from the main text of each chapter and placed in a box at the end. The topics are either of generic background interest or relate to specific material that is best treated in a discrete way.

Not surprisingly, in order to keep the text at a reasonable length and reasonable price, some material has had to be omitted. By trying to restrict omissions to material that is no longer relevant, either because scientific understanding has improved or because it is not mainstream to erosion control in practice, it is hoped that nothing vital has been lost. Reference to seminal work of the 1940s to 1970s has been retained, partly to give an important historical context but also to maintain an awareness of what has been achieved in the past so as to discourage others from attempting unnecessary repetition.

The text remains based on courses given on the Silsoe campus of Cranfield University and, again, contains material from research and advisory work carried out by myself, my colleagues and students. As before, the contributions of the last two groups are much appreciated. The text is intended for undergraduate and postgraduate students studying soil erosion and conservation as part of their courses in geography, environmental science, agriculture, agricultural engineering, hydrology, soil science, ecology and civil engineering. In addition, it provides an introduction to the subject for those working on soil erosion and conservation, either as consultants and advisers or at research and experimental stations.

I am grateful to two anonymous reviewers for their constructive comments on an earlier draft of the manuscript. My thanks also go to students and staff at Silsoe for encouraging me to produce a new edition and to Gillian, Richard and Gerald for their support.

R. P. C. Morgan
Silsoe

CHAPTER 1

Soil erosion: the global context

Soil erosion costs the US economy between US$30 billion (Uri & Lewis 1998) and US$44 billion (Pimental et al. 1993) annually. The annual cost in the UK is estimated at £90 million (Environment Agency 2002). In Indonesia, the cost is US$400 million per year in Java alone (Magrath & Arens 1989). These costs result from the effects of erosion both on- and off-site.

On-site effects are particularly important on agricultural land where the redistribution of soil within a field, the loss of soil from a field, the breakdown of soil structure and the decline in organic matter and nutrient result in a reduction of cultivable soil depth and a decline in soil fertility. Erosion also reduces available soil moisture, resulting in more drought-prone conditions. The net effect is a loss of productivity, which restricts what can be grown and results in increased expenditure on fertilizers to maintain yields. If fertilizers were used to compensate for loss of fertility arising from erosion in Zimbabwe, the cost would be equivalent to US$1500 million per year (Stocking 1986), a substantial hidden cost to that country's economy. The loss of soil fertility through erosion ultimately leads to the abandonment of land, with consequences for food production and food security and a substantial decline in land value.

Off-site problems arise from sedimentation downstream or downwind, which reduces the capacity of rivers and drainage ditches, enhances the risk of flooding, blocks irrigation canals and shortens the design life of reservoirs. Many hydroelectricity and irrigation projects have been ruined as a consequence of erosion. Sediment is also a pollutant in its own right and, through the chemicals adsorbed to it, can increase the levels of nitrogen and phosphorus in water bodies and result in eutrophication. Erosion leads to the breakdown of soil aggregates and clods into their primary particles of clay, silt and sand. Through this process, the carbon that is held within the clays and the soil organic content is released into the atmosphere as CO_2. Lal (1995) has estimated that global soil erosion releases $1.14\,PgC$ annually to the atmosphere, of which some $15\,TgC$ is derived from the USA. Erosion is therefore a contributor to climatic change, since increasing the carbon dioxide content of the atmosphere enhances the greenhouse effect.

The on-site costs of erosion are necessarily borne by the farmer, although they may be passed on in part to the community in terms of higher food prices as yields decline or land goes out of production. The farmer bears little of the off-site costs, which fall on local authorities for road clearance and maintenance, insurance companies and all the land holders in the local community affected by sedimentation and flooding. Off-site costs can be considerable. Erosive runoff from arable land in four catchments in the South Downs, England, in October 1987 caused damage equivalent to £660,000 (Robinson & Blackman 1990). Sedimentation ponds to trap sediment and runoff generated from arable land in an area of $5516\,km^2$ in central Belgium cost €38 million to construct and €1.5 million annually to maintain (Verstraeten & Poesen 1999).

Although soil erosion is a physical process with considerable variation globally in its severity and frequency, where and when erosion occurs is also strongly influenced by social, economic, political and institutional factors. Conventional wisdom favours explaining erosion as a response to increasing pressure on land brought about by a growing world population and the abandonment of large areas of formerly productive land as a result of erosion, salinization or alkalinization. In the loess plateau region of China, for example, annual soil loss has increased exponentially since about 220 BC in a simple relationship with total population (Wen 1993). Population pressure forces people to farm more marginal land, often unwisely, especially in the Himalaya, the Andes and many mountainous areas of the humid tropics. In other parts of the world, however, erosion can be seen as a direct response to abandonment of the land associated with rural depopulation. A dramatic example comes from the terraced mountain slopes of the Haraz in Yemen, where land abandonment occurred following droughts in the 1900s, the 1940s and between 1967 and 1973, and then increased markedly in the 1970s as people migrated to Saudi Arabia and the Gulf States. With fewer people on the land, terrace walls were allowed to collapse and erosion is now reducing the depth of the already shallow soil by 1–3 cm yr^{-1} (Vogel 1990). In much of Mediterranean Europe, policies to reduce the number of people employed in agriculture and to increase farm size and the level of mechanization have had a twofold effect. First, traditional terrace structures are left to decay. Second, the increase in farm size is often accompanied by large-scale earth moving and land levelling, which makes the soil more erodible. Almost everywhere that land consolidation programmes have been carried out, rates of soil erosion have increased.

The prevention of soil erosion, which means reducing the rate of soil loss to approximately that which would occur under natural conditions, relies on selecting appropriate strategies for soil conservation, and this, in turn, requires a thorough understanding of the processes of erosion. The factors that influence the rate of erosion may be considered under three headings: energy, resistance and protection. The energy group includes the potential ability of rainfall, runoff and wind to cause erosion. This ability is termed erosivity. Also included are those factors that directly affect the power of the erosive agents, such as the reduction in the length of runoff or wind blow through the construction of terraces and wind breaks respectively. Fundamental to the resistance group is the erodibility of the soil, which depends upon its mechanical and chemical properties. Factors that encourage the infiltration of water into the soil and thereby reduce runoff decrease erodibility, while any activity that pulverizes the soil increases it. Thus cultivation may decrease the erodibility of clay soils but increase that of sandy soils. The protection group focuses on factors relating to the plant cover. By intercepting rainfall and reducing the velocity of runoff and wind, plant cover can protect the soil from erosion. Different plant cover affords different degrees of protection, so that human influence, by determining land use, can control the rate of erosion to a considerable degree.

The rate of soil loss is normally expressed in units of mass or volume per unit area per unit of time. Under natural conditions, annual rates are of the order of 0.0045 t ha^{-1} for areas of moderate relief and 0.45 t ha^{-1} for steep relief. For comparison, rates from agricultural land are in the range of 45–450 t ha^{-1} (Young 1969). These differences have encouraged many researchers and practitioners to distinguish between 'natural' and 'accelerated' erosion, the latter being the result of human impact on the landscape. In practice, such a distinction is often unhelpful because it leads to a view that all unacceptably high rates of erosion must be accelerated, whereas the rates are actually dependent on local conditions. So-called accelerated rates of erosion in lowland England may, in fact, be an order of magnitude lower than the natural rates recorded in the

Himalaya, Karakoram or Andes. Theoretically, whether or not a rate of soil loss is severe may be judged relative to the rate of soil formation. If soil properties such as nutrient status, texture and thickness remain unchanged through time, it can usually be assumed that the rate of erosion balances the rate of soil formation. More practically, severity is better judged in relation to the damage caused and the costs of its amelioration.

1.1

Spatial variations

On a world scale, investigations of the relationship between soil loss and climate show that at annual precipitation totals below 450 mm, erosion increases as precipitation increases (Walling & Kleo 1979). But as precipitation increases so does the vegetation cover, resulting in better protection of the soil surface, so that for annual precipitation between 450 and 650 mm, soil loss decreases as precipitation increases. However, as seen in Fig. 1.1, further increases in precipitation are sufficient to overcome the protective effect and erosion then increases until, again, the vegetation responds by becoming sufficiently dense to provide additional protection, causing erosion to decrease. Above 1700 mm, the volume and intensity of the rain outweigh the protective effect of the vegetation and erosion increases with precipitation.

It should be stressed that the general trends described above are often masked by the high variability in erosion rates for any given quantity of precipitation as a result of differences in soil, slopes and land cover (Table 1.1). However, if the rates are grouped into categories of natural vegetation, cultivated land and bare soil, each group follows a broadly similar pattern, with the highest rates associated with semi-arid, semi-humid and tropical monsoon conditions. One exception to this is the humid tropics. Measurements of soil loss from hillslopes in West Africa (Roose 1971), ranging in steepness from 0.3 to 4°, yield mean annual rates of 0.15, 0.20 and 0.03 t ha^{-1} under natural conditions of open savanna grassland, dense savanna grassland and tropical rain forest respectively. Clearance of the land for agriculture increases the rates to 8, 26 and 90 t ha^{-1}, while

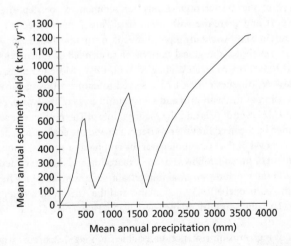

Fig. 1.1 Relationship between sediment yield and mean annual precipitation (after Walling & Kleo 1979).

Table 1.1 Annual rates of erosion in selected countries ($t ha^{-1}$)

	Natural	Cultivated	Bare soil
China	0.1–2	150–200	280–360
USA	0.03–3	5–170	4–9
Australia	0.0–64	0.1–150	44–87
Ivory Coast	0.03–0.2	0.1–90	10–750
Nigeria	0.5–1	0.1–35	3–150
India	0.5–5	0.3–40	10–185
Ethiopia	1–5	8–42	5–70
Belgium	0.1–0.5	3–30	7–82
UK	0.1–0.5	0.1–20	10–200

Sources: Browning et al. (1948), Roose (1971), Fournier (1972), Lal (1976), Bollinne (1978), Jiang et al. (1981), Singh et al. (1981), Morgan (1985a), Boardman (1990), Edwards (1993), Hurni (1993).

leaving the land as bare soil produces rates of 20, 30 and 170 $t ha^{-1}$ respectively. Thus, removal of the rain forest results in much greater rises in erosion rates than does removal of the savanna grassland. These measurements emphasize the high degree of protection afforded by the rain forest but also reflect the erosive capacity of the high rainfalls in the humid tropics when that protection is destroyed. The rates of removal of tropical rain forests over the past twenty years are therefore of major concern with respect to present and future erosion problems.

Many attempts have been made to produce maps of erosion at a global scale. Since some 70 per cent of the sediment delivered by the river systems to the oceans each year is carried in suspension, these maps are based largely on measurements of suspended sediment yields, with extrapolations to provide estimates in areas of sparse data. The results are subject to errors associated with inadequate extrapolation procedures, the different methods used to sample the sediment and process the data and differences between the river basins in their degree of human impact. In addition, suspended sediment yield is strongly influenced by the size of the catchment because of the greater opportunity for sediment to be deposited with increasing distance of transport and therefore with basin size. Thus a map based on data for drainage basins of 1000 km² in size would be very different from one based on data for basins of 10,000 km². Figure 1.2 shows the global pattern of suspended sediment yield for catchments between 1000 and 10,000 km² in area (Walling & Webb 1983). More recent assessments (Lvovich et al. 1991; Dedkov & Mozzherin 1996) have served to confirm this pattern, emphasizing the vulnerability to erosion of the semi-arid and semi-humid areas of the world, especially in China, India, the western USA, central Asia and the Mediterranean. The problem of soil erosion in these areas is compounded by the need for water conservation and the ecological sensitivity of the environment, so that removal of the vegetation cover for cropping or grazing results in rapid declines in the organic content of the soil, followed by soil exhaustion and the risk of desertification. Other areas of high erosion rates include mountainous terrain, such as much of the Andes, the Himalaya and the Karakoram, parts of the Rocky Mountains and the African Rift Valley; and areas of volcanic soils, such as Java, the South Island of New Zealand, Papua New Guinea and parts of Central America.

A further area of high erosion risk, not discernible from Fig. 1.2, occurs where the landforms and associated soils are relics of a previous climate. Over much of southern Africa, stratigraphi-

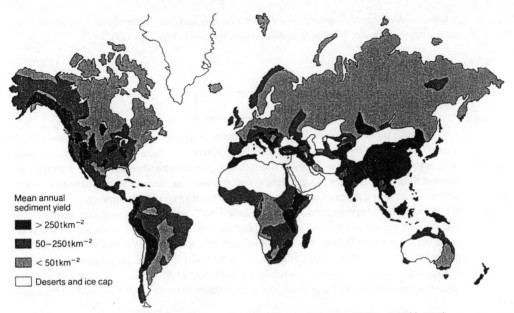

Fig. 1.2 A tentative map of global variations in suspended sediment yield (after Walling & Webb 1983).

cal evidence shows sequences of periods of comparative stability in the landscape, indicated by the development of humic layers and stone lines, and periods of instability, represented by colluvial sediments, often up to 5 m thick. Throughout much of Swaziland and Zimbabwe, present-day gully erosion is particularly severe on these colluvial deposits, which are often fine sandy or silty in nature and, therefore, inherently highly erodible (Shakesby & Whitlow 1991). Gullying is also extensive worldwide in areas of deeply weathered regoliths or saprolite, overlying granites and granodiorites. The deep weathered mantle was probably formed during a more humid tropical climate when the surface was protected from erosion by a dense vegetation cover. Clearance of the vegetation has led to an increase in runoff and erosion. Once the upper soil layers have been removed, the underlying, highly weathered, often very fine substrate is exposed. This offers limited resistance and rapidly becomes deeply dissected (Scholten 1997). Such conditions occur not only in southern Africa but also on the margins of the savanna lands in West Africa, Brazil and southern China.

Within relatively small areas, rainfall characteristics are reasonably uniform and erosion varies spatially in relation to soils, slopes and land use. Boardman (1990) found that between 1982 and 1987 in the area between Brighton and Lewes in the South Downs, England, most erosion occurred in fields on the sides of major dry valleys where the relief was greater than 100 m and the land was under winter cereals. Not all the sediment eroded from hillslopes finds its way into the river system. Some of it is deposited on footslopes and in flood plains, where it remains in temporary storage, sometimes until the next storm, or, at other times, as in the case of much colluvial and alluvial material, for millions of years. Larger drainage basins tend to have a larger proportion of these sediment sinks, which explains why erosion rates expressed per unit area are generally higher in small basins and decrease as the catchment becomes bigger. The proportion of the sediment eroded from the land surface that discharges into the river is known as the sediment delivery ratio. Research has shown that this can vary from about 3 to 90 per cent, decreasing with greater basin area and lower average slope (Walling 1983).

1.2

Temporal variations

Typically, data on erosion rates for individual events or years for given locations show a highly skewed distribution (Fig. 1.3), with a large number of very low magnitude events producing moderate amounts of soil loss and a small number of higher magnitude events. Over a long period of time, most erosion takes place during events of moderate frequency and magnitude simply because extreme or catastrophic events are too infrequent to contribute appreciably to the quantity of soil eroded. Experimental studies by Roose (1967) in Senegal showed that, between 1959 and 1963, 68 per cent of the soil loss took place in rain storms of 15–60 mm, events that occur about ten times a year. Studies in mid-Bedfordshire, England (Morgan et al. 1986) indicated that, in the period 1973–9, 80 per cent of the erosion occurred in 13 storms, the greatest soil loss, comprising 21 per cent of the erosion, resulting from a storm of 57.2 mm. These storms have a frequency of between two and four times a year. In contrast, Hudson (1981) emphasized the role of

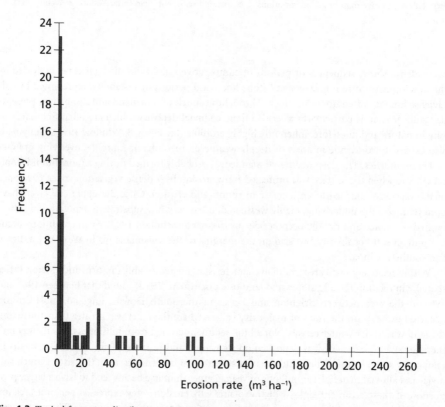

Fig. 1.3 Typical frequency distribution of annual erosion rates based on measurements at 270 field sites on arable land in England and Wales (after McHugh personal communication).

the more dramatic event. Quoting from research in Zimbabwe, he stated that 50 per cent of the annual soil loss occurs in only two storms and that, in one year, 75 per cent of the erosion took place in ten minutes. Moderate events also account for most of the erosion carried out by wind. Studies on coastal dunes at Cape Moreton, New South Wales, showed that most sand transport occurred in strong winds of about $14\,m\,s^{-1}$, with relatively little in winds of gale force and above because their greater competence was compensated for by their rarity (Chapman 1990). The frequency of the dominant erosion event may vary for different erosion processes. For example, for shallow debris slides and mudflows on cultivated fields and grassland in the Mgeta area of Tanzania the dominant event has a return period of once in five years (Temple & Rapp 1972).

The more dramatic events may become important where erosion is not a function of climate alone but depends on the frequency at which potentially erosive events coincide with ground conditions that favour erosion. Analysis of 28 years of data for nine small catchments under a four-year rotation of maize–wheat–grass–grass at Coshocton, Ohio (Edwards & Owens 1991) showed that the three largest storms, all with return periods of 100 years or more, accounted for 52 per cent of the erosion and that 92 per cent of the soil loss occurred in the years when the land was under maize. Extreme events may also produce landscape features that are both dramatic and long lasting. A slow-moving equatorial storm deposited 631 mm of rain on 28 December 1926 and 1194 mm between 26 and 29 December in the Kuantan area of Malaysia, resulting in extensive gully erosion and numerous landslides. The scars produced in the landscape were still visible 35 years later (Nossin 1964).

In addition to the variations in erosion associated with the frequency and magnitude of single storms, rates of erosion often follow a seasonal pattern. This is best illustrated with reference to a rainfall regime with a wet and dry season (Fig. 1.4). The vegetation growth follows a similar pattern but peaks later than the rainfall. The most vulnerable time for erosion is the early part of the wet season when the rainfall is high but the vegetation has not grown sufficiently to protect the soil. Thus the erosion peak precedes the rainfall peak.

Somewhat more complex seasonal patterns occur with less simple rainfall regimes or where the land is used for arable farming. Generally, the period between ploughing and the growth of the crop beyond the seedling stage contains an erosion risk if it coincides with heavy rainfall or strong winds. Thus, in western Europe, the period in spring before the crop cover reaches 20 per cent is often a peak time for erosion when rainfall degrades the bare soil surface, causing the development of a surface seal (Cerdan et al. 2002a).

Longer-term spatial variations in erosion occur in relation to changes in land cover. A typical sequence of events is described by Wolman (1967) for Maryland, where soil erosion rates increased with the conversion of woodland to cropland after AD 1700 (Fig. 1.5). They declined as the urban fringe extended across the area in the 1950s and the land reverted to scrub when the farmers sold out to speculators, before accelerating rapidly, reaching annual rates of $7000\,t\,ha^{-1}$, when the area was laid bare during housing construction. With the completion of urban development, runoff from concrete surfaces is concentrated into gutters and sewers, and annual soil loss falls below $4\,t\,ha^{-1}$.

Based on stratigraphical and archaeological evidence of valley floor deposits and archival material, Bork (1989) reconstructed the history of soil erosion in Niedersachsen, Germany. From the early Holocene, when soils developed under the natural woodlands, up to the early Middle Ages, erosion rates were extremely low. With the clearance of forest for agriculture between AD 940 and 1340, erosion increased and reached annual rates of about $10\,t\,ha^{-1}$. Between 1340 and 1350, annual erosion rates rose dramatically to $2250\,t\,ha^{-1}$ as a result of gully erosion induced by extreme climatic events such as that on 21 July 1342, when the largest flood ever

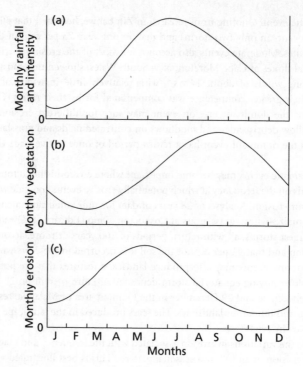

Fig. 1.4 Seasonal cycles of rainfall, vegetation cover and erosion in a semi-humid climate (after Kirkby 1980a).

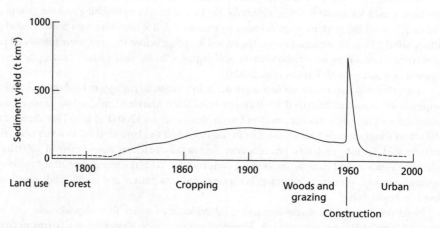

Fig. 1.5 Relationship between sediment yield and changing land use in the Piedmont region of Maryland, USA (after Wolman 1967).

recorded in central Europe occurred (Bork et al. 1998). Erosion declined afterwards, partly as a result of a decrease in the area under arable as land was abandoned due to impoverishment by erosion. The rate of erosion did not return to early mediaeval levels but remained at an average annual rate of around $25\,t\,ha^{-1}$. The higher rate reflected sheet erosion on the land remaining in arable as production went over to the three-field system, with one-third of the land in fallow at any one time. The period between 1750 and 1800 saw a second episode of gullying, with an average annual soil loss around $160\,t\,ha^{-1}$, in response to an increase in the frequency of heavy rainfall events. The soil loss did not reach mid-fourteenth-century levels, however, because of the establishment of terraces, the use of contour ploughing and grass strips and a higher proportion of the land under grass and trees. Since 1800 annual soil loss has averaged $20\,t\,ha^{-1}$ but it has increased in recent years following land consolidation, which has resulted in larger fields, removal of terraces and grass strips and land levelling. A similar history of fluctuating rates of soil erosion in relation to changes in land use has been reconstructed for the Wolfsgraben in northern Bavaria, Germany (Dotterweich et al. 2003). In periods when the land was under arable cultivation, annual erosion rates averaged $2.8\,t\,ha^{-1}$ and sedimentation occurred on the valley floors. In extreme rainfall events in the early fourteenth century and again in the late eighteenth century, these sediments were cut through by gullies, up to 5 m deep. Whenever land was taken out of cultivation and reverted to forest, erosion rates were very low and the gullies were infilled.

These historical studies indicate the complex nature of soil erosion. Although erosion is a natural process and, therefore, naturally variable with climate, soils and topography, human impact can make the landscape either more or less resilient to climatic events. Rates of erosion quickly accelerate to high levels whenever land is misused.

Box 1

Erosion, population and food supply

Only 22 per cent of the earth's land area of 14,900 million hectares is potentially productive (El-Swaify 1994). Since this has to provide 97 per cent of the food supply (3 per cent comes from oceans, rivers and lakes), it is under increasing pressure as world population numbers continue to grow. The fear is that meeting the greater demand for food through more intensive use of existing agricultural land and expansion of agriculture on to more marginal land will substantially increase erosion. Failure to control erosion will therefore seriously endanger global food security. Concern about the future is based upon:

■ very high rates of erosion measured from agricultural land, with annual rates often 20 to over $100\,t\,ha^{-1}$;
■ declines in the productivity of the soil by as much as 15–30 per cent annually;

■ the difficulty of restoring severely degraded land because of the loss of fertility;
■ an estimated loss of some 6 million hectares annually as a result of degradation by erosion and other causes (Pimental et al. 1993).

Unfortunately, it is impossible to know whether the above data represent a realistic picture because they ignore important issues. First, the data on erosion rates are highly selective and often based on short periods of measurement; it is statistically invalid to extrapolate them over large areas. Pimental et al. (1995) estimated that Europe was losing soil at an annual rate of $17\,t\,ha^{-1}$ but, according to Lomborg (2001), this figure is largely based on extrapolating measurements from a 0.1 hectare plot of land in Belgium. Second, most studies of productivity in relation to erosion come from low-input

Continued

---

agriculture and therefore ignore the effects of improved farming practices, including greater use of irrigation, pesticides and fertilizers. In much of Western Europe and the USA, annual increases of 1–2 per cent in productivity can more than offset the effects of erosion, which, locally, are generally in the 0.1–0.5 per cent range (Crosson 1995). In these areas, agricultural production has allowed increasing numbers of people to be fed despite the proportion of the population directly employed on the land falling to below 10 per cent.

In order to gain a better understanding of the global situation, more information is required on the status of the earth's land resource and how fast soil is being lost by erosion. An accurate assessment of land degradation is not straightforward. Statements on the area affected by erosion can be misleading unless supported by field observations. In order to provide a systematic method, UNEP co-sponsored a Global Assessment of Soil Degradation (GLASOD) using over 200 experts to assess the state of degradation in their own countries against clearly defined criteria. The results (Table B1.1; Oldeman 1994) indicated that soil erosion accounted for 82 per cent of human-induced soil degradation, affecting some 1643 million hectares, but only 0.5 per cent of this had reached an irreversible stage. It should be stressed that there is considerable uncertainty about these figures since there appears to have been no control over how the experts interpreted the various grades of land degradation. The grades were more often interpreted in relation to conditions within each country rather than to any consistent world standard. Nevertheless, the GLASOD survey represents the only global scale assessment available at present.

This analysis of the global situation would appear to indicate that soil erosion should not be a threat to the ability of the world to feed itself. The greater proportion of the world's arable land remains productive. Changes in farming practice can more than offset the effects of erosion and feed more people from a unit area of land. Studies in Nigeria and Kenya (Bridges & Oldeman 2001) indicate that, even in developing countries, high population densities can lead to higher productivity and better soil protection. Against this, there are many areas of the world where soil erosion presents major problems that need to be addressed. In addition, this global analysis ignores the environmental impacts of erosion with respect to water quality, flooding and carbon emission. There is, therefore, a clear need for soil protection, but the case for it needs to be made with reference to local on-site problems and off-site effects.

Table B1.1 Extent of human-induced soil degradation by erosion (million hectares)

	Light	Moderate	Strong	Extreme	Total
Water erosion	343	527	217	7	1094
Wind erosion	269	254	24	2	549
Total	612	781	241	9	1643

Light: somewhat reduced productivity which can be restored by local farming systems.
Moderate: greatly reduced productivity which can be restored by use of structural measures such as terracing and contour banks.
Strong: land cannot be reclaimed at farm level; restoration requires major engineering works.
Extreme: land is unreclaimable.
Source: after Oldeman (1994).

CHAPTER 2

Processes and mechanics of erosion

Soil erosion is a two-phase process consisting of the detachment of individual soil particles from the soil mass and their transport by erosive agents such as running water and wind. When sufficient energy is no longer available to transport the particles, a third phase, deposition, occurs.

Rainsplash is the most important detaching agent. As a result of raindrops striking a bare soil surface, soil particles may be thrown through the air over distances of several centimetres. Continuous exposure to intense rainstorms considerably weakens the soil. The soil is also broken up by weathering processes, both mechanical, by alternate wetting and drying, freezing and thawing and frost action, and biochemical. Soil is disturbed by tillage operations and by the trampling of people and livestock. Running water and wind are further contributors to the detachment of soil particles. All these processes loosen the soil so that it is easily removed by the agents of transport.

The transporting agents comprise those that act areally and contribute to the removal of a relatively uniform thickness of soil, and those that concentrate their action in channels. The first group consists of rainsplash, surface runoff in the form of shallow flows of infinite width, sometimes termed sheet flow but more correctly called overland flow, and wind. The second group covers water in small channels, known as rills, which can be obliterated by weathering and ploughing, or in the larger more permanent features of gullies and rivers. A distinction is commonly made for water erosion between rill erosion and erosion on the land between the rills by the combined action of raindrop impact and overland flow. This is termed interrill erosion. To these agents that act externally, picking up material from and carrying it over the ground surface, should be added transport by mass movements such as soil flows, slides and creep, in which water affects the soil internally, altering its strength.

The severity of erosion depends upon the quantity of material supplied by detachment over time and the capacity of the eroding agents to transport it. Where the agents have the capacity to transport more material than is supplied by detachment, the erosion is described as detachment-limited. Where more material is supplied than can be transported, the erosion is transport limited.

The energy available for erosion takes two forms: potential and kinetic. Potential energy (PE) results from the difference in height of one body with respect to another. It is the product of mass (m), height difference (h) and acceleration due to gravity (g), so that

$$PE = mhg \qquad (2.1)$$

Table 2.1 Efficiency of forms of water erosion

Form	Mass*	Typical velocity (m s^{-1})	Kinetic energy[†]	Energy for erosion[‡]	Observed sediment transport§ (g cm^{-1})
Raindrops	R	6.0	$18R$	$0.036R$	20
Overland flow	$0.5R$	0.01	$2.5 \times 10^{-5}R$	$7.5 \times 10^{-7}R$	400
Rill flow	$0.5R$	4¶	$4R$	$0.12R$	19,000

* Assumes rainfall mass of R of which 50 per cent contributes to runoff.
† Based on $\frac{1}{2}mv^2$.
‡ Assumes that 0.2 per cent of the kinetic energy of raindrops and 3 per cent of the kinetic energy of runoff is utilized in erosion.
§ Totals observed in mid-Bedfordshire, England, on an 11° slope, on sandy soil, over 900 days. Most of the energy of raindrops contributes to soil particle detachment rather than transport.
¶ Estimated using the Manning equation of flow velocity for a rill, 0.3 m wide and 0.2 m deep, on a slope of 11°, at bankfull, assuming a roughness coefficient of 0.02.

which, in units of kg, m and m s^{-2} respectively, yields a value in Joules. The potential energy for erosion is converted into kinetic energy (KE), the energy of motion. This is related to the mass and velocity (v) of the eroding agent in the expression

$$KE = \frac{1}{2}mv^2 \qquad (2.2)$$

which, in units of kg and m s^{-1}, also gives a value in Joules. Most of this energy is dissipated in friction with the surface over which the agent moves so that only 3–4 per cent of the energy of running water and 0.2 per cent of that of falling raindrops is expended in erosion (Pearce 1976). An indication of the relative efficiencies of the processes of water erosion can be obtained by applying these figures to calculations of kinetic energy, using eqn 2.2, based on typical velocities (Table 2.1). The concentration of running water in rills affords the most powerful erosive agent but raindrops are potentially more erosive than overland flow. Most of the raindrop energy is used in detachment, however, so that the amount available for transport is less than that from overland flow. This is illustrated by measurements of soil loss in a field in mid-Bedfordshire, England. Over a 900-day period on an 11° slope on a sandy soil, transport across a centimetre width of slope amounted to 19,000 g of sediment by rills, 400 g by overland flow and only 20 g by rainsplash (Morgan et al. 1986).

2.1

Hydrological basis of erosion

The processes of water erosion are closely related to the pathways taken by water in its movement through the vegetation cover and over the ground surface. During a rainstorm, part of the water falls directly on the land, either because there is no vegetation or because it passes through gaps in the plant canopy. This component of the rainfall is known as direct throughfall. Part of the

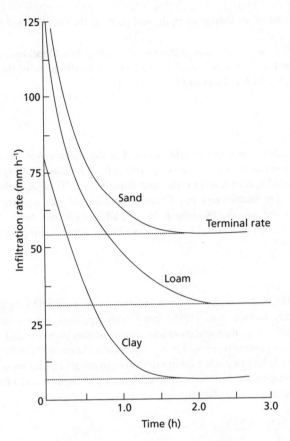

Fig. 2.1 Typical infiltration rates for various soils (after Withers & Vipond 1974).

rain is intercepted by the canopy, from where it either returns to the atmosphere by evaporation or finds its way to the ground by dripping from the leaves, a component termed leaf drainage, or by running down the plant stems as stemflow. The action of direct throughfall and leaf drainage produces rainsplash erosion. The rain that reaches the ground may be stored in small depressions or hollows on the surface or it may infiltrate the soil, contributing to soil moisture storage, to lateral movement downslope within the soil as subsurface or interflow or, by percolating deeper, to groundwater. When the soil is unable to take in more water, the excess contributes to runoff on the surface, resulting in erosion by overland flow or by rills and gullies.

The rate at which water passes into the soil is known as the infiltration rate and this exerts a major control over the generation of surface runoff. Water is drawn into the soil by gravity and by capillary forces, whereby it is attracted to and held as a thin molecular film around the soil particles. During a rainstorm, the spaces between the soil particles become filled with water and the capillary forces decrease so that the infiltration rate starts high at the beginning of a storm and declines to a level that represents the maximum sustained rate at which water can pass through the soil to lower levels (Fig. 2.1). This level, the infiltration capacity or terminal infiltration rate, corresponds theoretically to the saturated hydraulic conductivity of the soil. In practice, however, the infiltration capacity is often lower than the saturated hydraulic conductivity because of air entrapped in the soil pores as the wetting front passes downwards through the soil.

Various attempts have been made to describe the change in infiltration rate over time mathematically. One of the most widely used equations is the modification of the Green and Ampt (1911) equation proposed by Mein and Larson (1973):

$$i = A + \frac{B}{t} \tag{2.3}$$

where i is the instantaneous rate of infiltration, A is the transmission constant or saturated hydraulic conductivity of the soil, B is the sorptivity, defined by Talsma (1969) as the slope of the line when i is plotted against t, and t is the time elapsed since the onset of the rain. This equation has been found to describe well the infiltration behaviour of soils in southern Spain (Scoging & Thornes 1979) and Arizona (Scoging et al. 1992) but Bork and Rohdenburg (1981), also working in southern Spain, obtained better results with the equation proposed by Philip (1957):

$$i = A + \frac{B}{\sqrt{t}} \tag{2.4}$$

while Gifford (1976) found neither equation satisfactory for semi-arid rangelands in northern Australia and Utah. Kutílek et al. (1988) tested both equations against field measurements obtained with double-ring infiltrometers and found that neither fitted the data well, giving errors of between 10 and 59 per cent when used to estimate saturated hydraulic conductivity. One reason for the error is the failure to predict infiltration correctly under conditions of surface ponding when the soil develops a viscous resistance to air flow. Morel-Seytoux and Khanji (1974) developed the following equation to allow for this:

$$i = \frac{k_s}{\beta}\left(1 + \frac{(\theta_t - \theta_i)(H_0 + \Delta\psi)}{I}\right) \tag{2.5}$$

where k_s is the saturated hydraulic conductivity; β is a viscous correction factor, which varies in value between 1.1 and 1.7, depending on soil type and ponding depth but averages 1.4; θ_i is the initial soil moisture content by volume; θ_t is the actual volumetric moisture content of soil in the zone between the ground surface and the wetting front; H_0 is the depth of ponded water; $\Delta\psi$ is the change in ψ between the soil surface and the wetting front; ψ is the difference in pressure between the pore-water and the atmosphere; and I is the total amount of water already infiltrated. As a result of including the viscous correction factor, eqn 2.5 predicts lower infiltration rates than either eqn 2.3 or eqn 2.4.

Infiltration rates depend upon the characteristics of the soil. Generally, coarse-textured soils such as sands and sandy loams have higher infiltration rates than clay soils because of the larger spaces between the pores. Infiltration capacities may range from more than 200 mm h^{-1} for sands to less than 5 mm h^{-1} for tight clays (Fig. 2.1). In addition to the role played by the inter-particle spacing or micropores, the larger cracks or macropores exert an important influence over infiltration. They can transmit considerable quantities of water so that clays with well defined structures can have infiltration rates that are much higher than would be expected from their texture alone. Infiltration behaviour on many soils is also rather complex because the soil profiles are characterized by two or more layers of differing hydraulic conductivities; most agricultural soils, for example, consist of a disturbed plough layer and an undisturbed subsoil. Many soils on construction sites comprise a heavily compacted subsoil covered by a thinner and less compacted topsoil. Local variability in infiltration rates can be quite high because of differences in the structure, compaction, initial moisture content and profile form of the soil and in vegetation density. Field determinations of average infiltration capacity using infiltrometers may have coefficients of variation of 70–75 per cent. Eyles (1967) measured infiltration capacity on soils of the Melaka Series near Temerloh, Malaysia, and obtained values ranging from 15 to 420 mm h^{-1}, with a mean of 147 mm h^{-1}.

According to Horton (1945), if rainfall intensity is less than the infiltration capacity of the soil, no surface runoff occurs and the infiltration rate equals the rainfall intensity. If the rainfall intensity exceeds the infiltration capacity, the infiltration rate equals the infiltration capacity and the excess rain forms surface runoff. As a mechanism for generating runoff, however, this comparison of rainfall intensity and infiltration capacity does not always hold. Studies in Bedfordshire, England (Morgan et al. 1986) on a sandy soil show that measured infiltration capacity is greater than 400 mm h^{-1} and that rainfall intensities rarely exceed 40 mm h^{-1}. Thus no surface runoff would be expected, whereas, in fact, the mean annual runoff is about 55 mm from a mean annual rainfall of 550 mm. The reason runoff occurs is that these soils are prone to the development of a surface crust. Two types of crust can be distinguished. Where a crust forms *in situ* on the soil, it is termed a structural crust; where it results from the deposition of fine particles in puddles, it is called a depositional crust (Boiffin 1985). As shown by studies on loamy soils in north-east France, crusting can reduce the infiltration capacity from 45–60 to about 6 mm h^{-1} with a structural crust and 1 mm h^{-1} with a depositional crust (Boiffin & Monnier 1985; Martin et al. 1997). Reductions in infiltration of 50 (Hoogmoed & Stroosnijder 1984) to 100 per cent (Torri et al. 1999) can occur in a single storm. The importance of crusting and sealing was also emphasized by Poesen (1984), who found that infiltration rates were higher on steeper slopes where the higher erosion rate prevented the seal from forming.

The presence of stones or rock fragments on the surface of a soil also influences infiltration rates but in a rather complex way depending on whether the stones are resting on top of the surface or are embedded within the soil. Generally, rock fragments protect the soil against physical destruction and the formation of a crust, so that infiltration rates are higher than on a comparable stone-free bare soil. However, on soils that are subject to crusting, a high percentage stone cover can produce a worse situation; a 75 per cent cover of rock fragments embedded in a crusted surface on a silt-loam soil reduced infiltration rates to 50 per cent of those on a stone-free soil (Poesen & Ingelmo-Sanchez 1992).

The important control for runoff production on many soils is not infiltration capacity but a limiting moisture content. When the actual moisture content is below this value, pore water pressure in the soil is less than atmospheric pressure and water is held in capillary form under tensile stress or suction. When the limiting moisture content is reached and all the pores are full of water, pore water pressure equates to atmospheric pressure, suction reduces to zero and surface ponding occurs. This explains why sands that have low levels of capillary storage can produce runoff very quickly even though their infiltration capacity is not exceeded by the rainfall intensity. Since hydraulic conductivity is a flux partly controlled by rainfall intensity, increases in intensity can cause conductivity to rise so that, although runoff may have formed rapidly at a relatively low intensity, higher rainfall intensities do not always produce greater runoff. This mechanism explains why infiltration rates sometimes increase with rainfall intensity (Nassif & Wilson 1975). Bowyer-Bower (1993) found that, for a given soil, infiltration capacity was higher with higher rainfall intensities because of their ability to disrupt surface seals and crusts that would otherwise keep the infiltration rate low.

Once water starts to pond on the surface, it is held in depressions or hollows and runoff does not begin until the storage capacity of these is satisfied. On agricultural land, depression

Table 2.2 Surface roughness (*RFR*) for different tillage implements compared to other expressions of random roughness

Implement	Roughness (*RFR*) (cm m^{-1})	Random roughness (*RR*) (mm)
Moldboard plough	30–33	33–48
Chisel plough	24–28	17–26
Cultivator	15–23	6–15
Tandem disc	25–28	18–26
Offset disc	32–35	38–51
Paraplough	32–35	10
Spike tooth harrow	17–23	8–15
Spring tooth harrow	25	18
Rotary hoe	21–22	12–13
Rototiller	23	15
Drill	20–21	10–12
Row planter	13–22	5–13

Note: The term, *RFR*, is essentially an index of the tortuosity of the soil surface. An alternative and widely used descriptor of the roughness of the soil surface is random roughness (*RR*, mm), defined as the standard deviation of a series of surface height measurements (Currence & Lovely 1970). There is a good correlation between *RR* and *RFR* which can be expressed by (Auerswald personal communication):

$\ln RR = 0.29 + 0.099 RFR$, $r = 0.995$, $n = 27$

$RFR = -1.77 + 9.25 \ln RR$, $r = 0.912$, $n = 27$

Surface roughness, expressed by *RR*, declines over time as a function of cumulative rainfall (*R_c*):

$RR(t) = RR(0)e^{-\alpha R_c}$

where *RR(t)* is the random roughness at time (*t*), *RR*(0) is the original random roughness after tillage and $\alpha = 2.8 \times 0.35_i$, where S_i is the silt content of the soil (0–1) (if $\alpha \geq 0$, α is set to -1) (Alberts et al. 1989).

Source: after Auerswald, personal communication.

storage varies seasonally depending on the type of cultivation that has been carried out and the time since cultivation for the roughness to be reduced by weathering and raindrop impact. Table 2.2 gives typical values of depression storage (*DS*; mm) for surfaces produced by different tillage implements, based on their roughness index (*RFR*; cm m^{-1}) (Auerswald, personal communication):

$$DS = 0.14e^{0.04 RFR} \tag{2.6}$$

$$RFR = \frac{L_A - L_0}{L_A} \times 100 \tag{2.7}$$

where L_0 is the straight-line distance between two points along a transect of the soil surface and L_A is the actual distance measured over all the microtopographic irregularities.

Surface roughness and therefore depression storage decline over time through weathering and raindrop impact. Auerswald (personal communication) developed the following relationship to express the decline in roughness as a function of the cumulative kinetic energy of rainfall:

$$RFR(t) = RFR_0 e^{-0.7\sqrt{KE(t)}} \tag{2.8}$$

where $RFR(t)$ is the roughness at a certain time, RFR_0 is the initial roughness and $KE(t)$ is the accumulated kinetic energy of the rain at time (t). Depression storage also varies with the soil with clay soils having 1.6–2.3 times the storage volume of sandy soils. The roughness values given in Table 2.2 relate to soils with about 20 per cent clay. These base values (RFR_{base}) can be adjusted to give roughness for different clay contents (RFR_{CC}) using the relationship:

$$RFR_{cc} = RFR_{base}(0.4 + 0.025CC) \tag{2.9}$$

where CC is the percentage clay content of the soil. This relationship is valid for clay contents up to 25 per cent; for higher clay contents it is recommended to use the 25 per cent value.

2.2

Rainsplash erosion

The action of raindrops on soil particles is most easily understood by considering the momentum of a single raindrop falling on a sloping surface. The downslope component of this momentum is transferred in full to the soil surface but only a small proportion of the component normal to the surface is transferred, the remainder being reflected. The transfer of momentum to the soil particles has two effects. First, it provides a consolidating force, compacting the soil; second, it produces a disruptive force as the water rapidly disperses from and returns to the point of impact in laterally flowing jets. Whereas the impact velocity of falling raindrops striking the soil surface varies from about $4\,\mathrm{m\,s^{-1}}$ for a 1 mm diameter drop to $9\,\mathrm{m\,s^{-1}}$ for a 5 mm diameter drop, the local velocities of these jets are about twice these (Huang et al. 1982). These fast-moving water jets impart a velocity to some of the soil particles and launch them into the air, entrained within water droplets that are themselves formed by the break-up of the raindrop on contact with the ground (Mutchler & Young 1975). Thus, raindrops are agents of both consolidation and dispersion.

The consolidation effect is best seen in the formation of a surface crust, usually only a few millimetres thick, which results from clogging of the pores by soil compaction and by the infilling of surface pore spaces by fine particles detached from soil aggregates by the raindrop impact. Studies of crust development under simulated rainfall show that crusts have a dense surface skin or seal, about 0.1 mm thick, with well oriented clay particles. Beneath this is a layer, 1–3 mm thick, where the larger pore spaces are filled by finer washed-in material (Tackett & Pearson 1965). That raindrop impact is the critical process was shown by Farres (1978), who found that, after a rainstorm, most aggregates on the soil surface were destroyed, while those in the lower layer of the crust remained intact, even though completely saturated. A tap of these aggregates, however, caused their instant breakdown. This evidence indicates that although saturation reduces the internal strength of soil aggregates, they do not disintegrate until struck by raindrops.

The actual response of a soil to a given rainfall depends upon its moisture content and, therefore, its structural state and the intensity of the rain. Le Bissonnais (1990) describes three possible responses:

■ If the soil is dry and the rainfall intensity is high, the soil aggregates break down quickly by slaking. This is the breakdown by compression of air ahead of the wetting front. Infiltration capacity reduces rapidly and on very smooth surfaces runoff can be generated after only a few millimetres of rain. With rougher surfaces, depression storage is greater and runoff takes longer to form.

■ If the aggregates are initially partially wetted or the rainfall intensity is low, microcracking occurs and the aggregates break down into smaller aggregates. Surface roughness thus decreases but infiltration remains high because of the large pore spaces between the microaggregates.

■ If the aggregates are initially saturated, infiltration capacity depends on the saturated hydraulic conductivity of the soil and large quantities of rain are required to seal the surface. Nevertheless, soils with less than 15 per cent clay content are vulnerable to sealing if the intensity of the rain is high.

Over time, the percentage area of the soil surface affected by crust development increases exponentially with cumulative rainfall energy (Govers & Poesen 1985), which, in turn, brings about an exponential decrease in infiltration capacity (Boiffin & Monnier 1985). Crustability decreases with increasing contents of clay and organic matter since these provide greater strength to the soil. Thus loams and sandy loams are the most vulnerable to crust formation.

Studies of the kinetic energy required to detach one kilogram of sediment by raindrop impact show that minimal energy is needed for soils with a geometric mean particle size of 0.125 mm and that soils with geometric mean particle size between 0.063 and 0.250 mm are the most vulnerable to detachment (Fig. 2.2; Poesen 1985). Coarser soils are resistant to detachment because of the weight of the larger particles. Finer soils are resistant because the raindrop energy has to overcome the adhesive or chemical bonding forces that link the minerals comprising the clay particles. The wide range in energy required to detach clay particles is a function of different levels of resistance in relation to the type of clay minerals and the relative amounts of calcium, magnesium and sodium ions in the water passing through the pores (Arulanandan & Heinzen 1977). Overall, silt loams, loams, fine sands and sandy loams are the most detachable. Selective removal of particles by rainsplash can cause variations in soil texture downslope. Splash erosion on stony loamy soils in the Luxembourg Ardennes has resulted in soils on the valley sides becoming deficient in clay and silt particles and high in gravel and stone content, whereas the colluvial soils at the base of the slopes are enriched by the splashed-out material (Kwaad 1977). Selective erosion can affect soil aggregates as well as primary particles. Rainfall simulation experiments on clay soils in Italy show that splashed-out material is enriched in soil aggregates of 0.063–0.50 mm in size (Torri & Sfalanga 1986).

The detachability of soil depends not only on its texture but also on top soil shear strength (Cruse & Larsen 1977), a finding that has prompted attempts to understand splash erosion in terms of shear. The detachment of soil particles represents a failure of the soil by the combined mechanism of compression and shear under raindrop impact, an event that is most likely to occur under saturated conditions when the shear strength of the soil is lowest (Al-Durrah & Bradford 1982). Generally, detachment decreases exponentially with increasing shear strength. Broadly linear relationships have been obtained, however, between the quantity of soil particles detached by raindrop impact and the ratio of the kinetic energy of the rainfall to soil shear strength (Al-Durrah & Bradford 1981; Torri et al. 1987b; Bradford et al. 1992).

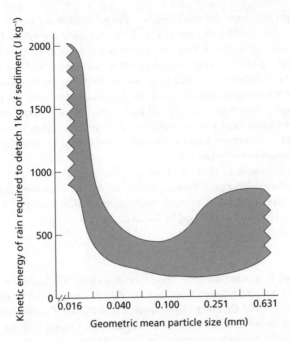

Fig. 2.2 Relation between geometric mean particle size of the soil and the rainfall energy required to detach 1 kg of sediment. Shaded area shows range of experimental values (after Poesen 1992).

Rain does not always fall on to a dry surface. During a storm it may fall on surface water in the form of puddles or overland flow. Studies by Palmer (1964) show that as the thickness of the surface water layer increases, so does splash erosion. This is believed to be due to the turbulence that impacting raindrops impart to the water. No increase in splash erosion with water depth has been observed, however, on sandy soils (Ghadiri & Payne 1979; Poesen 1981). There is, however, a critical water depth beyond which erosion decreases exponentially with increasing water depth because more of the rainfall energy is dissipated in the water and does not affect the soil surface. Laboratory experiments have shown that the critical depth is approximately equal to the diameter of the raindrops (Palmer 1964) or to one-fifth (Torri & Sfalanga 1986) or one-third (Mutchler & Young 1975) of the diameter. These differences in the value of critical depth are due to the different experimental conditions used in the experiments, particularly the soils, which ranged from clays to silt loams, loams and sandy loams.

Experimental studies show that the rate of detachment of soil particles with rainsplash varies with the 1.0 power of the instantaneous kinetic energy of the rain (Free 1960; Quansah 1981) and with the square of the instantaneous rainfall intensity (Meyer 1981). The detachment rate (D_r) on bare soil can be expressed by equations of the form:

$$D_r \propto I^a s^c \tag{2.10}$$

$$D_r \propto KE^b s^c e^{-dh} \tag{2.11}$$

where I is the rainfall intensity (mm h^{-1}), s is the slope expressed in m m^{-1} or as a sine of the slope angle, KE is the kinetic energy of the rain (J m^{-2}) and h is the depth of surface water (m). Although

Processes and mechanics of erosion

2.0 is a convenient value for a, the value may be adjusted to allow for variations in soil texture using the term $a = 2.0 - (0.01 \times \%\ \text{clay})$ (Meyer 1981). Similarly, the value of 1.0 for b may be varied from 0.8 for sandy soils to 1.8 for clays (Bubenzer & Jones 1971). Values for c are in the range of 0.2–0.3 (Quansah 1981; Torri & Sfalanga 1986), also varying with the texture of the soil (Torri & Poesen 1992). It should be remembered that the slope term in this equation refers to the local slope for a distance equivalent to only a few drop diameters from the point of raindrop impact – for example, that on the side of a soil clod – and not the average ground slope. Thus, for practical purposes, the slope term is often omitted from calculations of soil particle detachment. A value of 2.0 is convenient for d as representative of a range of values between 0.9 and 3.1 for different soil textures (Torri et al. 1987b).

In contrast, average ground slope is important when considering the overall transport of splashed particles. On a sloping surface more particles are thrown downslope than upslope during the detachment process, resulting in a net movement of material downslope. Splash transport per unit width of slope (T_r) can be expressed by the relationship:

$$T_r \propto I^j S^f \tag{2.12}$$

where $j = 1.0$ (Meyer & Wischmeier 1969) and $f = 1.0$ (Quansah 1981; Savat 1981). There is some evidence to suggest that the value of f decreases on steeper slopes; Mosley (1973) gives a value of 0.8 and Moeyersons and De Ploey (1976) a value of 0.75 where slope angles rise to 20 and 25° respectively. Foster and Martin (1969) and Bryan (1979) found that splash transport increases with slope angle to reach a maximum at about 18° and that on steeper slopes f becomes negative.

These relationships for detachment and transport of soil particles by rainsplash ignore the role of wind. Windspeed imparts a horizontal force to a falling raindrop until its horizontal velocity component equals the velocity of the wind. As a result, the kinetic energy of the raindrop is increased. Not surprisingly, detachment of soil particles by impacting wind-driven raindrops can be some 1.5–3 times greater than that resulting from rains of the same intensity without wind (Disrud & Krauss 1971; Lyles et al. 1974a). Wind also causes raindrops to strike the surface at an angle from vertical. This affects the relative proportions of upslope versus downslope splash. Moeyersons (1983) shows that where the angle between the falling raindrop and the vertical is 20°, net splash transport is reduced to zero for slopes of 17–19° and has a net upslope component for gentler slopes. Where the angle between the falling raindrop and the vertical is 5°, zero splash occurs on a slope of 3°.

Since splash erosion acts uniformly over the land surface its effects are seen only where stones or tree roots selectively protect the underlying soil and splash pedestals or soil pillars are formed. Such features frequently indicate the severity of erosion. Splash erosion is most important for detaching the soil particles that are subsequently eroded by running water. However, on the upper parts of hillslopes, particularly those of convex form, splash transport may be the dominant erosion process. In Calabria, southern Italy, under forest and under scattered herb and shrub vegetation, splash erosion accounts for 30–95 per cent of the total transport of material by water erosion (van Asch 1983). In Bedfordshire, England, splash accounts for 15–52 per cent of total soil transport on land under cereals and grass but only 3–10 per cent on bare ground (Morgan et al. 1986). As runoff and soil loss increase, the importance of splash transport declines, although very low contributions of splash to total transport were also measured in Bedfordshire under woodland because of the protective effect of a dense litter layer. Govers and Poesen (1988) found that although raindrop impact detached 152 t ha^{-1} of soil over one year on a bare loam soil on a 14° slope in Belgium, splash transport accounted for only 0.2 t ha^{-1} of the soil loss. The most

Thus, while, as seen by eqns 2.21 and 2.22, the detachment capacity of flow is reduced by raindrop impact, its transporting capacity is enhanced (Savat 1979; Guy & Dickinson 1990; Proffitt & Rose 1992). The degree of enhancement depends on the resistance of the soil, the diameter of the raindrops and the depth and velocity of the flow. Govers (1989) found that high sediment concentrations could increase velocity by up to 40 per cent, especially at low discharges and flow depths. His experiments, however, were carried out for flow without rain, whereas Guy et al. (1990) found that the impact of rain decreased flow velocity by about 12 per cent.

Govers (1990) investigated three different types of equations, based on grain shear velocity, effective stream power and unit stream power for describing the transport capacity of overland flow, defined as the maximum sediment concentration that can be carried. For ease of use, the relationships based on unit stream power were preferred since this is simply the product of slope and flow velocity. He found that:

$$C_{max} = a(sv - 0.4)^b \qquad (2.28)$$

where a and b are empirical coefficients dependent on grain size. Everaert (1991) confirmed the above equation for flows without simultaneous rainfall, obtaining values of b from 1.5 to 3.5 for particles with median grain diameters (D_{50}) of 33 and 390 μm respectively. The impact of rainfall had a negligible influence on the relationship for fine particles but reduced the exponent for coarser particles to 1.5, indicating that rainfall diminishes the ability of overland flow to transport coarse material.

Instead of trying to define transport capacity only in terms of flow properties, some researchers have attempted to relate transport capacity to the maximum sediment concentration that a flow can carry when a balanced condition exists between detachment and deposition (Rose et al. 1983; Styczen & Nielsen 1989). The rate of deposition (D_p) is:

$$D_p = v_s \cdot C \qquad (2.29)$$

where v_s is the settling velocity of the particles (Proffitt et al. 1991). Torri and Borselli (1991) took data from the experiments of Govers (1990) and obtained a good agreement between the transport capacity of the flow estimated from a balance-based approach and that estimated from eqn 2.28, indicating that the latter is a reasonable expression of the transport capacity of flow.

Given the rather shallow depths of overland flow, the considerable role played by surface roughness and the generally low Reynolds and Froude numbers, it can be proposed that most of the sediment transported is derived by raindrop impact and that, except on steep slopes or on smooth bare soil surfaces, grain shear velocity rarely attains the level necessary to detach soil particles. Since, as seen earlier, particles between 0.063 and 0.250 mm in size are the most detachable by raindrop impact and, from Figs 2.2 and 2.3, it can be seen that the most detachable particles by flow are within the 0.1–0.3 mm range, the sediment carried in overland flow is deficient in particles larger than 1 mm and enriched in finer material. Thus, over time, areas of erosion on a hillside will become progressively sandier and areas of deposition, particularly in valley floors, will be enriched with clay particles.

The trend towards increasing sandiness in eroded areas is also brought about by another mechanism. Most of the sediment splashed into the flow is moved only relatively short distances before being deposited. Since deposition is a particle-size selective process, with the coarser particles being deposited first, the deposited layer becomes progressively coarser (Proffitt et al. 1991) and, as seen in section 2.2, may develop into a depositional crust. Less of the finer material is then exposed to erosion. This mechanism can take place even within an individual storm so that detachment is highest at the beginning of the storm and transport capacity is reached very quickly.

Plots of the relationship between sediment transport by overland flow and discharge, as measured in the field, do not always conform with those expected from the research described above. Work in Bedfordshire, England (Jackson 1984; Morgan et al. 1986), and in southern Italy (van Asch 1983) shows that sediment transport varies with discharge raised by a power of 0.6–0.8. The similarity of this value to that in equations for bed load transport in rivers implies that the transport process is dominantly one of rolling of the particles over the soil surface as bed load. Kinnell (1990) considers that the sediment component contributed to overland flow by raindrop impact is moved as bed load and the component contributed through detachment by the flow itself is moved as suspended load. This implies that sediment transport may be better expressed by eqn 2.20 on low slopes where soil particle detachment is solely by rainsplash but by eqn 2.27 on steeper slopes or with higher flow velocities when particle detachment by flow also takes place. It is also likely that the process is extremely dynamic, so that the most relevant equation for describing sediment transport continually changes through time.

2.3.4 Spatial distribution

The dynamic nature of the process is even more apparent when the spatial extent and distribution of overland flow over a hillside is considered. Horton (1945) described overland flow as covering two-thirds or more of the hillslopes in a drainage basin during the peak period of the storm. He viewed overland flow as being the result of the rainfall intensity exceeding the infiltration capacity of the soil, with the following pattern of distribution over land surface. At the top of the slope is a zone without flow, which forms a belt of no erosion. At a critical distance from the crest sufficient water accumulates on the surface for flow to begin. Moving further downslope, the depth of flow increases with distance from the crest until, at a further critical distance, the flow becomes concentrated into fewer and deeper flow paths, which occupy a progressively smaller proportion of the hillslope (Parsons et al. 1990). Hydraulic efficiency improves, allowing the increased discharge to be accommodated by a higher flow velocity. Nevertheless, the hydraulic characteristics of the flow vary greatly over very short distances because of the influence of bed roughness associated with vegetation and stones. As a result, erosion is often localized and after a rainstorm the surface of a hillside displays a pattern of alternating scours and sediment fans (Moss & Walker 1978). Eventually, the flow breaks up into rills. That overland flow occurs in such a widespread fashion has been questioned, particularly in well vegetated areas where such flow occurs infrequently and covers only that 10–30 per cent of the area of a drainage basin closest to the stream sources (Kirkby 1969a). Under these conditions its occurrence is more closely related to the saturation of the soil and the fact that moisture storage capacity is exceeded, rather than infiltration capacity. Although, as illustrated by the detailed studies of Dunne and Black (1970) in a small forested catchment in Vermont, the saturated area expands and contracts, being sensitive to heavy rain and snow melt, rarely can erosion by overland flow affect more than a small part of the hillslopes.

Since most of the observations testifying to the power of overland flow relate to semi-arid areas or to cultivated land with sparse plant cover, it would appear that vegetation is the critical factor. Some form of continuum exists, ranging from well vegetated areas where overland flow occurs rarely and is mainly of the saturation type, to bare soil where it frequently occurs and is of the Hortonian type. Removal of the plant cover can therefore enhance erosion by overland flow. The change from one type of overland flow to another results from more rain reaching the ground surface, less being intercepted by the vegetation and decreased infiltration as rainbeat and deposition of material from the flow cause a surface crust to develop. Exceptions to this trend occur in areas where high rainfall intensities are recorded. Hortonian overland flow is widespread in the tropical rain forests near Babinda, northern Queensland, where six-minute rainfall intensities of $60-100\,mm\,h^{-1}$ are common, especially in the summer, and the saturated hydraulic conductivity of the soil at 200 mm depth is only $13\,mm\,h^{-1}$. As a result, a temporary perched water table develops in the soil soon after the onset of rain, subsurface flow commences and this quickly emerges on the soil surface (Bonell & Gilmour 1978).

Where runoff rates are relatively high over most of the hillside, overland flow, or, more strictly, the combined action of overland flow and raindrop impact as interrill erosion, can be the dominant erosion process on the upper and middle slopes, with deposition of material as colluvium on the footslopes. This appears to be true for many agricultural areas on non-cohesive soils. On loose, freshly ploughed soils on colluvial deposits on 18–22° slopes in Calabria, Italy, van Asch (1983) found that overland flow accounted for 80–95 per cent of the sediment transport. On unvegetated sandy soils in Bedfordshire with an 11° slope, it accounts for 50–80 per cent (Morgan et al. 1986). On a loam soil on a 14° slope in northern Belgium it accounts for 22–46 per cent of the total soil loss, with rates ranging from 24 to $100\,t\,ha^{-1}$ (Govers & Poesen 1988). Interrill processes can also be the main agent of erosion on well vegetated slopes if the rainfall is very high.

2.4

Subsurface flow

The lateral movement of water downslope through the soil is known as interflow. Where it takes place as concentrated flow in tunnels or subsurface pipes its erosive effects through tunnel collapse and gully formation are well known. Less is known about the eroding ability of water moving through the pore spaces in the soil, although it has been suggested that fine particles may be washed out by this process. Pilgrim and Huff (1983) measured sediment concentrations as high as $1\,g\,l^{-1}$ in subsurface flow through a silt-loam soil on a 17° slope under grass in California in storms of $10\,mm\,h^{-1}$ intensity or less. The material, uniformly fine with particles ranging from 4 to $8\,\mu m$ in diameter, was being detached by raindrop impact at the surface and then moved by the flow through the macropores in the soil. Under tropical rain forest on 8–14° slopes at Pasoh, Negeri Sembilan, Malaysia, where erosion rates are very low, the material removed as dissolved solids in the subsurface flow can amount to 15–23 per cent of the total sediment transport (Leigh 1982).

Subsurface flow is enhanced where subsurface drainage systems have been installed, which can then serve as important pathways for sediment movement. On a 30.6 ha catchment with silty clay loam soils at Rosemaund, Herefordshire, flows through tile drains account for up to 50 per cent of the annual sediment loss of $0.8\,t\,ha^{-1}$ (Russell et al. 2001). More important than the sediment concentrations, however, are the concentrations of base minerals, which can be twice those found in overland flow. Essential plant nutrients, particularly those added by fertilizers, can be removed, thereby impoverishing the soil and reducing its resistance to erosion. In the Syvbroek catchment, Denmark, 58 per cent of the total phosphorus delivered annually to the water course comes from subsurface drains (Hansen 1990).

Processes and mechanics of erosion

2.5

Rill erosion

As indicated earlier, it is widely accepted that rills are initiated at a critical distance downslope, where overland flow becomes channelled. The break-up of overland flow into small channels or microrills was examined by Moss et al. (1982). They found that, in addition to the main flow path downslope, secondary flow paths developed with a lateral component. Where these converged, the increase in discharge intensified particle movement and small channels or trenches were cut by scouring. Studies of the hydraulic characteristics of the flow show that the change from overland flow to rill flow passes through four stages: unconcentrated overland flow; overland flow with concentrated flow paths; microchannels without headcuts; and microchannels with headcuts (Merritt 1984). The greatest differences exist between the first and second stages, suggesting that the flow concentrations within the overland flow should strictly be treated as part of an incipient rill system. In the second stage, small vortices appear in the flow and, in the third stage, develop into localized spots of turbulent flow characterized by roll waves (Rauws 1987) and eddies (Savat & De Ploey 1982). At the point of rill initiation, flow conditions change from subcritical to supercritical (Savat 1979). The overall change in flow conditions through the four stages seems to take place smoothly as the Froude number increases from about 0.8 to 1.2, rather than occurring when a threshold value is reached (Torri et al. 1987b; Slattery & Bryan 1992). For this reason, attempts to explain the onset of rilling through the exceedance of a critical Froude number have not been successful and additional factors have had to be included when defining its value. Examples are the particle size of the material (Savat 1979) and the sediment concentration in the flow (Boon & Savat 1981).

Greater success has been achieved relating rill initiation to the exceedance of a critical shear velocity of the runoff (eqn 2.17). Govers (1985) found that on smooth or plane surfaces, where all the shear velocity is exerted on the soil particles, the sediment concentration in the flow increased with shear velocity more rapidly once a critical value of about 3.0–3.5 cm s^{-1} was reached. At this point, the erosion becomes non-selective regarding particle size, so that coarser grains can be as easily entrained in the flow and removed as finer grains. A value of about 3.5 cm s^{-1} for the critical shear velocity only applies to non-cohesive soils or soils, which, because they are highly sensitive to dispersal or to liquefaction, resemble loose sediments. Rauws and Govers (1988) proposed that, except for soils with high clay contents, the critical shear velocity for rill initiation (u_{*crit}) is linearly related to the shear strength of the soil (τ_s) as measured at saturation with a torvane:

$$u_{*crit} = 0.89 + 0.56\tau_s \tag{2.30}$$

Using a shear vane, the equivalent equation is (Brunori et al. 1989):

$$u_{*crit} = 0.9 + 0.3\tau_s \tag{2.31}$$

An alternative but similarly conceived approach relates rill initiation to a critical value of the ratio between the shear stress exerted by the flow (τ) and the shear strength of the soil (τ_s) measured with a shear vane. When $\tau/\tau_s > 0.0001–0.0005$, rills will form (Torri et al. 1987a). In all these relationships, it should be stressed that the shear velocity or shear stress is applied wholly to the soil particles and should be strictly known as the grain shear velocity or the grain shear stress.

Once rills have been formed, their migration upslope occurs by the retreat of the headcut at the top of the channel. The rate of retreat is controlled by the cohesiveness of the soil, the height and angle of the headwall, the discharge and the velocity of the flow (De Ploey 1989a). Downslope extension of the rill is controlled by the shear stress exerted by the flow and the strength of the soil (Savat 1979). Shear stress also determines the rate of detachment of soil particles by flow within the rill, which can be broadly described by an equation of the type (Foster 1982):

$$Df = K_r(\tau - \tau_c) \tag{2.32}$$

where K_r is a measure of the detachability of the soil and τ_c is the critical shear stress for the soil.

The transport capacity of rill flow can be approximately represented by eqns 2.27 or 2.28. Govers (1992) found experimentally that flow velocity in rills could be related to the discharge by the relationship:

$$v = 3.52Q^{0.294} \tag{2.33}$$

and that this gave better predictions than the Manning equation (eqn 2.18). This was because over a range from 2 to 8° slope had no effect on flow velocity; neither did the grain roughness or the surface form of the soil. Govers (1992) therefore modified eqn 2.28 by replacing the velocity term with eqn 2.33 to read:

$$C_{max} = a(3.52Q^{0.294} - 0.0074)Q \tag{2.34}$$

where a is dependent upon the grain size of the sediment and the value of 0.0074 is interpreted as the critical value of unit stream power. Although this equation expressed the maximum sediment concentration that can be carried by runoff in a rill, the actual sediment concentration and, therefore, the erosion may vary considerably from this. This is because the supply of sediment to the rill is not solely dependent upon the detachment of soil particles by the flow. Instead, the rill has to adjust continually for pulse influxes of sediment due to wash-in from interrill flow on the surrounding land, erosion and collapse of the head wall and collapse of the side walls. Mass failure of the side walls can contribute more than half of the sediment removed in rills, particularly when heavy rains follow a long dry period during which cracks have developed in the soil (Govers & Poesen 1988).

Since raindrop impact increases the transport capacity of the flow and, through the detachment of soil particles, causes higher sediment concentrations, Savat (1979) argued that the interaction of rainfall with the flow would enhance the probability of rilling. Quansah (1982) and Dunne and Aubry (1986), however, found that the particles detached by rain filled in the microchannels as fast as they could form, so that rilling was inhibited. It appears that the two sets of processes compete so that either the microchannels are short-lived because they drain away the overland flow, become laterally isolated and fill in, or the concentration of flow increases its erosive power and the channels deepen, widen and migrate both upslope and downslope.

Since rill flow is non-selective in the particle size it can carry, large grains, even rock fragments up to 9 cm in diameter (Poesen 1987), can be moved. Meyer et al. (1975) found that 15 per cent of the particles carried in rills on a 3.5° slope of tilled silt loam were larger than 1 mm in size and that 3 per cent were larger than 5 mm. On a 4.5° slope of bare untilled silt loam, 80 per cent of the sediment transported in rills was between 0.21 and 2.0 mm in size and most of the clay particles were removed as aggregates within this size range (Alberts et al. 1980).

As expected from its considerable erosive power, rill erosion may account for the bulk of the sediment removed from a hillside, depending on the spacing of the rills and the extent of the area affected. Govers and Poesen (1988) found that the material transported in rills accounted for 54–78 per cent of the total erosion. In a cloud burst on 23 May 1958 near Banská Bystrica, in the Czech Republic, rills accounted for 70 per cent of the total sediment eroded (Zachar 1982). These figures contrast with the situation in mid-Bedfordshire, England, where rills contributed only 20–50 per cent of the total erosion (Morgan et al. 1986) and in the Hammerveld-1 catchment in the Belgian loess, where rills accounted for one-third of the erosion over a three-year period (Vandaele & Poesen 1995).

2.6

Gully erosion

Gullies are relatively permanent steep-sided water courses that experience ephemeral flows during rainstorms. Compared with stable river channels, which have a relatively smooth, concave-upwards long profile, gullies are characterized by a headcut and various steps or knick-points along their course. These rapid changes in slope alternate with sections of very gentle gradient, either straight or slightly convex in long profile. Gullies also have relatively greater depth and smaller width than stable channels, carry larger sediment loads and display very erratic behaviour, so that relationships between sediment discharge and runoff are frequently poor (Heede 1975a). A widely recognized definition used to separate gullies from rills is that gullies have a cross-sectional area greater than $1\,m^2$ (actually $929\,cm^2$) (Poesen 1993). Gullies are almost always associated with accelerated erosion and therefore with instability in the landscape.

2.6.1 Gully formation

At one time it was thought that gullies developed as enlarged rills but studies of the gullies or *arroyos* of the southwest USA revealed that their initiation is a more complex process. In the first stage small depressions or knicks form on a hillside as a result of localized weakening of the vegetation cover by grazing or fire. Water concentrates in these depressions and enlarges them until several depressions coalesce and an incipient channel is formed. Erosion is concentrated at the heads of the depressions, where near-vertical scarps develop over which supercritical flow occurs. Some soil particles are detached from the scarp itself but most erosion is associated with scouring at the base of the scarp, which results in deepening of the channel and the undermining of the headwall, leading to collapse and retreat of the scarp upslope. Sediment is also produced further down the gully by bank erosion. This occurs partly by the scouring action of running water and the sediment it contains and partly by slumping of the banks. Between flows sediment is made available for erosion by weathering and bank collapse. This sequence of gully formation, described by Leopold et al. (1964) in New Mexico, is shown in Fig. 2.4.

Not all gullies develop purely by surface erosion, however. Berry and Ruxton (1960), investigating gullies in Hong Kong that formed following clearance of natural forest cover, found that most water was removed from the hillsides by subsurface flow in natural pipes or tunnels, and when heavy rain provided sufficient flow to flush out the soil in these, the ground surface subsided, exposing the pipe network as gullies. Numerous studies record the formation of gullies by pipe collapse in many different materials and climatic environments. The essential requirements are steep hydraulic gradients in a soil of high infiltration capacity through macropores but

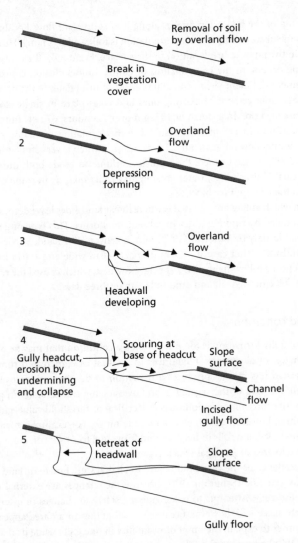

Fig. 2.4 Stages in the surface development of gullies on a hillside (after Leopold et al. 1964).

low intrinsic permeability, so that water does not move readily into the matrix (Crouch 1976; Bryan & Yair 1982). Suitable soils include those prone to cracking as a result of high sodium absorption, shrinkage on drying or release of pressure following unloading of overlying material.

Tunnel erosion has been widely reported in many hilly and rolling areas of Australia, where it is associated with duplex soils. These are soils characterized by a sharp increase in clay content between the A and B horizons so that the upper layer, 0.03–0.6 m in depth, varies from a loamy sand to a clay loam and the lower layer ranges from a light to heavy clay. According to Downes (1946), overgrazing and removal of vegetation cover cause crusting of the surface soil, resulting in greater runoff. This passes into the soil through small depressions, cracks and macropores but,

Processes and mechanics of erosion

on reaching the top of the B horizon, moves along it as subsurface flow. Localized dispersion of the clays in areas of subsurface moisture accumulation is followed by piping. Heavy summer rains cause the water in the pipes to break out on to the surface. Eventually, the roofs of the pipes collapse and gullying occurs. In the Loess Plateau of China, tunnel erosion contributes 25–30 per cent of the catchment sediment yield (Zhu 2003). Most of this relates to the development of new tunnels rather than enlargement of existing ones and takes place in single storms with return periods of 50 years or more. In a storm of 107 mm over 7.5 hours, 67 new tunnels were created; another storm of 37 mm in 115 minutes resulted in the formation of 123 new inlets. Sediment concentration in tunnel flow ranges from 8.2 to 893.2 g l⁻¹, which is very similar to that in channel flow (Zhu et al. 2002). Once formed, the tunnels continue to erode both during and between storms through earth falls, slumps, water erosion and roof collapse; in some cases, slumps and earth falls lead to their temporary blockage.

A third way in which gullies are initiated is where linear landslides leave deep, steep-sided scars, which may be occupied by running water in subsequent storms. This type of gully development has been described in Italy by Vittorini (1972) and in central Värmland, Sweden, by Fredén and Furuholm (1978). In the latter case, a 3–20 m deep, 20–40 m wide and 100 m long gully formed in glaciofluvial deposits following the 1977 spring snowmelt, which caused the removal, as a mass flow, of 20,000 m³ of saturated silt and sand in less than three days.

2.6.2 Threshold conditions

The main cause of gully formation is too much water, a condition that may be brought about by either climatic change or alterations in land use. In the first case, increased runoff may occur if rainfall increases or if less rainfall produces a reduction in the vegetation cover. In the second case, deforestation, burning of the vegetation and overgrazing can all result in greater runoff. If the velocity or tractive force of the runoff exceeds a critical or threshold value, gullying will occur. The threshold values, however, can show a very wide range. For example, τ_c varies from 3.3 to 32.2 Pa on cultivated silt-loam soils in Belgium and from 16.8 to 74.4 Pa on cultivated stony sandy loams in the Alentejo area of Portugal (Nachtergaele 2001).

Where the exceedance of the threshold relates to changes in climate or land use, the threshold is described as extrinsic (Schumm 1979) because the changes are external to the processes operating within the gully. Attempts to relate gullying solely to changes in external factors have not proved entirely successful, however, because not all gullies in an area appear to respond in the same way. In order to explain the onset of instability in one gully while its neighbours remain stable, Schumm (1979) examined the role of intrinsic thresholds, which are related to the internal working of the gully. From a review of studies in Wyoming, Colorado, New Mexico and Arizona, a discriminant function was established between stable and unstable conditions in terms of the size of the catchment area (a), which controls discharge, and the channel slope (s), which controls the velocity of runoff. When, for a given catchment area, channel slope exceeds a critical value, incision occurs, creating a channel characterized by one or more head scarps. Subsequent scouring causes the gully to become very active: the channel widens, deepens and extends headwards. Over time, the channel slope is reduced, promoting a consolidation phase as the gully stabilizes, the channel fills in, the sides and head wall become flatter and vegetation regrows. Deposition steepens the slope again and triggers a new phase of gullying. Thus gullies pass through successive cycles of erosion and deposition. It is not uncommon for the head of a gully to be extremely active while the lower section of the gully is stabilizing or for gullies to contain a sequence of alternating stable and unstable sections.

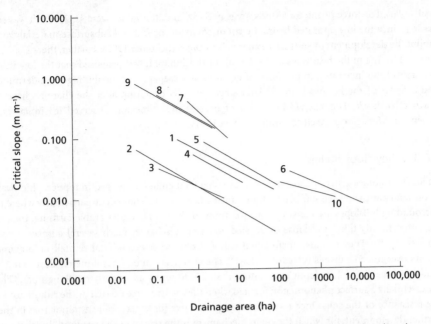

1, Central Belgium; 2, Central Belgium; 3, Portugal; 4, France; 5, United Kingdom (South Downs); 6, Colorado, USA; 7, Sierra Nevada, USA; 8, California, USA; 9, Oregon, USA; 10, New South Wales, Australia

Fig. 2.5 Relationship between critical slope and drainage area for development of gullies (after Poesen et al. 2003).

Begin and Schumm (1979) and Moore et al. (1988) have established critical s–a relationships to the effect that gullies form when:

$$sa^b > t \tag{2.35}$$

where t is the threshold value. Threshold values are higher for non-cultivated land than for cropland and also vary with the type of vegetation cover, differences in soil structure and soil moisture and type of tillage (Poesen et al. 2003). The threshold values also depend on the methodology. The lines plotted in Fig. 2.5 are best-fit regression lines passing through data obtained for gullied catchments. As a result, some gullied catchments will plot below the lines. Begin and Schumm (1979), however, proposed defining the line below which gullied catchments did not occur and which could therefore be interpreted as defining the condition at which all valley floors were stable; there will, however, be some ungullied catchments, which will plot above the line. More recently, Morgan and Mngomezulu (2003) showed that discriminant functions relating s to a could be used to separate gullied from ungullied catchments in four areas of Swaziland. Out of a sample of 201 catchments, only 8 per cent were incorrectly classified by this method.

The value of b in eqn 2.35 is generally interpreted in relation to the processes operating in the catchment. Values >0.2 are associated with erosion by surface runoff and those <0.2 as indicating subsurface processes and mass movement (Montgomery & Dietrich 1994; Vandekerckhove et al. 2000). For three of the four study areas in Swaziland, the values were <0.2, which was surprising, since the low saturated hydraulic conductivity of the soil and subsurface material would inhibit the development of subsurface channels or pipes (Scholten 1997). Further, there is no evidence of piping in the headwalls and sidewalls of the gullies. It was proposed that the low values may reflect the increasing importance of groundwater seepage in contributing to undermining and collapse of the headwalls in the later stages of gully evolution after the channels have cut down to bedrock. This change in dominant process over time was observed in similar gully systems in Madagascar (Wells & Andriamihaja 1991).

2.6.3 Valley floor gullies

Valley floor gullies generally take the form of ephemeral gullies developed in topographic swales in the landscape where runoff concentrates during heavy rains. They occur particularly where the surrounding hillslopes are convexo-concave, most of the land is under arable farming, the soils are either freshly tilled and loose or crusted and peak discharges reach several cumecs ($m^3 s^{-1}$) (Poesen 1989). They can also form when runoff from either exceptional rainfall or snowmelt occurs over frozen subsoils (Øygarden 2003). The channels can be several metres deep but generally they are limited in depth to no more than 25–30 cm by an underlying plough pan. They are essentially a surface phenomenon formed when the tractive force exerted by the runoff exceeds the resistance of the soil. Once formed, however, tractive force plays only a minor role in their further development, for which the main mechanism is the retreat of the headwall. In loess soils, this occurs mainly by slab failure as the wall is undermined by plunge-pool erosion and basal sapping, and vertical tension cracks develop on the slope above. Although head wall collapse may be the most important source of sediment, as in the gullies in western Iowa (Bradford & Piest 1980), measurements southwest of Sydney, Australia, show that erosion of the sidewalls subsequent to retreat of the headcut is much more important in contributing sediment to the gullies than the process of incision by the headcut itself (Blong et al. 1982).

Even though flow along the floor of the gully may contribute very little sediment compared to head wall and side wall processes, it is still important in removing sediment from the gully, without which the gully would fill in and stabilize. Stabilization can often be temporary, however, with the channels reforming in the same place in subsequent storms. Provided the material is flushed out, the gully can continue to retreat headwards until the area upslope contributing runoff to the gully head decreases sufficiently for the s–a threshold to fall below the critical value (Nachtergaele et al. 2002).

2.6.4 Valley side gullies

Valley side gullies develop more or less at right angles to the main valley line where local concentrations of surface runoff cut into the hillside, subsurface pipes collapse or local mass movements create a linear depression in the landscape. Valley side gullies may be continuous – that is, they discharge into the river at the bottom of the slope – or discontinuous, fading out into a depositional zone and not reaching the valley floor. Once formed, they can grow upslope by headward retreat and downslope by incision of the channel floor.

Among the most spectacular valley side gullies are the *lavakas* found in the deeply weathered hills of the central part of Madagascar. They have developed on convex hillsides with basal slopes of 30–45° adjacent to wide flat-floored valleys. Under the average rainfall of 1000–2000 mm per year and maximum mean monthly temperatures of 18–26°C, the underlying metamorphic rocks have been chemically weathered to form a 5–25 m deep cover of saprolite. According to Wells and

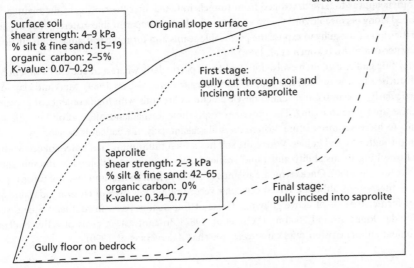

Headward retreat ceases
when upslope catchment
area becomes too small

Surface soil
shear strength: 4–9 kPa
% silt & fine sand: 15–19
organic carbon: 2–5%
K-value: 0.07–0.29

Original slope surface

First stage:
gully cut through soil and
incising into saprolite

Saprolite
shear strength: 2–3 kPa
% silt & fine sand: 42–65
organic carbon: 0%
K-value: 0.34–0.77

Final stage:
gully incised into saprolite

Gully floor on bedrock

Fig. 2.6 Gully formation on deeply weathered saprolite in Swaziland. Note the high erodibility of the saprolite compared to the surface soil.

Andriamihaja (1991), gullies form where the vegetation is destroyed, probably by grazing, or along paths and tracks. Infiltration capacities in these areas are reduced to <50 mm h^{-1} compared with >300 mm h^{-1} on the surrounding land. With ten-minute rainfall intensities of 80–100 mm h^{-1}, sometimes 200–500 mm h^{-1}, runoff is quickly produced with sufficient force to develop rills within about 20–30 m of the crest, just where the slope steepens rapidly into the convexity. The rills enlarge over time, develop head scarps and cut down into the underlying saprolite. Once this material is reached, the rate of downcutting, particularly in the head area, is very rapid until the channel reaches the groundwater zone above the bedrock. Thereupon, downcutting ceases but the gully erosion continues to expand by sapping, undermining and collapse of the headwall and side walls (Wells et al. 1991). Since this occurs in the area of greatest relief, the gully develops its typical pear-shaped plan form, broadest towards the top of the slope and with a narrow outlet downslope. In this stage, upslope runoff contributes little, if anything, to the gully growth, whereas localized concentrations of groundwater enhance headward bifurcation and cause the perimeter of the gully to take on a feather-edged appearance when viewed in plan. Figure 2.6 shows the development of similar gully systems in Swaziland. Compared to the surface soil, the underlying saprolite has a high content of silt and very fine sand, no organic matter and a low shear strength (Scholten 1997), which means it has a much higher erodibility. These characteristics, combined with the depth of material into which downcutting can occur, explain why gully erosion is so rapid once the more resistant surface soil has been breached. Similar patterns of valley side gully development have been observed in the Appalachian Piedmont, southeast USA (Ireland et al. 1939), southeast Brazil (de Meis & de Moura 1984) and Zimbabwe (Whitlow & Bullock 1986).

In the examples just described, there is no evidence of piping. Yet piping is often associated with valley side gullies, even though the relationship can be complex. At Springdale in the

Riverina area of New South Wales, some gullies clearly owe their origin to the collapse of tunnels, as evidenced by the remnants of the soil forming bridges where the collapse has been incomplete, while others appear to have developed from tunnels initiated after the cutting of the main gully. Developed along permeable zones between layers of relatively impermeable material, they depend on the incision of the gully to expose the permeable seams and form the hydraulic gradient along which water can move (Crouch et al. 1986).

Valley side gullies can often evolve into badlands, particularly where a fragile environment has been disturbed by unwise land use (López-Bermúdez & Romero-Diaz 1989). Marl and clay soils are particularly susceptible to badland development, as are soils with high contents of gypsum, since these are prone to piping. The processes responsible for badlands depend on how the soil responds to increases in moisture. Where the soil seals quickly, the badlands develop by surface erosion through rills and gullies. Where the soil has a high infiltration rate, water collects within the soil, resulting in instability and rapid shallow mass movements (Bouma & Imeson 2000; Moretti & Rodolfi 2000). Once formed, badlands take a long time to heal and many of those found in the Mediterranean, although very spectacular, record rather low rates of erosion today and are relics of former periods of active gullying. Those of the Guadix Basin in southeast Spain were probably developed around 2000 BC (Wise et al. 1982). In contrast, in areas of active badland development annual erosion rates can exceed $260\,t\,ha^{-1}$ (Cervera et al. 1990).

2.6.5 Erosion rates

Although gullies can remove vast quantities of soil, gully densities are not usually greater than $10\,km\,km^{-2}$ and the surface area covered by gullies is rarely more than 15 per cent of the total area (Zachar 1982). This results in a considerable contrast between the erosion rate for an individual gully and its contribution to the overall soil loss of an area. Rates of headwall extension can be very rapid for relatively short periods of time. Measurements on the Mbothoma Gully system, Swaziland, showed very rapid retreats of between 2.5 and $6.3\,m\,yr^{-1}$ from 1947 to 1961, followed by a slowing down to $0.13–0.51\,m\,yr^{-1}$ for the period 1961–1980 (WMS Associates 1988). A new gully opposite Mbothoma developed in the 1960s and up to 1990 eroded headwards at a rate of $14\,m\,yr^{-1}$; the rate then decreased to $5\,m\,yr^{-1}$ between 1990 and 1998 (Sidorchuk et al. 2003). Erosion from a gully developed on arable land near Cromer, Norfolk, England in 1975 was estimated at $195\,t\,ha^{-1}$ (Evans & Nortcliff 1978) and erosion from a winter runoff event in southern Norway in January 1990 exceeded $100\,t\,ha^{-1}$ in many gullies (Øygarden 2003). In gullies near Bathurst, New South Wales, soil loss from the side walls alone amounted to $1100\,t\,ha^{-1}$ over a three-year period beginning April 1984 (Crouch 1990a). These figures contrast with annual rates for whole catchments for which typical figures include $3–5\,t\,ha^{-1}$ for the gullied watersheds at Treynor, Iowa (Bradford & Piest 1980), $3–16\,t\,ha^{-1}$ for the *lavakas* in Madagascar (Wells & Andriamihaja 1991) and $6.4\,t\,ha^{-1}$ for a watershed near Gilgranda, New South Wales, of which 60 per cent came from gully heads (Crouch 1990b).

In a review of worldwide data, Poesen et al. (2003) showed that gully erosion can contribute between 10 and 94 per cent of overall soil loss from an area, with values between 30 and 75 per cent being typical. The contribution of gullies to total erosion is therefore not easily predictable. It depends on the characteristics of the storm, the topography of the catchment and the land cover at the time the storm occurs. In the 94-hectare Blosseville catchment, Normandy, France, a rain storm of 60 mm on 26 December 1999 with a maximum six-minute intensity of $55\,mm\,h^{-1}$ caused erosion of $10\,t\,ha^{-1}$. Some 93 per cent of the catchment had less than 20 per cent vegetation cover. Ephemeral gullies contributed 24 per cent of the erosion. A 60 mm storm with a maximum six-minute intensity of $105\,mm\,h^{-1}$ in the same catchment on 9 May 2000 produced a much lower erosion of $1\,t\,ha^{-1}$ because some 73 per cent of the catchment had a vegetation cover greater than 60 per cent. However, 83 per cent of the erosion occurred in ephemeral gullies (Cerdan et al. 2002a).

2.7

Mass movements

Although mass movement has been widely studied by geologists, geomorphologists and engineers, it is generally neglected in the context of soil erosion. Yet Temple and Rapp (1972) found that in the western Uluguru Mountains, Tanzania, landslides and mudflows are the dominant erosion processes. They occur in small numbers about once every ten years. The quantity of sediment moved from the hillsides into rivers by mass movement is far in excess of that contributed by gullies, rills and overland flow. Further, less than 1 per cent of the slide scars are in areas of woodland, 47 per cent being on the cultivated plots and another 47 per cent on land lying fallow. The association of erosion with woodland clearance for agriculture is thus very clear. Further evidence for this is provided by Rogers and Selby (1980) in respect of shallow debris slides on clay and silty clay soils derived from greywackes in the Hapuakohe Range, south Auckland, New Zealand. Clearance of the forest for pasture causes a decline in the shear strength of the soil over the five- to ten-year period needed for the tree roots to decay. As a result, landslides under pasture are triggered by a storm with a return period of 30 years, whereas a storm with a 100-year return period is required to produce slides beneath the forest. The close relationship between mass movement and water erosion is illustrated by the studies of the 'bottle slides' in the Uluguru Mountains (Temple & Rapp 1972; Lundgren & Rapp 1974), which develop in areas of large subsurface pipes as a result of the flushing out of a muddy viscous mass of debris and subsequent ground collapse.

The stability of the soil mass on a hillslope in respect of mass movement can be assessed by a safety factor (F), defined as the ratio between the total shear strength (σ) of the soil material along a given shear surface and the amount of stress (τ) developed along that surface. Thus

$$F = \frac{\sigma}{\tau} \tag{2.36}$$

The slope is stable if $F > 1$ and failure occurs if $F \leq 1$. For the simple case of shallow or translational slides, F can be defined as:

$$F = \frac{c' + (\gamma z - \gamma_w h) \cos^2 \theta \tan \phi'}{\gamma z \sin \theta \cos \theta} \tag{2.37}$$

where c' is the effective cohesion of the soil, γ is the unit weight of soil above the slide plane, z is the vertical depth of soil above the slide plane, γ_w is the unit weight of water, h is the height of the groundwater (piezometric) surface above the slide plane, θ is the slope angle and ϕ' is the effective angle of internal friction. Applications of the equation to slides in the Hapuakohe Range show that the value of F is particularly sensitive to changes in c' and z. Thus, control measures should be directed at influencing these (Rogers & Selby 1980).

Mass movements, in the varied forms of creep, slides, rock falls and mudflows, are given detailed treatment in numerous books (Sharpe 1938; Zaruba & Mencl 1969; Brunsden & Prior 1984; Anderson & Richards 1987), together with details of how eqn 2.37 can be modified for application to different types of slope failure.

Wind erosion

The main factor in wind erosion is the velocity of moving air. Because of the roughness imparted by soil, stones, vegetation and other obstacles, wind speeds are lowest near the ground surface. A plane of zero wind velocity can be defined at some height (z_0) above the mean aerodynamic surface. Above z_0, windspeed increases exponentially with height so that velocity values plot as a straight line on a graph against the logarithmic values of the height (Fig. 2.7). The change in velocity with height is expressed by the relationship (Bagnold 1941):

$$\bar{v}_z = \frac{2.3}{k} u_* \log\left(\frac{z}{z_0}\right)$$

(2.38)

where $\bar{v}$ is the mean velocity at height z, k is the von Kármán universal constant for turbulent flow, assumed to equal 0.4 for clear fluids, and u_* is the drag or shear velocity.

Although the movement of soil particles can be related to a critical wind velocity, many workers have attempted to define the conditions more precisely in terms of a critical value of the dimensionless shear stress. Using eqn 2.15 and substituting ρ_a (the density of air) for ρ_w (the density of water), the critical value of Θ, the Shields parameter, for initiating soil particle movement approximates 0.01 (Bagnold 1941), which is much lower than that obtained for water. The difference may be due to the very great difference in the density of the fluids relative to the particle density. A sand grain in air is about 2000 times more massive than the surrounding fluid, whereas it is only about 2.6 times more massive than water (Bagnold 1979). As a result, much higher shear velocities are required to move particles by air and the initial particle motion is more violent. This violence is rapidly transmitted to neighbouring particles, causing a general chain reaction of motion. At this point, grains are dislodged and entrained in the flow relatively easily and therefore at relatively low Shields numbers (Iversen 1985).

Although the Shields coefficient has been successfully applied to particle movement in air (Bagnold 1951; Iversen 1985), for most soil conservation work it is sufficient to relate the detachment of soil particles by wind to a critical value of the shear velocity, using it as a surrogate measure of the drag force exerted by the flow. Shear velocity is directly proportional to the rate of increase in wind velocity with the logarithm of height and is therefore the slope of the line in Fig 2.7(d). Its value can be determined by measuring the wind speed at two heights but, in practice, by assuming $v = 0$ at height z_0, it can be obtained by measuring the speed at one height and applying the formula derived from eqn 2.37:

$$u_* = \frac{k}{2.3} \cdot \frac{\bar{v}_z}{\log(z/z_0)}$$

(2.39)

For use in estimates of sediment transport by wind, the velocities should be measured within 0.2 m of the ground surface (Rasmussen et al. 1985; Mulligan 1988).

Bagnold (1937) identifies two threshold velocities required to initiate grain movement. The static or fluid threshold applies to the direct action of the wind. The dynamic or impact threshold allows for the bombardment of the soil by grains already in motion. Impact thresholds are about 80 per cent of the fluid threshold velocities in value. In addition to detaching soil particles at a lower threshold velocity, the detachment potential of sediment-laden air is further enhanced by increases of 8–18 per cent in shear velocity close to the ground surface (Sørensen

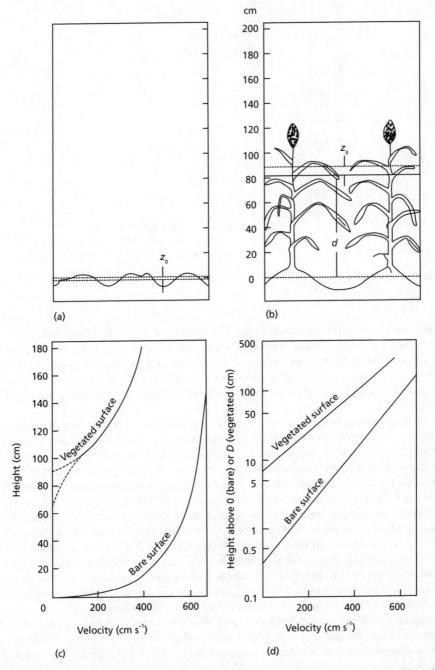

Fig. 2.7 Wind velocity near a soil surface. (a) Zero wind velocity occurs at a height z_0, which lies above the height of the mean aerodynamic surface and below the high points. (b) A crop cover raises the height of the mean aerodynamic surface by a distance d and also increases the value of z_0 so that the plane of zero wind velocity ($d + z_0$) occurs at a height that is equal to about 70 per cent of the height of the plants. (c) Wind velocity profiles above a bare surface and a vegetated surface plotted with linear scales. (d) Wind velocity profiles plotted with a logarithmic scale for height (after Troeh et al. 1980).

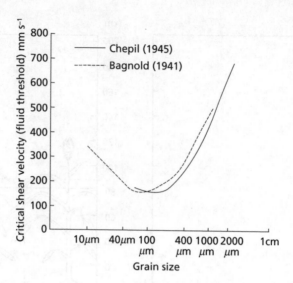

Fig. 2.8 Critical shear velocities for wind for erosion as a function of particle size (after Savat 1982).

1985; Dong et al. 2002). This arises through the addition of grain-borne shear stress to the air flow. The critical shear velocities vary with the grain size of the material, being least for particles of 0.10–0.15 mm in diameter, and increase with both increasing and decreasing grain size (Chepil 1945). The resistance of the larger particles results from their size and weight. That of the finer particles is due to their cohesiveness and the protection afforded by surrounding coarser grains (Fig. 2.8).

Once in motion, the transport of soil and sand particles by wind takes place in suspension, surface creep and saltation. Suspension describes the movement of fine particles, usually less than 0.2 mm in diameter, high in the air and over long distances. Surface creep is the rolling of coarse grains along the ground surface. Saltation is the process of grain movement in a series of jumps. Initially, drag and lift forces cause a particle to rise with a vertical ejection velocity that is about twice the shear velocity of the air (Willetts & Rice 1985). Drag with the surrounding air quickly reduces the vertical velocity, which is also opposed by the settling velocity of the particle. Once in the air the particle takes on a horizontal velocity, imparted by the wind, so that, while settling back to the ground, it is blown forwards. The result is an overall particle movement or jump comprising a vertical rise to a maximum height at which the settling velocity exceeds the vertical velocity, followed by a falling path at an angle of 6–12° to the horizontal. Individual jump lengths for coarse grains vary from about 60 to 400 mm, increasing with the shear velocity (Sørensen 1985; Willetts & Rice 1988). On striking the ground, the impact energy of the saltating grain is distributed into a disruptive part that causes disintegration of the soil and a dispersive part that imparts a velocity to other soil particles and launches them into the air (Smalley 1970). In a soil blow, between 55 and 72 per cent of the moving particles are carried in saltation.

The rate at which soil particles are dislodged from a bare soil surface was found in wind tunnel experiments (Sørensen 1985; Jensen & Sørensen 1986) to follow the relationship:

$$D_{fa} \propto u_*^{2.8}$$

(2.40)

where D_{fa} is the rate of detachment of soil particles in moving air. Although Willetts and Rice (1988) produce comparable data from their experiments, they caution that the overall data base is too small to determine the value of the exponent in eqn 2.39. It probably lies, however, between 2.0 and 3.0.

In contrast, general agreement exists among researchers that the transport capacity of moving air varies with the cube of the shear velocity. Based on this relationship and considering the transport of grains as representing a transfer of momentum from the air to the moving particles, Bagnold (1941) developed the following equation for determining the maximum sediment discharge per unit width (T_{fa}):

$$T_{fa} = C(d/D)^{1/2}(\rho_a/g)u_*^3 \tag{2.41}$$

where C is an experimentally derived parameter relating to grain size, d is the average diameter of the material, D is a standard grain diameter of 0.25 mm, ρ_a is the density of the air and g is the gravitational constant. Among the more commonly used expressions of eqn 2.40 is that proposed by Lettau and Lettau (1978):

$$T_{fa} = C(\rho_a/g)u_*^3(1 - u_t/u_z) \tag{2.42}$$

where T_{fa} is in $g\,m^{-1}\,s^{-1}$, C is a constant related to grain size and typically is about 6.5 in value, u_t is the impact threshold velocity for particle movement ($m\,s^{-1}$) and u_z is the wind velocity at height z.

It should be stressed that, at best, eqns 2.41 and 2.42 provide estimates of the maximum sediment transport rate (transport capacity) that can occur. They will not necessarily predict the actual rate of sediment transport. Where the soil surface is very resistant, the surface has become armoured by rock fragments or a vegetation cover is present, the equations will overpredict. Under these conditions, sediment transport will vary with the shear velocity of the wind raised to a power of less than 2.0. Further, sediment transport is a highly dynamic process reflecting the considerable short-term fluctuations in wind velocity (Sterk et al. 1998). Periods of sediment transport alternate with periods of no transport, even over a duration as short as ten minutes. During individual wind storms, activity may occur for only 16–21 per cent of the time. The fluctuations in sediment transport most probably relate to short-term changes in the turbulent structure of the storm.

Wind erosion impoverishes the soil and also buries the soil and crops on surrounding land. Although, as already seen, the most erodible particles are 0.10–0.15 mm in size, particles between 0.05 and 0.5 mm are generally selectively removed by the wind. Chepil (1946) found that areas of wind-blown deposits are enriched by particles in the 0.30 and 0.42 mm range. Resistance to wind erosion increases rapidly when primary particles and aggregates larger than 1 mm predominate. If erosion results in armouring of the surface so that more than 60 per cent of the surface material is of this size, the soil is almost totally resistant to wind erosion. Although saltation is the most important process from the viewpoint of soil erosion, wind erosion through the movement of dust particles in suspension can give rise to additional effects of contamination of food and water, aggravation of respiratory diseases, health risks associated with the transport of pathogens and interference with switches in machinery. Wind erosion also creates problems of visibility for road, rail and air transport.

Box 2

Initiation of soil particle movement

The initial movement of a soil particle by running water or wind takes place when the forces tending to move the particle exceed the forces resisting movement. In simple terms, there are three forces acting on the particle: the vertical or weight force moving the particle vertically downwards towards the ground surface, the lift force tending to make the particle rise vertically and the drag force exerted by the flow moving the particle horizontally along the surface (Fig. B2.1). The relative proportions of the downward, lift and drag forces depend upon the slope angle. On a horizontal surface the weight force predominates and on a steeply sloping surface the lift and drag forces are dominant.

Considering initially only the weight and drag forces, it is possible to define the critical condition at which a particle will move forwards, rotating about its point of contact with its neighbouring particle in a downslope or downwind direction (point A in Fig. B2.1). The condition is defined as the point at which the two forces are equal. Thus:

$$\text{weight force} = \text{drag force} \tag{B2.1}$$

$$g(\rho_s - \rho)\frac{\pi}{6}d^3 \tan\phi \cdot \frac{d}{2}\sin\phi = \tau_c \cdot \frac{d}{2}\sin\phi \tag{B2.2}$$

where g is the gravity force, ρ_s is sediment density, ρ is the fluid density, d is the particle diameter, ϕ is the angle of repose of the soil particles and τ_c is the critical value of the drag force (τ) for particle movement. For running water, $\tau = \gamma rs$ in which γ is the specific weight of the fluid, r is the hydraulic radius and s is the slope. For air flow, $\tau = \rho_a u_*^2$, in which ρ_a is the density of the air and u_* is the shear velocity of the flow, defined as the rate of change in velocity with height. Rearranging the equa-

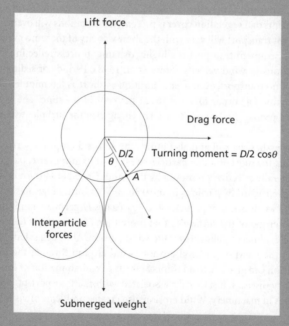

Fig. B2.1 Forces involved in the initiation of particle movement. When movement begins the uppermost particle will move against the underlying particle, turning at contact point A.

tion to determine τ_c and considering the degree of packing of the grains as an additional factor gives:

$$\tau_c = \eta g(\rho_s - \rho)\frac{\pi}{6}d\tan\phi \qquad (B2.3)$$

where η is a packing index.

The lift force arises from two components:
- differences in flow velocity between the top and bottom surfaces of the particle leading to a pressure gradient which encourages the particle to rise, which is known as the Bernoulli effect;
- turbulent eddies within the flow producing localized flow velocities close to the ground surface acting in an upwards direction.

The Bernoulli effect occurs in the following way. After a grain has been rolling along the ground for a short distance, the velocity of the air at any point near the grain is made up of two components, one due to the wind and the other to the spinning of the grain. On the upper side of the grain both components have the same direction but on the lower side they are in opposite directions. As a result of greater velocity on the top surface of the grain, the pressure there is reduced, while pressure at the lower surface increases. This difference in static pressure produces a lifting force and when this is sufficient to overcome the weight of the grain, the grain rises vertically.

Once a soil particle has been detached from the soil mass and entrained in the flow, it is carried forwards until the flow velocity falls below the settling velocity of the particle. The settling velocity (v_s) can be calculated from Stokes' Law:

$$v_s = \frac{2gr^2(\rho_s - \rho)}{9\mu} \qquad (B2.4)$$

where r is the radius of the particle assuming it to be spherical and μ is the viscosity of the fluid.

Figure B2.2 shows the relationship between the critical velocity for entrainment and the fall or settling velocity for particles of different sizes based on studies made in rivers (Hjulström 1935). This shows that a soil particle of 0.01 mm requires a flow of 60 cm s⁻¹ to detach it but it is not deposited until the flow velocity falls below 0.1 cm s⁻¹. In contrast, a coarse particle of 1.0 mm diameter needs a velocity of 25 cm s⁻¹ to detach it and it will be deposited when the velocity falls below 7 cm s⁻¹. The difference between the erosion and fall velocities is much less for the coarser particles, which means that, once entrained, they are carried for only relatively short distances.

Since settling velocity is related to particle size, during deposition the coarser particles are deposited first, with progressively finer

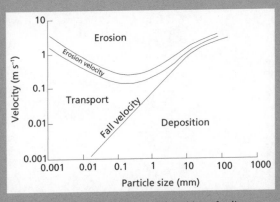

Fig. B2.2 Critical water velocities for erosion, transport and deposition of sediment as a function of particle size (after Hjulström 1935).

Continued

grains dropping out of the flow as the flow velocity continues to decline. As a result, deposits from slope wash on concave slopes at the foot of hillsides tend to comprise mainly sands, grading into silts and clays with increasing slope length and decreasing slope angle. On many slopes, silts and clays can be transported across the footslope and into the adjacent river. However, these relationships can sometimes be offset by the redetachment of the deposits and the trapping of fine particles in the wake of coarser ones, with the result that the transport of coarse material over fines is enhanced (Beuselinck et al. 2002). Deposits of wind-blown material comprise mainly silts and sands, moved in saltation, whereas the clay particles are carried long distances in suspension in the atmosphere.

CHAPTER 3

Factors influencing erosion

The factors controlling soil erosion are the erosivity of the eroding agent, the erodibility of the soil, the slope of the land and the nature of the plant cover.

3.1

Erosivity

3.1.1 Rainfall

Soil loss is closely related to rainfall partly through the detaching power of raindrops striking the soil surface and partly through the contribution of rain to runoff. This applies particularly to erosion by overland flow and rills, for which intensity is generally considered to be the most important rainfall characteristic. The effect of rainfall intensity is illustrated by the data for 183 rain events at Zanesville, Ohio, between 1934 and 1942, which show that average soil loss per rain event increases with the intensity of the storm (Table 3.1; Fournier 1972).

The role of intensity is not always so obvious, however, as indicated by studies of erosion in mid-Bedfordshire, England, taking data for the ten most erosive storms between May 1973 and October 1975. While intense storms, such as the one of 34.9 mm on 6 July 1973, in which 17.7 mm fell at intensities greater than 10 mm h^{-1}, produce erosion, so do storms of long duration and low intensity, like the one of 19 June 1973 when 39.6 mm of rain fell in over 23 hours (Morgan et al. 1986). It appears that erosion is related to two types of rain event, the short-lived intense storm where the infiltration capacity of the soil is exceeded, and the prolonged storm of low intensity that saturates the soil.

The response of the soil to rainfall may also be determined by previous meteorological conditions. This can again be demonstrated by data for Zanesville, Ohio (Fournier 1972). A storm of 19.3 mm on 9 June 1940 fell on dry ground and, despite the quantity, only 25 per cent went into runoff, most of the water soaking into the soil. On the following day, in a storm of 13.7 mm, 66 per cent of the rain ran off and soil loss almost trebled. The control in this case is the closeness of the soil to saturation, which is dependent on how much rain has fallen in the previous few days. The pattern of low soil loss in the first and high loss in the second of a series of storms is reversed, however, when, between erosive storms, weathering and light rainfall loosen the soil surface. Most of the loose material is removed during the first runoff event, leaving little for erosion in subsequent storms. This sequence is illustrated by studies in the Alkali Creek

Table 3.1 Relationship between rainfall intensity and soil loss

Maximum 5-min intensity (mm h^{-1})	Number of falls of rain	Average soil loss per rainfall (t ha^{-1})
0–25.4	40	3.7
25.5–50.8	61	6.0
50.9–76.2	40	11.8
76.3–101.6	19	11.4
101.7–127.0	13	34.2
127.1–152.4	4	36.3
152.5–177.8	5	38.7
177.9–254.0	1	47.9

Data for Zanesville, Ohio, 1934–42 (after Fournier 1972).

watershed, Colorado (Heede 1975b), where, following a year without runoff, a sediment discharge peak of 143 kg s^{-1} was observed on 15 April 1964 in a runoff from snowmelt of 2.21 m^3 s^{-1}. Next day, peak runoff increased to 3.0 m^3 s^{-1} but the sediment discharge fell to 107 kg s^{-1}. Although this type of evidence clearly points to the importance of antecedent events in conditioning erosion, no relationship was obtained between soil loss and antecedent precipitation in mid-Bedfordshire, England (Morgan et al. 1986).

The question arises of how much rain is required to induce significant erosion. Hudson (1981) gives a figure, based on his studies in Zimbabwe, of 25 mm h^{-1}, a value that has also been found appropriate in Tanzania (Rapp et al. 1972a) and Malaysia (Morgan 1974). It is too high for western Europe, however, where it is only rarely exceeded. Arbitrary thresholds of 10, 6 and 1.0 mm h^{-1} have been used in England (Morgan 1980a), Germany (Richter & Negendank 1977) and Belgium (Bollinne 1977) respectively.

Threshold values vary with the erosion process. The figures quoted above are typical for erosion by overland flow, rills and mass movements, which, as seen in Chapter 1, is characteristic of moderate events, whereas higher magnitude events are required for the initiation of fresh gullies. The distinction between these types of event can be blurred, however, in those areas that, by world standards, regularly experience what would be described elsewhere as extreme events. Starkel (1972) stresses the importance of regular gully erosion in the Assam Uplands where monthly rainfall may total 2000–5000 mm and in the Darjeeling Hills where over 50 mm of rain falls on an average of 12 days each year and rainfall intensities are often highest at the end of the rain event. In this very active landscape, overland flow and slope wash can start during rain storms of 50 mm with intensities greater than 30 mm h^{-1}; slides and slumps can occur after daily rains of 100 to 150 mm or a rainfall total of 200 mm in two or three days; and debris flows and mudflows are generated when 500 and 1000 mm of rain fall within two or three days (Froehlich & Starkel 1993). The effects of an extreme event may be long lasting and give rise to high soil losses for a number of years. The length of time required for an area to recover from a severe rainstorm, flooding and gullying has not been fully investigated but, in a review of somewhat sparse evidence, Thornes (1976) quotes figures up to 50 years.

3.1.2 Rainfall erosivity indices

The most suitable expression of the erosivity of rainfall is an index based on the kinetic energy of the rain. Thus the erosivity of a rainstorm is a function of its intensity and duration, and of the mass, diameter and velocity of the raindrops. To compute erosivity requires an analysis of the drop-size distributions of rain. Laws and Parsons (1943), based on studies of rain in the eastern USA, show that the drop-size characteristics vary with the intensity of the rain, with the median drop diameter by volume (d_{50}) increasing with rainfall intensity. Studies of tropical rainfall (Hudson 1963) indicate that this relationship holds only for rainfall intensities up to 100 mm h^{-1}. At higher intensities, median drop size decreases with increasing intensity, presumably because greater turbulence makes larger drop sizes unstable. However, at intensities above 200 mm h^{-1}, coalescence of smaller drops takes place so that the median drop diameter begins to increase again (Carter et al. 1974). Considerable variability exists because the relationship between median drop size and intensity is not constant; both median drop size and drop-size distribution vary for rains of the same intensity but different origins (Mason & Andrews 1960; Carter et al. 1974; Kinnell 1981; McIsaac 1990). The drop-size characteristics of convectional and frontal rain differ, as do those of rain formed at the warm and cold fronts of a temperate depression system.

Despite the difficulties posed by these variations, it is possible to derive general relationships between kinetic energy and rainfall intensity. Based on the work of Laws and Parsons (1943), Wischmeier and Smith (1958) obtained the equation:

$$KE = 0.0119 + 0.0873 \log_{10} I \tag{3.1}$$

where I is the rainfall intensity (mm h^{-1}) and KE is the kinetic energy (MJ ha^{-1} mm^{-1}). Many researchers (Mason & Ramandham 1953; Carte 1971; Houze et al. 1979; Styczen & Høgh-Schmidt 1988) consider the drop-size distribution of rainfall described by Marshall and Palmer (1948) as representative of a wide range of environments. The equivalent formula for calculating kinetic energy is:

$$KE = 0.0895 + 0.0844 \log_{10} I \tag{3.2}$$

For tropical rainfall, Hudson (1965) gives the equation:

$$KE = 0.298 \left(1 - \frac{4.29}{I}\right) \tag{3.3}$$

based on measurements of rainfall properties in Zimbabwe.

Given the variability of rainfall characteristics across the globe, it is not surprising that a large number of relationships have been established by different workers in different countries (Table 3.2). Many of these studies show that at intensities greater than 75 mm h^{-1}, the kinetic energy levels off at a value of about 0.29 MJ ha^{-1} mm^{-1} (Kinnell 1987). However, much higher values of 0.34–0.38 have been obtained in northern Nigeria (Kowal & Kassam 1976; Osuji 1989), Tuscany, Italy (Zanchi & Torri 1980), Okinawa, Japan (Onaga et al. 1988), Cévennes, France (Sempere-Torres et al. 1992), Portugal (Coutinho & Tomás 1995), Hong Kong (Jayawardena & Rezaur 2000) and Spain (Cerro et al. 1998). Carter et al. (1974) found that in the southern USA the kinetic energy increased to a maximum value at about 75 mm h^{-1}, decreased with further increases in intensity up to about 175 mm h^{-1} and then increased again at still higher intensities. In contrast, other studies in Japan (Mihara 1951) and in the Marshall Islands (McIsaac 1990) have recorded rainfall energies some 6–20 per cent lower than those calculated from eqn 3.1. Rainfall energy also varies with the density of the air raised to the 0.9 power; as a result, energy increases with altitude. Tracy et al. (1984) found that the kinetic energy of rainfall at 900–1800 m above sea level in Arizona was about 15 per cent higher than that predicted by eqn 3.1. Based on a review of previous research, van Dijk et al. (2002) proposed the following as a general equation:

Factors influencing erosion

Table 3.2 Relationship between kinetic energy of rain (KE, MJ ha^{-1} mm^{-1}) and rainfall intensity (I, mm h^{-1})

Equation	Source
$E = 0.0119 + 0.0873 \log_{10} I$	Used in Universal Soil Loss Equation (Wischmeier & Smith 1978); based on drop-size distributions of rainfall measured by Laws and Parsons (1943)
$E = 0.29(1 - 0.72e^{-I/20})$	Used in Revised Universal Soil Loss Equation; Brown and Foster (1987)
$E = 0.0895 + 0.0844 \log_{10} I$	Based on drop-size distributions of rainfall measured by Marshall and Palmer (1948)
$E = 0.0981 + 0.1125 \log_{10} I$	Zanchi and Torri (1980) for Toscana, Italy
$E = 0.359(1 - 0.56e^{-0.034I})$	Coutinho and Tomás (1995) for Portugal
$E = 0.0981 + 0.106 \log_{10} I$	Onaga et al. (1988) for Okinawa, Japan
$E = 0.298(1 - 4.29/I)$	Hudson (1965) for Zimbabwe
$E = 0.29(1 - 0.6e^{-0.04I})$	Rosewell (1986) for New South Wales, Australia
$E = 0.26(1 - 0.7e^{-0.035I})$	Rosewell (1986) for southern Queensland, Australia
$E = 0.1132 + 0.0055I - 0.005 \times 10^{-2} I^2 + 0.00126 \times 10^{-4} I^3$	Carter et al. (1974) for south-central USA
$E = 0.384(1 - 0.54e^{-0.029I})$	Cerro et al. (1998) for Barcelona, Spain
$E = 0.369(1 - 0.69e^{-0.038I})$	Jayawardena and Rezaur (2000) for Hong Kong
$E = 0.283(1 - 0.52e^{-0.042I})$	Proposed by van Dijk et al. (2002) as a universal relationship

$$KE = 0.283(1 - 0.52^{-0.042I})$$

(3.4)

This relationship generally provides estimates to within 10 per cent of measured values, although it overpredicts in climates with a strong coastal influence and underpredicts in semi-arid and sub-humid areas.

To compute the kinetic energy of a storm, a trace of the rainfall from an automatically record-ing rain gauge is analysed and the storm divided into small time increments of uniform inten-sity. For each time period, knowing the intensity of the rain, the kinetic energy of the rain at that intensity is estimated using an equation from Table 3.2 and this, multiplied by the amount of rain received, gives the kinetic energy for that time period. The sum of the kinetic energy values for all the time periods gives the total kinetic energy for the storm (Table 3.3).

To be valid as an index of potential erosion, an erosivity index must be significantly correlated with soil loss. Wischmeier and Smith (1958) found that soil loss by splash, overland flow and rill erosion is related to a compound index of kinetic energy (E) and the maximum 30-minute inten-sity (I_{30}). This index, known as EI_{30}, is open to criticism. First, being based on estimates of kinetic energy using eqn 3.1, it is of suspect validity for tropical rains of high intensity as well as for high altitudes and for oceanic areas like the Marshall Islands, where rainfall energies are rather low. Second, it assumes that erosion occurs even with light intensity rain, whereas Hudson (1965) showed that erosion is almost entirely caused by rain falling at intensities greater than 25 mm h^{-1}. The inclusion of I_{30} in the index is an attempt to correct for overestimating the importance of light intensity rain but it is not entirely successful because the ratio of intense erosive rain to non-erosive rain is not well correlated with I_{30} (Hudson, personal communication). In fact, there is no obvious reason why the maximum 30-minute intensity is the most appropriate parameter to choose. Stocking and Elwell (1973a) recommend its use only for bare soil conditions. With sparse

Table 3.3 Calculation of erosivity

Time from start of storm (min)	Rainfall (mm)	Intensity $(mm\,h^{-1})$	Kinetic energy $(MJ\,ha^{-1}mm^{-1})$	Total kinetic energy (col 2 × col 4) $(MJ\,ha^{-1})$
0–14	1.52	6.08	0.0877	0.1333
15–29	14.22	56.88	0.2755	3.9180
30–44	26.16	104.64	0.2858	7.4761
45–59	31.50	126.00	0.2879	9.0674
60–74	8.38	33.52	0.2599	2.1776
75–89	0.25	1.00	0	0

Erosivity indices

Wischmeier index (EI_{30})

Maximum 30-minute rainfall   = 26.16 + 31.50 mm
                             = 57.66 mm

Maximum 30-minute intensity = 57.66 × 2
                             = 115.32 mm h^{-1}

Total kinetic energy   = total of column 5
                             = 22.7724 MJ ha^{-1}

EI_{30} = 22.7724 × 115.32
                             = 2262.12 MJ mm ha^{-1} h^{-1}

Hudson index (KE > 25)
Total kinetic energy for rainfall = total of lines 2, 3, 4 and 5 in column 5
intensity ≥25 mm h^{-1}
                             = 22.64 MJ ha^{-1}

and dense plant covers they obtain better correlations with soil loss using the maximum 15- and 5-minute intensities respectively. The maximum 5-minute intensity was also found superior to I_{30} for short duration storms in Mediterranean countries (Usón & Ramos 2001). In order to overcome the likelihood of overestimating soil loss from high-intensity rainfall, the recommended practice with the EI_{30} index is to use a maximum value of 0.28 MJ ha^{-1} mm^{-1} for the E component for all rains above 76.2 mm h^{-1} and a maximum value of 63.5 mm h^{-1} for the I_{30} term (Wischmeier & Smith 1978).

As an alternative erosivity index, Hudson (1965) proposed $KE > 25$, which, to compute for a single storm, means summing the kinetic energy received in those time increments when the rainfall intensity equals or exceeds 25 mm h^{-1} (Table 3.3). When applied to data from Zimbabwe, a better correlation was obtained between this index and soil loss than between soil loss and EI_{30}. Stocking and Elwell (1973a) reworked Hudson's data, taking account of more recent data, and suggested that EI_{30} was the better index after all. But, since they computed EI_{30} only for storms yielding 12.5 mm or more of rain and with a maximum 5-minute intensity greater than 25 mm h^{-1}, they removed most of the objections to the original EI_{30} index and produced an index that is philosophically very close to $KE > 25$. Hudson's index has the advantage of simplicity and less stringent data requirements. Although somewhat limiting for temperate latitudes, it can be modified by using a lower threshold value such as $KE > 10$.

By calculating erosivity values for individual storms over a period of 20–25 years, mean monthly and mean annual data can be obtained. Since the EI_{30} and $KE > 25$ indices yield vastly different values because of the inclusion of I_{30} in the former, the two indices cannot be substituted for each other.

Factors influencing erosion

3.1.3 Wind erosivity

The kinetic energy (KE_a; $\mathrm{J\,m^{-2}\,s^{-1}}$) of wind can be calculated from:

$$KE_a = \frac{\gamma_a u^2}{2g} \tag{3.5}$$

where u is the wind velocity ($\mathrm{m\,s^{-1}}$) and γ_a is the specific weight of air defined in terms of temperature (T) in °C and barometric pressure (P) in kPa by the relationship (Zachar 1982):

$$\gamma_a = \frac{1.293}{1+0.00367T} \cdot \frac{P}{101.3} \tag{3.6}$$

For $T = 15$ °C and $P = 101.3\,\mathrm{kPa}$, $\gamma_a = 0.0625\,u^2\,\mathrm{kg\,m^{-2}}$, which converts to $227\,u^2\,\mathrm{J\,m^{-2}\,s^{-1}}$. Energy values for wind storms can be obtained by summing the energies for the different velocities weighted by their duration.

In practice, kinetic energy is rarely used as an index of wind erosivity and, instead, a simpler index based on the velocity and duration of the wind (Skidmore & Woodruff 1968) has been developed. The erosivity of wind blowing in vector j is obtained from:

$$EW_j = \sum_{i=1}^{n} \overline{V}t_{ij}^3 f_{ij} \tag{3.7}$$

where EW_j is the wind erosivity value for vector j, $\overline{V}t$ is the mean velocity of wind in the ith speed for vector j above a threshold velocity, taken as $19\,\mathrm{km\,h^{-1}}$, and f_{ij} is the duration of the wind for vector j at the ith speed. Expanding this equation for total wind erosivity (EW) over all vectors yields:

$$EW = \sum_{j=0}^{15} \sum_{i=1}^{n} \overline{V}t_{ij}^3 f_{ij} \tag{3.8}$$

where vectors $j = 0$ to 15 represent the 16 principal compass directions beginning with $j = 0 = E$ and working anticlockwise so that $j = 1 = ENE$ and so on.

3.2

Erodibility

Erodibility defines the resistance of the soil to both detachment and transport. Although a soil's resistance to erosion depends in part on topographic position, slope steepness and the amount of disturbance, such as during tillage, the properties of the soil are the most important determinants. Erodibility varies with soil texture, aggregate stability, shear strength, infiltration capacity and organic and chemical content.

The role of soil texture has been indicated in Chapter 2, where it was shown that large particles are resistant to transport because of the greater force required to entrain them and that fine particles are resistant to detachment because of their cohesiveness. The least resistant particles are silts and fine sands. Thus soils with a silt content above 40 per cent are highly erodible (Richter & Negendank 1977). Evans (1980) prefers to examine erodibility in terms of clay content, indicating that soils with a clay content between 9 and 30 per cent are the most susceptible to erosion.

The use of the clay content as an indicator of erodibility is theoretically more satisfying because the clay particles combine with organic matter to form soil aggregates or clods and it is the stability of these that determines the resistance of the soil.

Soils with a high content of base minerals are generally more stable, as these contribute to the chemical bonding of the aggregates. Wetting of the soil weakens the aggregates because it lowers their cohesiveness, softens the cements and causes swelling as water is adsorbed on the clay particles. Rapid wetting can also cause collapse of the aggregates through slaking. The wetting-up of initially dry soils results in greater aggregate breakdown than if the soil is already moist because, in the latter case, less air becomes trapped in the soil (Truman et al. 1990). Aggregate stability also depends on the type of clay mineral present. Soils containing kaolinite, halloysite, chlorite or fine-grained micas, all of which are resistant to expansion on wetting, have a low level of erodibility, whereas soils with smectite or vermiculite swell on wetting and therefore have a high erodibility; soils with illite are in an intermediate position.

In detail, however, the interactions between the moisture content of the soil and the chemical composition of both the clay particles and the soil water are rather complex. This makes it difficult to predict how clays, particularly those susceptible to swelling, will behave. The erodibility of clay soils is highly variable (Chisci et al. 1989). Although most clays lose strength when first wetted because the free water releases bonds between the particles, some clays, under moist but unsaturated conditions, regain strength over time. This process, known as thixotropic behaviour, occurs because the hydration of clay minerals and the adsorption of free water promote hydrogen bonding (Grissinger & Asmussen 1963). Strength can also be regained if swelling brings about a reorientation of the soil particles from an alignment parallel to the eroding water to a more random orientation (Grissinger 1966). The strength of smectitic clays is largely dependent upon the sodium adsorption ration (SAR). As this increases, i.e. the replacement of calcium and magnesium ions by sodium increases, so do water uptake and the likelihood of swelling and aggregate collapse. High salt concentrations in the soil water, however, can partly offset this effect so that aggregate stability is maintained at higher levels of SAR (Arulanandan et al. 1975). Sodic and saline-sodic soils, where the exchangeable sodium percentage (ESP) exceeds $15\,cmol\,kg^{-1}$ or the SAR of the pore water exceeds 13, are highly erodible.

The shear strength of the soil is a measure of its cohesiveness and resistance to shearing forces exerted by gravity, moving fluids and mechanical loads. Its strength is derived from the frictional resistance met by its constituent particles when they are forced to slide over one another or to move out of interlocking positions, the extent to which stresses or forces are absorbed by solid-to-solid contact among the particles, cohesive forces related to chemical bonding of the clay minerals and surface tension forces within the moisture films in unsaturated soils. These controls over shear strength are only understood qualitatively, so that, for practical purposes, shear strength is expressed by an empirical equation:

$$\tau = c + \sigma \tan \phi \tag{3.9}$$

where τ is the shear stress required for failure to take place, c is a measure of cohesion, σ is the stress normal to the shear plane (all in units of force per unit area) and ϕ is the angle of internal friction. Both c and ϕ are best regarded as empirical parameters rather than as physical properties of the soil.

Increases in the moisture content of a soil decrease its shear strength and bring about changes in its behaviour. At low moisture contents the soil behaves as a solid and fractures under stress, but with increasing moisture content it becomes plastic and yields by flow without fracture. The point of change in behaviour is termed the plastic limit. With further wetting, the soil will reach its liquid limit and start to flow under its own weight. The behaviour of a compressible soil when saturated depends on whether the water can drain. If drainage cannot take place and the soil is subjected to further loading, pressure will increase in the soil water, the compaction load will not be supported by the particles and the soil will deform, behaving as a plastic material. If drainage can occur, more of the load will be supported and the soil is more likely to remain below the plastic limit and retain a higher shear strength. As seen in Chapter 2, shear strength is used as a basis for understanding the detachability of soil particles by raindrop impact. Since the soils are usually saturated and the process is virtually instantaneous, there is no time for drainage and undrained failure occurs. Bradford et al. (1992) found that soil strength measured with a drop-cone penetrometer after one hour of rainfall was a good indicator of a soil's resistance to splash erosion. The drop-cone apparatus simulates the same kind of failure mechanism, in terms of compression and shear, as the impact of a falling raindrop.

The mechanism of soil particle detachment by surface flow involves different failure stresses on the soil surface compared with those generated by raindrop impact. Rauws and Govers (1988) show that these can be represented by measurements of the strength of the soil at saturation, made with a torvane. Equations 2.30 and 2.31 predict the critical shear velocity (u_{*c}; cm s^{-1}) for rill initiation on a smooth bare soil surface as a function of the strength or apparent cohesion of the soil measured at saturation by a torvane and a laboratory shear vane respectively.

Infiltration capacity, the maximum sustained rate at which soil can absorb water, is influenced by pore size, pore stability and the form of the soil profile. Soils with stable aggregates maintain their pore spaces better, while soils with swelling clays or minerals that are unstable in water tend to have low infiltration capacities. Although estimates of the infiltration capacity can be obtained in the field using infiltrometers (Hills 1970), it was seen in Chapter 2 that actual capacities during storms are often much less than those indicated by field tests. Where soil properties vary with profile depth, it is the horizon with the lowest infiltration capacity that is critical. For sandy and loamy soils, the critical horizon is often the surface where, as described in Chapter 2, a crust of 2 mm thickness may be sufficient to decrease infiltration capacity enough to cause runoff, even though the underlying soil may be dry.

The organic and chemical constituents of the soil are important because of their influence on aggregate stability. Soils with less than 2 per cent organic carbon, equivalent to about 3.5 per cent organic content, can be considered erodible (Evans 1980). Most soils contain less than 15 per cent organic content and many sands and sandy loams have less than 2 per cent. Voroney et al. (1981) suggest that soil erodibility decreases linearly with increasing organic content over the range of 0–10 per cent, whereas Ekwue (1990) found that soil detachment by raindrop impact decreased exponentially with increasing organic content over a 0–12 per cent range. These relationships cannot be extrapolated, however, because some soils with very high organic contents, particularly peats, are highly erodible by wind and water, whereas others with very low organic content can become very hard and therefore stronger under dry conditions. The role played by organic material depends on its origin. While organic material from grass leys and farmyard manure contributes to the stability of the soil aggregates, peat and undecomposed haulm merely protect the soil by acting like a mulch and do little to increase aggregate strength (Ekwue et al. 1993). Thus peat soils have very low aggregate stability.

Chemically, the most important control over erodibility is the proportion of easily dispersible clays in the soil. As seen above, a high proportion of exchangeable sodium can cause rapid deterioration in a soil's structure on wetting, with consequent loss of strength, followed by the formation of a surface crust and a decline in infiltration as the detached clay particles fill the pore spaces in the soil. The addition of sodium-containing fertilizers to support crops such as tobacco can sometimes lead to quite small increases in exchangeable sodium yet result in very marked structural deterioration of a previously stable soil (Miller & Sumner 1988). Excess calcium carbonate within the clay and silt fractions of the soil also leads to high erodibility (Barahona et al. 1990; Merzouk & Blake 1991).

Many attempts have been made to devise a simple index of erodibility based on either the properties of the soil as determined in the laboratory or the field, or the response of the soil to rainfall and wind (Table 3.4). In a review of the indices related to water erosion, Bryan (1968) favoured aggregate stability as the most efficient index. Unfortunately, there is no agreement between researchers on the most appropriate method to evaluate aggregate stability. Indices like the instability index (*Is*) and the pseudo-textural aggregation index (*Ipta*) (Table 3.4) are based on breaking up the aggregates by wet-sieving a sample of the soil. But some researchers consider that wet-sieving does not adequately simulate the processes of breakdown as they occur in the field and prefer to measure the proportion of aggregates that can be destroyed by the impact of water drops (Bruce-Okine & Lal 1975). Different researchers also follow different methods for the duration and speed of oscillation of the sieves in wet-sieving tests, and for the size and height of fall in water-drop tests. Further work on developing an appropriate test probably needs to take account of the factors that contribute to the stability of the aggregates. These are, respectively: for aggregates >10 mm in size, the binding and adhesive effects of plant roots; for aggregates of 2–10 mm, the calcium carbonate and organic matter content; for aggregates of 1–2 mm, the network of roots and hyphae; for aggregates of 0.105–1.0 mm, organic matter, roots and hyphae; and for aggregates <0.105 mm, clay mineralogy and cementing agents derived from microbiological activity (Boix-Fayos et al. 2001).

Given the above, it is not surprising that attempts have been made to develop a more universally accepted index. The one most commonly used is the *K* value which represents the soil loss per unit of EI_{30} as measured in the field on a standard bare soil plot, 22 m long and at 5° slope. Estimates of the *K* value may be made if the grain-size distribution, organic content, structure and permeability of the soil are known (Wischmeier et al. 1971; Fig. 3.1). Soil erodibility has been satisfactorily described by the *K* value for many agricultural soils in the USA (Wischmeier & Smith 1978) and for ferrallitic and ferruginous soils in West Africa (Roose 1977). Where *K* values have been determined from field measurements of erosion, they are valid. Difficulties arise, however, with attempts to predict the values from the nomograph (Fig. 3.1). Where it is applied to soils with similar characteristics to those in the USA, a close correlation exists between predicted and measured values, but poorer predictions are obtained where it is necessary to extrapolate the nomograph values. This applies to soils with organic contents above 4 per cent, swelling clays and those where resistance to erosion is a function of aggregate stability rather than primary particle size.

The resistance of the soil to wind erosion depends upon dry rather than wet aggregate stability and on the moisture content, wet soil being less erodible than dry soil, but is otherwise related to much the same properties as affect its resistance to water erosion. Chepil (1950), using wind tunnel experiments, related wind erodibility of soils to various indices of dry aggregate structure, but little of the work was tested in field conditions. Nevertheless, the results were extrapolated to give an index in $t\,ha^{-1}\,yr^{-1}$ based on climatic data for Garden City, Kansas. Values range from 84–126 for non-calcareous silty clay loams, silt loams and loams to 356–694 for sands. Similar research by Dolgilevich et al. (1973) in western Siberia and northern Kazakhstan yields values in $t\,ha^{-1}\,h^{-1}$ (Table 3.5). Both indices are closely related to the percentage of dry stable aggregates larger than 0.84 mm.

The indices described above treat soil erodibility as constant over time. They thus ignore seasonal variations on agricultural land associated with tillage operations, which alter the bulk density and hydraulic conductivity of the soil. Erodibility is four times higher in summer than in

Table 3.4 Indices of soil erodibility for water erosion

Static laboratory tests

Index	Formula	Reference
Dispersion ratio	$\dfrac{\%\text{silt} + \%\text{clay in undispersed soil}}{\%\text{silt} + \%\text{clay after dispersal of the soil in water}}$	Middleton (1930)
Clay ratio	$\dfrac{\%\text{sand} + \%\text{silt}}{\%\text{clay}}$	Bouyoucos (1935)
Surface aggregation ratio	$\dfrac{\text{surface area of particles} > 0.05\text{mm}}{(\%\text{silt} + \%\text{clay in dispersed soil}) - (\%\text{silt} + \%\text{clay in undispersed soil})}$	André and Anderson (1961)
Erosion ratio	$\dfrac{\text{dispersion ratio}}{\text{colloid content/moisture equivalent ratio}}$	Lugo-Lopez (1969)
Instability index (Is)	$\dfrac{\%\text{silt} + \%\text{clay}}{Ag_{air} + Ag_{alc} + Ag_{benz}}$ where Ag is the % aggregates >0.2mm after wet sieving for no pretreatment and pretreatment of the soil by alcohol and benzine respectively	Hénin et al. (1958)
Instability index (Is)	$\dfrac{\%\text{silt} + \%\text{clay}}{(\%\text{aggregates} > 0.2\text{mm after wet sieving}) - 0.9(\%\text{coarse sand})}$	Combeau and Monnier (1961)
Pseudo-textural aggregation index ($Ipta$)	$\dfrac{MWDw - MWDt}{X - MWDt} \times 100$ where $MWDw$ is the mean weight diameter of the wet-sieving grain-size distribution (mm), $MWDt$ is the mean weight diameter of the primary particle grain-size distribution (mm) and X is the maximum average grain-size diameter of the particles in the given grain-size distribution	Chischi et al. (1989)

Static field tests

Erodibility index	$\dfrac{1}{\text{mean shearing resistance} \times \text{permeabilty}}$	Chorley (1959)
Soil cohesion	direct measure of soil cohesion at saturation using a torvane	Rauws and Govers (1988)

Dynamic laboratory tests

Simulated rainfall test	Comparison of erosion of different soils subjected to a standard storm	Woodburn and Kozachyn (1956)
Water-stable aggregate (WSA) content	%WSA > 0.5 mm after subjecting the soil to rainfall simulation	Bryan (1968)
Water drop test	% aggregates destroyed by a pre-selected number of impacts by a standard raindrop (e.g. 5.5 mm diameter, 0.1 g from a height of 1 m)	Bruce-Okine and Lal (1975)
Erosion index	$\dfrac{dh}{a}$ where d is an index of dispersion (ratio of % particles >0.05 mm without dispersion to % particles >0.05 mm after dispersion of the soil by sodium chloride); h is an index of water-retaining capacity (water retention of soil relative to that of 1 g of colloids); and a is an index of aggregation (% aggregates >0.25 mm after subjecting the soil to a water flow of 100 cm min^{-1} for 1 h)	Voznesensky and Artsruui (1940)

Dynamic field tests

Erodibility index (K)	mean annual soil loss per unit of EI_{30}	Wischmeier and Mannering (1969)

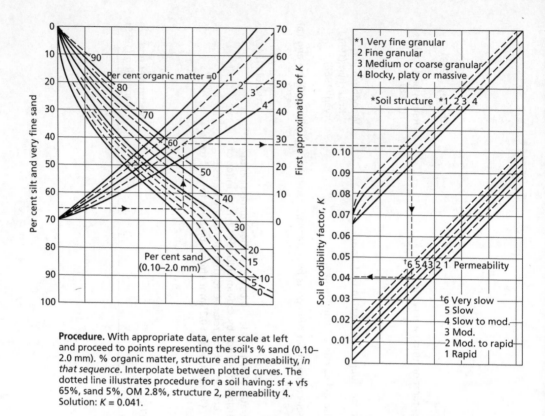

Procedure. With appropriate data, enter scale at left and proceed to points representing the soil's % sand (0.10–2.0 mm). % organic matter, structure and permeability, *in that sequence.* Interpolate between plotted curves. The dotted line illustrates procedure for a soil having: sf + vfs 65%, sand 5%, OM 2.8%, structure 2, permeability 4. Solution: K = 0.041.

Fig. 3.1 Nomograph for computing the K value (metric units) of soil erodibility for use in the Universal Soil Loss Equation (after Wischmeier et al. 1971). Divide values by 0.13 to obtain K values in the original American units.

Table 3.5 Assessments of soil erodibility by wind

% dry stable aggregates >0.84 mm	>80	70–80	50–70	20–50	<20
Erodibility $(t ha^{-1} h^{-1})$*	<0.5	0.5–1.5	1.5–5	5–15	>15
Erodibility $(t ha^{-1} yr^{-1})$†	<4	4–84	84–166	166–220	>220

* After Dolgilevich et al. (1973) for windspeeds of 20–25 m s⁻¹.
† After Chepil (1960) for Garden City, Kansas.

winter on bare, uncultivated sandy soil in Bedfordshire, England (Martin & Morgan 1980) and two times higher on silts and silt loam soils in Limbourg, The Netherlands (Kwaad 1991). There are also more frequent changes in erodibility related to changes in moisture content during and between rainstorms. While the expectation is that most soils become more erodible when they are wet because of aggregate destruction during the wetting-up process and the loss of cohesion, some soils are also very erodible when dry and more susceptible to detachment by raindrop impact (Martínez-Mena et al. 1998) and rilling (Govers 1991). A key factor for coarser soils is their tendency to become hydrophobic when dry, which then leads to an increase in runoff and, until the depth of water becomes sufficient to absorb their impact, an increase in soil particle detachment by raindrops (Terry & Shakesby 1993; Doerr et al. 2000). Freezing and thawing also alter the erodibility of the soil. Conditions of low bulk density and high soil moisture during periods of thaw produce a surface that is highly erodible. The erodibility of agricultural soils in Ontario, Canada, is 15 times higher in winter thaw conditions than in summer (Coote et al. 1988).

Under natural conditions, the seasonal activity of burrowing animals is important, giving rise to considerable disturbance of the soil. Earthworms bring to the surface as casts as much as 2–5.8 t ha^{-1} of material on agricultural land (Evans 1948). Rates of 15 t ha^{-1} have been measured in temperate woodland in Luxembourg (Hazelhoff et al. 1981) and 50 t ha^{-1} in tropical forest in the Ivory Coast (Roose 1976). Other animals and their annual rates of production of sediment at the surface include: ants, with 4 to 10 t ha^{-1} observed in Utah (Thorp 1949); termites, with 1.2 t ha^{-1} in tropical forest in the Ivory Coast (Roose 1976); voles and moles, with 19 t ha^{-1} in temperate woodland in Luxembourg (Imeson 1976) and 6 t ha^{-1} from Pyrenean mountain voles in the Spanish Pyrenees (Borghi et al. 1990); and isopods and porcupines, with 0.3 to 0.7 t ha^{-1} on stony land in the Negev Desert, Israel (Yair & Rutin 1981). On coastal sand dunes in the western Netherlands, rabbits displaced locally between 0.9 and 5.1 t ha^{-1} of sediment from their burrows (Rutin 1992). In many cases, the material brought to the surface comprises loose sediment with low bulk density and cohesion, which is rapidly broken down by splash erosion. The material contained in earthworm casts, however, consists of soil aggregates that are more stable under raindrop impact than the surrounding top soil, probably as a result of their higher organic content and secretions from the gut of the worm. Thus, earthworms have a positive effect on the stability of soil aggregates and the hydraulic conductivity of the soil (Glasstetter & Prasuhn 1992).

3.3
Effect of slope

Erosion would normally be expected to increase with increases in slope steepness and slope length as a result of respective increases in velocity and volume of surface runoff. Further, while on a flat surface raindrops splash soil particles randomly in all directions, on sloping ground more soil is splashed downslope than upslope, the proportion increasing as the slope steepens. The relationship between erosion and slope can be expressed by the equation:

$$E \propto \tan^m \theta L^n \qquad (3.10)$$

where E is soil loss per unit area, θ is the slope angle and L is slope length. Zingg (1940), in a study of data from five experimental stations of the United States Soil Conservation Service, found that the relationship had the form:

$$E \propto \tan^{1.4} \theta L^{0.6} \qquad (3.11)$$

To express E proportional to distance downslope, the value of n must be increased by 1.0. Since the values for the exponents have been confirmed in respect of m by Musgrave (1947) and m and n by Kirkby (1969b), there is some evidence to suggest that eqn 3.11 has general validity. Other studies, however, show that the values are sensitive to the interaction of other factors.

3.3.1 Exponents for slope steepness

Working with data from experimental stations in Zimbabwe, Hudson and Jackson (1959) found that m was close to 2.0 in value, indicating that the effect of slope is stronger under tropical conditions where rainfall is heavier. The effect of soil is illustrated by the laboratory experiments of Gabriels et al. (1975), who showed that m increases in value with the grain size of the material, from 0.6 for particles of 0.05 mm diameter to 1.7 for particles of 1.0 mm. The value also changes with slope, decreasing from 1.6 on slopes of 0–2.5° to 0.7 on slopes between 3 and 6.5°, to 0.4 on slopes over 6.5° (Horváth & Erődi 1962). On steeper slopes, the value may be expected to decrease further as soil-covered slopes give way to rock surfaces and soil supply becomes a limiting factor. In a detailed study of soil loss from 33 road-cut slopes on the Benin–Lagos Highway in Nigeria, Odermerho (1986) found values of $m = 1.09$ for slopes between 1.4 and 6°, 1.80 for slopes between 6 and 8.5°, −2.18 for slopes between 8.5 and 11° and −1.39 for slopes between 11 and 26.5°. Combining the results of these studies suggests a curvilinear relationship between soil loss and slope steepness, with erosion initially increasing rapidly as slope increases from gentle to moderate, reaching a maximum on slopes of about 8–10° and then decreasing with further increases in slope. Such a relationship would apply only to erosion by rainsplash and surface runoff; it would not apply to landslides, piping or gully erosion by pipe collapse.

The exponents in eqn 3.11 also vary in value with slope shape. D'Souza and Morgan (1976) obtained values of $m = 0.5$ on convex slopes, 0.4 on straight slopes and 0.14 on concave slopes. No studies have been made of the effect of changes of slope in plan, but Jackson (1984) found from erosion surveys and laboratory experiments that discharge varies with an index of contour curvature to the power of 5.5. If soil loss is assumed to vary with the square of the discharge, the value of m becomes 3.5. Contour curvature is here defined as the proportion of a circle centred on a point on a hillside that lies at a higher altitude than that point. The index ranges from 0 to 1 in value with values <0.5 indicating diverging slopes, a value of 0.5 a straight slope and values >0.5 converging slopes.

Few studies have looked at the effect of variations in plant cover. Quinn et al. (1980) investigated the change in the value of m for soil loss from 1.2 m long plots with slopes of 5–30° under simulated rainfall in relation to decreasing grass cover brought about by human trampling. They found that m was 0.7 for fully grassed slopes, rose to 1.9 in the early stages of trampling and fell to 1.1 when only about 25 per cent of the grass cover remained. Lal (1976) obtained a value of $m = 1.1$ for both bare fallow and maize on erosion plots in Nigeria; use of a mulch, however, reduced m to 0.5.

The exponent values also vary with the process of erosion: $m = 1.0$ for soil creep, ranges between 1.0 and 2.0 for splash erosion and between 1.3 and 2.0 for erosion by overland flow, and may be as high as 3.0 for rivers (Kirkby 1971). It was shown in Chapter 2 that m was about 0.3–1.0 in value for rainsplash and about 0.7 and 1.7–2.0 respectively for detachment and transport of soil particles by overland flow.

Increases in slope steepness may also cause an increase in the intensity of wind erosion on windward slopes and on the crests of knolls. Data from Chepil et al. (1964) and Stredňanský (1977) show that $m = 0.4$ for slopes up to 2° and 1.2 for slopes from 2 to 15°. The increase in value is attributed to increases in wind speed, shear and turbulence close to the ground as the air moves upslope (Livingstone and Warren, 1996).

3.3.2 Exponents for slope length

The value of 0.6 for exponent n applies only to overland flow on slopes about 10–20 m long, with steepnesses greater than 3°. Wischmeier and Smith (1978) propose values of $n = 0.4$ for slopes of

3°, 0.3 for slopes of 2°, 0.2 for slopes of 1° and 0.1 for slopes of less than 1°. Kirkby (1971) suggests that $n = 0$ for soil creep and splash erosion, ranges between 0.3 and 0.7 for overland flow and rises to between 1.0 and 2.0 if rilling occurs. This implies that the value of n will vary with distance along a hillside as, for example, soil creep close to the summit gives way first to overland flow and then to rill flow. Without rills, n may become negative on slopes longer than about 10 m. The increasing depth of overland flow downslope protects the soil from raindrop impact so that, even though the transporting capacity of the overland flow increases, erosion becomes limited by the rate of detachment, which is decreasing with slope length (Gilley et al. 1985). Once rills form, soil loss will either increase with slope length (Meyer et al. 1975), particularly if the density of rills is very high, or decrease because, as the flow becomes concentrated, there is no longer sufficient flow on the interrill areas to remove all the material detached by rainsplash (Abrahams et al. 1991). Erosion may also decrease with increasing slope length if, as the slope steepens, the soil becomes less prone to crusting and infiltration rates remain higher than on the gentler-sloping land at the top of the slope (Poesen 1984). Similarly, if the slope declines in angle as length increases, soil loss may decrease as a result of deposition. Clearly, with such a great range of possible conditions, a single relationship between soil loss and slope length cannot exist.

3.4

Effect of plant cover

Vegetation acts as a protective layer or buffer between the atmosphere and the soil. The above-ground components, such as leaves and stems, absorb some of the energy of falling raindrops, running water and wind, so that less is directed at the soil, while the below-ground components, comprising the root system, contribute to the mechanical strength of the soil.

The importance of plant cover in reducing erosion is demonstrated by the mosquito gauze experiment of Hudson and Jackson (1959), in which soil loss was measured from two identical bare plots on a clay loam soil. Over one plot was suspended a fine wire gauze, which had the effect of breaking up the force of the raindrops, absorbing their impact and allowing the water to fall to the ground from a low height as a fine spray. The mean annual soil loss over a ten-year period was 126.6 t ha^{-1} for the open plot and 0.9 t ha^{-1} for the plot covered by gauze.

Although numerous measurements have been made of erosion under different plant covers for comparison with that from bare ground, there is little agreement on the nature of the relationship between soil loss and changes in the extent of cover. Elwell (1981) favoured an exponential decrease in soil loss with increasing percentage interception of rainfall energy and, therefore, increasing percentage cover. Such a relationship was suggested by Wischmeier (1975) as applicable to covers in direct contact with the soil surface and has been verified experimentally for crop residues (Laflen & Colvin 1981; Hussein & Laflen 1982) and grass covers (Lang & McCaffrey 1984; Morgan et al. 1997a). The relationship can be described by the equation:

$$SLR = e^{-j.PC} \qquad (3.12)$$

where SLR is the ratio between the soil loss with the plant cover and that from bare ground, PC is the percentage cover and j varies in value from 0.025 to 0.06, with 0.035 taken as typical (Fig. 3.2). Foster (1982) attributes the exponential form of the relationship for covers in proximity to the ground to the ponding of water behind the plant elements, which reduces the

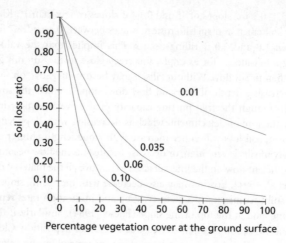

Fig. 3.2 Relationship between soil loss ratio (*SLR*) value and percentage vegetation cover at the ground surface. Curves represent different values of *j* in eqn 3.12.

effectiveness of the raindrop impact. For plant covers where the leaves and stems are not in contact with the soil but form a canopy at some height above the soil surface, the soil loss ratio is conventionally considered to reduce in a linear relationship with increasing percentage cover (Wischmeier & Smith 1978; Dissmeyer & Foster 1981) but more research is needed to confirm this. Overall, it is generally recognized that for adequate protection at least 70 per cent of the ground surface must be covered (Elwell & Stocking 1976) although reasonable protection can sometimes be achieved with between 30 and 40 per cent cover. The effects of vegetation, however, are far from straightforward and, under certain conditions, a plant cover can exacerbate erosion depending on how it interacts with the erosion processes.

3.4.1 Effect on rainfall

The effectiveness of a plant cover in reducing erosion by raindrop impact depends upon the height and continuity of the canopy, and the density of the ground cover. The height of the canopy is important because water drops falling from 7 m may attain over 90 per cent of their terminal velocity. Further, raindrops intercepted by the canopy may coalesce on the leaves to form larger drops, which are more erosive. Brandt (1989) showed that, for a wide range of plant types, leaf drips have a mean volume drop diameter between 4.5 and 4.9 mm, which is about twice that of natural rainfall. In contrast, Hall and Calder (1993) found that *Pinus caribaea* and *Eucalyptus camaldulensis* forest in southern India produced median volume drop diameters of only 2.3 and 2.8 mm, very similar to those of natural rainfall. For many types of vegetation canopy, raindrop-size distributions are characteristically bimodal with peaks around 2 and 4.8 mm, corresponding to the direct throughfall and the leaf drainage respectively. The effects of these changes in drop-size distribution have been studied mainly in relation to forest canopies. Chapman (1948) under pine forest in the USA, Wiersum et al. (1979) under *Acacia* forest in Indonesia, Mosley (1982) under beech forest in New Zealand and Vis (1986) under tropical rain forest in Colombia all show that while interception by the canopy reduces the volume of rain reaching the ground surface, it does not significantly alter its kinetic energy, which may even be increased compared with that in open ground. As a result, rates of soil particle detachment by rainsplash under the forest canopies can be between 1.2 and 3.1 times those in open ground, unless the soil surface is protected by a litter layer. It is possible that kinetic energy underplays the importance of leaf drips in contributing to soil detachment because it emphasizes drop velocity over drop size. Indices such as the square of the rainfall momentum (Styczen & Høgh-Schmidt 1988) or the product of momentum and drop diameter (Salles et al. 2000), which assign greater importance to drop size, may describe the process better.

Fewer investigations have been made to assess the effects of lower growing canopies. McGregor and Mutchler (1978) found that while cotton reduced the kinetic energy of rainfall by 95 per cent under the canopy and 75 per cent overall, it was locally increased between the rows where the leaf drips were concentrated. Armstrong and Mitchell (1987) found that the detachment under soya bean was about 94 per cent of that in the open despite the very low canopy height. Somewhat lower detachment rates were measured in the field by Morgan (1985b), who found that, with a 90 per cent cover, detachment under soya bean was 0.2 times that in open ground for a 100 mm h^{-1} rainfall intensity and 0.6 times for a 50 mm h^{-1} intensity. Finney (1984) showed in a laboratory study that detachment rates from leaf drip were 1.7 and 1.3 times those in open ground for Brussels sprouts and sugar beet at 23 and 16 per cent cover respectively. In another laboratory study under Brussels sprouts, Noble and Morgan (1983) found that the average detachment rate was the same as that in open ground. In a field study under maize with 88 per cent canopy cover at a height of 2 m, detachment was 14 times greater than that in open ground for a rainfall intensity of 100 mm h^{-1} and 2.4 times greater for an intensity of 50 mmh (Morgan 1985b).

In addition to modifying the drop-size distribution of the rainfall, a plant canopy changes its spatial distribution at the ground surface. Concentrations of water at leaf drip points can result in very high localized rainfall intensities, which can considerably exceed infiltration capacities and play an important role in the generation of runoff. Under mature soya bean, Amstrong and Mitchell (1987) found that half of the rainfall reaching the ground in a storm of 25 mm h^{-1} did so at intensities greater than those in open ground and that 10 per cent of the rain had an intensity of 385 mm h^{-1}. Stemflow also concentrates rain at the ground surface. De Ploey (1982) found that the effective intensity of stemflow beneath a canopy of tussocky grass was 150–200 per cent greater than the rainfall intensity in open ground. Herwitz (1986) recorded local stemflows of between 830 and 18,878 mm h^{-1} in a rainstorm of 118 mm h^{-1} over a six-minute period in a tropical rain forest in northern Queensland.

3.4.2 Effect on runoff

A plant cover dissipates the energy of running water by imparting roughness to the flow, thereby reducing its velocity. In most soil conservation work, the roughness is expressed as a value of Manning's n, which represents the summation of roughness imparted by the soil particles, surface microtopography (form roughness) and vegetation, acting independently of each other. Typical values of Manning's n are given in Table 3.6 (Petryk & Bosmajian 1975; Temple 1982; Engman 1986). The level of roughness with different forms of vegetation depends upon the morphology and the density of the plants, as well as their height in relation to the depth of flow. When the flow depth is shallow, as with overland flow, the vegetation stands relatively rigid and imparts a high degree of roughness, represented for grasses by n values of 0.25 to 0.3. As flow depths increase, the grass stems begin to oscillate, disturbing the flow and causing n values to increase to around 0.4. With further increases in flow depth, the vegetation is submerged; the plants tend to lie down in the flow and offer little resistance, so that n values decrease rapidly (Ree 1949).

Table 3.6 Guide values for Manning's *n*

Land use or cover	Manning's *n*
Bare soil	
Roughness depth <25 mm	0.010–0.030
Roughness depth 25–50 mm	0.014–0.033
Roughness depth 50–100 mm	0.023–0.038
Roughness depth >100 mm	0.045–0.049
Bermuda grass – sparse to good cover	
Very short (<50 mm)	0.015–0.040
Short (50–100 mm)	0.030–0.060
Medium (150–200 mm)	0.030–0.085
Long (250–600 mm)	0.040–0.150
Very long (>600 mm)	0.060–0.200
Bermuda grass – dense cover	0.300–0.480
Other dense sod-forming grasses	0.390–0.630
Dense bunch grasses	0.150
Kudzu	0.070–0.230
Lespedeza	0.100
Natural rangeland	0.100–0.320
Clipped rangeland	0.020–0.240
Wheat straw mulch	
2.5 t ha^{-1}	0.050–0.060
5.0 t ha^{-1}	0.075–0.150
7.5 t ha^{-1}	0.100–0.200
10.0 t ha^{-1}	0.130–0.250
Chopped maize stalks	
2.5 t ha^{-1}	0.012–0.050
5.0 t ha^{-1}	0.020–0.075
10.0 t ha^{-1}	0.023–0.130
Cotton	0.070–0.090
Wheat	0.100–0.300
Sorghum	0.040–0.110
Concrete or asphalt	0.010–0.013
Gravelled surface	0.012–0.030
Chisel-ploughed soil	
<0.6 t ha^{-1} residue	0.006–0.170
0.6–2.5 t ha^{-1} residue	0.070–0.340
2.5–7.5 t ha^{-1} residue	0.190–0.470
Disc-harrowed soil	
<0.6 t ha^{-1} residue	0.008–0.140
0.6–2.5 t ha^{-1} residue	0.100–0.250
2.5–7.5 t ha^{-1} residue	0.140–0.530
No tillage	
<0.6 t ha^{-1} residue	0.030–0.070
0.6–2.5 t ha^{-1} residue	0.010–0.130
2.5–7.5 t ha^{-1} residue	0.160–0.470
Bare mouldboard-ploughed soil	0.020–0.100
Bare soil tilled with coulter	0.050–0.130

Source: after Petryk and Bosmajian (1975), Temple (1982), Engman (1986).

Greatest reductions in velocity occur with dense, spatially uniform, vegetation covers. Clumpy, tussocky vegetation is less effective and may even lead to concentrations in flow with localized high velocities between the clumps. When flow separates around a clump of vegetation, the pressure exerted by the flow is higher on the upstream face than it is downstream, and eddying and turbulence occur immediately downstream of the vegetation. Vortex erosion is induced both upstream and downstream (Babaji 1987). Detailed observations during laboratory experiments on overland flow (De Ploey 1981) show that for slopes above about 8°, erosion under grass is higher than that from an identical plot without grass until the percentage grass cover reaches a critical value. Beyond this point, the grass has the expected protective effect.

3.4.3 Effect on air flow

Vegetation reduces the shear velocity of wind by imparting roughness to the air flow. It increases the roughness length, z_0, and raises the height of the mean aerodynamic surface by a distance, d, known as the zero plane displacement (Fig. 2.7). Estimates of d and z_0 can be obtained from the relationships:

$$D = HF \tag{3.13}$$

$$z_0 = 0.13(H - d) \tag{3.14}$$

where H is the average height of the roughness elements and F is the fraction of the total surface area covered by those elements (Abtew et al. 1989). From this, it follows that the key plant parameter is the lateral cover (L_c), defined as:

$$L_c = \frac{NS}{A} \tag{3.15}$$

where N is the number of roughness elements per unit area (A), and S is the mean frontal silhouette area of the plants, i.e. the cross-sectional area of the plant facing the wind (Musick & Gillette 1990). An increase in the value of L_c results in an exponential decrease in the proportion of the shear velocity of the wind exerted on the soil surface (Wolfe & Nickling 1996). This, in turn, causes an exponential decrease in sediment transport. Al-Awadhi and Willetts (1999) showed from wind tunnel experiments with cylinders that sediment transport levels off to very low levels when L_c exceeds 0.18 in value. When L_c reaches 0.5, sediment transport ceases (Gillette & Stockton 1989; Nickling & McKenna Neuman 1995). However, low densities of vegetation can sometimes increase the rate of erosion over bare ground through the development of turbulent eddies in the flow between individual plants (Logie 1982).

The effect of the vegetation can be described by a frictional drag coefficient (C_d) exerted by the plant layer in bulk and computed from:

$$C_d = \frac{2u_*^2}{u^2} \tag{3.16}$$

where u is the mean velocity measured at a height z, which equals 1.6 times the average height of the roughness elements. The coefficient generally decreases in value from about 0.1 in light winds to 0.01 in strong winds for a wide range of crops (Ripley & Redman 1976; Uchijima 1976; Morgan & Finney 1987) but both Randall (1969) in apple orchards and Bache (1986) with cotton canopies found that C_d could also increase with windspeed. When C_d exceeds 0.0104 in value, no regional scale wind erosion occurs (Lyles et al. 1974b).

Instead of considering these bulk drag coefficients, more insight can be gained by examining conditions close to the ground surface. Drag coefficients within the plant layer (C_d') can be calculated from:

$$C_d' = \frac{2u_*^2}{\int_0^h u^2 A(z)dz} \tag{3.17}$$

where h is the height of the vegetation, $A(z)$ is the leaf area per unit volume for the vegetation at height z and dz is the difference in height between z and the ground surface. For a wide range of crops, values of C_d' within the lowest 0.5 m of the plant layer decrease from about 0.1 in low windspeeds to about 0.001 in high windspeeds. However, when the wind is moderate to strong, consistent over time, and the crops are at an early stage of growth, the drag coefficient is found to increase with windspeed (Morgan & Finney 1987). This is probably due to the waving of the leaves in the wind, which disturbs the surrounding air, creating a wall effect that acts as a barrier to the air flow. The result is that the windspeed is reduced close to the ground surface but remains the same or even increases at the canopy level, thereby increasing the drag or shear velocity and enhancing the risk of erosion. The effect is particularly marked in crops of young sugar beet and onions. Similar increases in the drag coefficient with windspeed within a crop have been reported for maize (Wright & Brown 1967).

3.4.4 Effect on slope stability

It was shown in Chapter 2 that forest covers generally help to protect the land against mass movements partly through the cohesive effect of the tree roots. The fine roots, 1–20 mm in diameter, interact with the soil to form a composite material in which root fibres of relatively high tensile strength reinforce a soil matrix of lower tensile strength. In addition, soil strength is increased by the adhesion of soil particles to the roots. Roots can make significant contributions to the cohesion of a soil, even at low root densities and in materials of low shear strength. Increases in cohesion in forest soils due to roots can range from 1.0 to 17.5 kPa (Greenway 1987), although local variability in this may be as high as 30 per cent (Wu 1995). Grasses, legumes and small shrubs can reinforce a soil down to depths of 0.75–1.0 m and trees can enhance soil strength to depths of 3 m or more. The magnitude of the effect depends upon the angle at which tree roots cross the potential slip plane, being greatest for those at right angles, and whether the strain exerted on the slope is sufficient to mobilize fully the tensile strength of the roots. The effect is limited where roots fail by pull-out because of insufficient bonding with the soil, as can occur in stony materials, or where the soil is forced into compression instead of tension, as can occur at the bottom of a hillslope, and the roots fail by buckling.

Following observations on the forested slopes of the Serra do Mar, east of Santos, Brazil, De Ploey (1981) proposed that trees could sometimes induce landslides through an increase in loading (surcharge) brought about by their weight and an increase in infiltration which allows more water to penetrate the soil, lowering its shear strength. Bishop and Stevens (1964) showed that large trees can increase the normal stress on a slope by up to 5 kPa but that less than half of this contributes to an increase in shear stress and the remainder has the beneficial effect of increasing the frictional resistance of the soil. While, generally, surcharge enhances slope stability, under certain circumstances it can be detrimental. Trees planted only at the top of a slope can reduce stability, as can trees planted on steep slopes with shallow soils characterized by low angles of internal friction. In the Serra do Mar, the landslides occurred in a soil with an angle of internal friction of less than 20°, on slopes greater than 20°, after two days on which respectively 260 and 420 mm of rain fell.

A vegetation cover should theoretically contribute to slope stability as a result of evapo-transpiration producing a drier soil environment so that a higher intensity and longer duration rainfall are required to induce a slope failure compared with an unvegetated slope (Greenway 1987). Further, since soil moisture depletion can affect depths well below those reached by the roots, increases in slope stability should extend some 4–6 m below ground level. In practice, however, as found by Terwilliger (1990) in southern California, soil moisture levels after a few storms reach similar levels in both vegetated and unvegetated soils, so that under the conditions when the risk of mass movement is highest, the drying effect of vegetation is unlikely to play a role. Despite this, overall increases in the factor of safety arising from vegetation are generally in the range of 20–30 per cent (Greenway 1987; Wu 1995).

Box 3

Scale and erosion processes

Erosion processes and the factors that influence them vary according to the scale at which they are studied.

Micro-scale (mm² to 1 m²)

At this scale, erosion is controlled largely by the stability of the soil aggregates. Soil moisture, organic matter content and the activity of soil fauna, particularly earthworms, are major influences. Since aggregate breakdown is largely a result of raindrop impact, the frequency and erosivity of individual rainstorms control the rate of erosion through the rate of soil particle detachment. Soil type, slope and land cover are reasonably uniform over areas of this size so that differences in these can be used to demarcate different micro-scale units. For example, Cammeraat (2002) distinguished between areas of bare crusted soils and areas covered by tussocks of *Stipa tenacissima* when describing erosion processes at this scale in the Guadalentín Basin, Spain. In woodland in the Luxembourg Ardennes, he distinguished between bare soil areas where earthworms remove and digest the freshly fallen litter, and litter-covered areas with lower biotic activity.

Plot-scale (1 m² to 100 m²)

Erosion at the plot scale is controlled by the processes that generate surface runoff. These include the infiltration characteristics of the soil and changes in the microtopography of the surface related to crust development and surface roughness. The spatial distribution of crusted and uncrusted areas or vegetated and bare soil areas determines the locations of runoff and the patterns of flow and sediment movement over the soil surface. Depending on the size of the plot, interrill erosion will dominate, but, on steep slopes or in areas with highly erodible soils, the flow may exceed the critical conditions for rills to develop. It is at this scale that rock fragments can increase rates of soil erosion compared with bare ground if the surface between the stones becomes sealed (Poesen et al. 1994).

Field scale (100 m² to 10,000 m²)

At the field scale, there is usually a reasonably well defined spatial pattern of runoff pathways in locations such as swales and valley bottoms, separated by areas of either interrill erosion or no erosion. The extent of interrill erosion depends on the severity of the rainfall event so that the size of the area contributing

Continued

Factors influencing erosion

runoff is quite dynamic. The direction of runoff pathways is often controlled by tillage. There may be spatial variations in soil erodibility and slope within a field. The normally expected increases in runoff and erosion downslope can sometimes be offset by soils with greater infiltration rates at the bottom of the field, resulting in a decrease in runoff, or by gentler slopes, resulting in deposition of sediment.

Catchment scale (>10,000 m²)

Depending on the spatial distribution of soils, slope and land cover in a catchment, it is possible to recognize different types of process-domains (De Ploey 1989b):

I: areas dominated by rainsplash erosion and infiltration, typical of soils with high infiltration rates and high aggregate stability.
II: areas dominated by interrill erosion but with sediment deposition on the lower slopes.
III: areas dominated by rill erosion, again with deposition on the lower slopes.
IV: areas of ephemeral gullies, particularly along the valley floors.

The pathways of runoff and sediment movement through the catchment are influenced by the patterns of field boundaries, gateways, tracks and roads, as well as the natural topography. The effectiveness of these pathways and the spatial extent of the process-domains depend on the magnitude of the erosion event. In low magnitude events, erosion is generally limited to local slope wash, but with higher rainfall, runoff pathways develop over the whole hillside with local discharges into

the river; with more extreme events, overland flow and slope wash may be widespread, with many points of discharge into the river channel (Święchowicz 2002). Thus, in order to understand the sources of sediment and associated pollutants in water bodies, the dynamic nature of the process-domains and their connectivity must be investigated.

The way in which the effectiveness of erosion events depends on the scale of analysis was demonstrated by Carver and Schreier (1995) for the Jhikhu Khola watershed, Nepal (Table B3.1). They examined three storms, one of 49.5 mm occurring in the pre-monsoon period, one of 35.8 mm during the monsoon and an extreme event of 90.6 mm in the transition period just before the onset of the monsoon season. On a 70 m² area of a terraced hillslope, erosion in the pre-monsoon storm when the cultivated land was bare was higher than that from the monsoon storm when the land was protected by a crop cover; the extreme event was only half as damaging as the pre-monsoon event. In a sub-watershed of 540 ha, the pre-monsoon and monsoon storms had less overall effect but the extreme event became very important as a result of erosion of the stream bed. At the scale of the whole watershed, 11,141 ha, the differences between the pre-monsoon and monsoon periods disappeared and the effect of the extreme event was much reduced because the heavy rain did not extend beyond the area of the sub-watershed. Thus, what was an extreme event at a small watershed scale was no larger than the annual event at the larger watershed scale.

Table B3.1 Effectiveness of erosion events $(t\,ha^{-1})$ with changes in scale

Season	Rainfall total (mm)	Peak rainfall intensity (mm h⁻¹)	Terraced hillside (70 m²)	Sub-watershed (540 ha)	Jhikhu Khola watershed (11,141 ha)
Pre-monsoon	49.5	109	20	2	0.1
Monsoon	35.8	63	0.02	0.4	0.1
Extreme event	90.6	103	10	40	2

Source: after Carver and Schreier (1995).

Chapter 3

CHAPTER 4

Erosion hazard assessment

Erosion hazard assessment aims to identify those areas of land where the maximum sustained productivity from a given land use is threatened by excessive soil loss or the off-site damage arising from erosion is unacceptable. A distinction is made between potential erosion risk, reflecting the local conditions of soil, climate and slope, and actual erosion risk, which additionally takes account of land cover. It is therefore possible to recognize areas of high potential risk but low actual risk as a result of the protection afforded by vegetation.

4.1
Generalized assessments

4.1.1 Erosivity indices

Erosivity data can be used as an indicator of regional variations in erosion potential. Stocking and Elwell (1976) present a generalized picture of erosion risk in Zimbabwe, based on mean annual erosivity values, showing that high-risk areas are in the Eastern Districts, the region east of Masvingo and the High and Middle Veld north and east of Harare (Fig. 4.1). The area south and west of Bulawayo, by contrast, has a much lower risk. Temporal variations in erosion risk are revealed by the mean monthly erosivity values. Hudson (1981; Fig. 4.2) contrasts the erosivity patterns of Bulawayo and Harare. At Bulawayo, erosivity is low at the beginning of the wet season and increases as the season progresses. By the time the maximum values are experienced, the plant cover has had a chance to become established, giving protection against erosion and lowering the risk. At Harare, however, erosivity is highest at the time of minimal vegetation cover. The seasonal pattern of erosivity in Kenya was analysed in by Rowntree (1983) for the Katumani Research Station, near Machakos. The mean annual erosivity of $164,352\,J\,mm\,m^{-2}h^{-1}$ is much higher than the values recorded in Zimbabwe. The highest mean monthly value, $69,688\,J\,mm\,m^{-2}h^{-1}$, occurs in April, the beginning of the period when the potential soil moisture deficit is severe and the vegetation cover poorly established. Cultivation of the land at this time creates a great risk of erosion. May has a much lower erosivity, only $20,961\,J\,mm\,m^{-2}h^{-1}$, and is also a month with good vegetation cover. Erosivity is very low during the summer with mean monthly values of less than 1000 but rises again in the short rainy season, reaching 15,306 in November before falling to mean monthly values of about 10,000 during the winter.

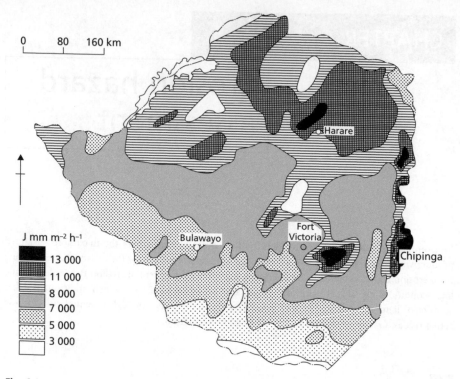

Fig. 4.1 Mean annual erosivity in Zimbabwe (after Stocking & Elwell 1976).

Maps of erosivity using the rainfall erosion index, R, have been produced for the USA (Wischmeier & Smith 1978). Originally R was calculated as mean annual $EI_{30}/100$ with E being the rainfall energy in foot-tons per acre and I_{30} being the maximum 30-minute rainfall intensity in inches per hour; this can be described as an R value in American units. With E calculated in $MJ\,ha^{-1}$ and I_{30} in $mm\,h^{-1}$, an R can be derived in metric units as mean annual $EI_{30}/100$. The conversion is 1 unit of metric $R = 17.3$ units of American R. Since maps for many parts of the world exist in both units, care should be taken to ensure which one is being used. The metric version of the map for the USA is shown in Fig. 4.3. Values range from about 50 in the arid west to over 1000 along the Gulf Coast. These are rather low compared to those of more tropical climates. Values in India range from about 250 in western Rajasthan to over 1250 along the coasts of Maharashtra and Karnataka (Singh et al. 1981; Fig. 4.4). Those in Zambia range from 500 in the south to about 800–900 in the Copper Belt and over 1000 in the western part of the Northern Province (Lenvain et al. 1988). In South Africa (Smithen & Schulze 1982) values are lowest in the southwest of Cape Province where they are about 50–100 and rise eastwards reaching over 400 in eastern Transvaal and the midlands of Natal (Fig. 4.5). Very low values are characteristic of northern Europe. Those in France range from 50 in the north and west to 340 in the southern Massif Central and Languedoc (Pihan 1979); those in Belgium and the Netherlands from 60 to 75 but rising to 100–150 in the Ardennes (Laurant & Bollinne 1978).

In many countries insufficient rainfall records from autographic gauges are available to calculate erosivity nationwide. In such cases, an attempt is made to find a more widely available

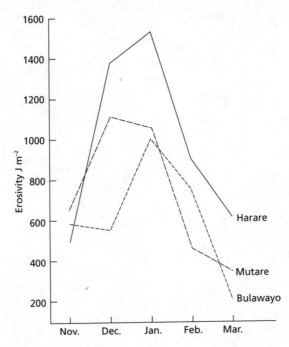

Fig. 4.2 Mean monthly erosivity for three towns in Zimbabwe (after Hudson 1965).

Fig. 4.3 Mean annual values of the rainfall erosion index R ($10\,MJ\,ha^{-1}$) in the USA (after Wischmeier & Smith 1978). Divide values by 1.73 to obtain R values in original American units (10^2 foot-tons per acre).

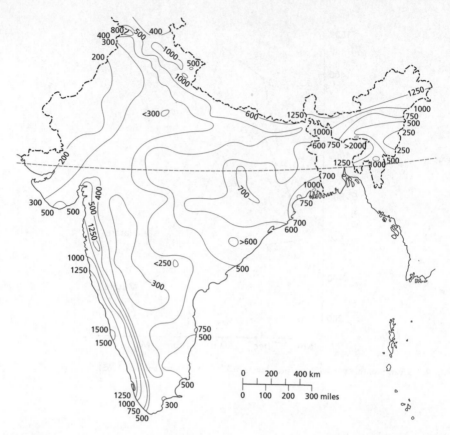

Fig. 4.4 Mean annual values of the rainfall erosion index R ($10\,MJ\,ha^{-1}$) in India (after Singh et al. 1981).

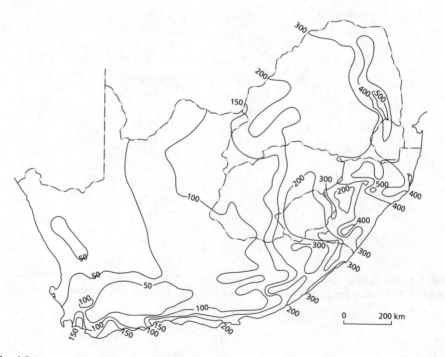

Fig. 4.5 Mean annual values of the rainfall erosion index R ($10\,MJ\,ha^{-1}$) in southern Africa (after Smithen & Schulze 1982).

rainfall parameter which significantly correlates with erosivity and from which erosivity values might be predicted using a best-fit regression equation. Considerable caution should be taken when estimating erosivity in this way because the results are only valid if the rainfall parameters used are themselves significantly correlated with soil loss; otherwise the exercise is pointless (Hudson 1981). Roose (1975) found that in the Ivory Coast and Burkina Faso, the mean annual R value (in American units) could be approximated by the mean annual rainfall total (mm) multiplied by 0.5. This relationship was used to produce a map of R values for west Africa. The map of mean annual erosivity for Peninsular Malaysia (Fig. 4.6; Morgan 1974) is based on a

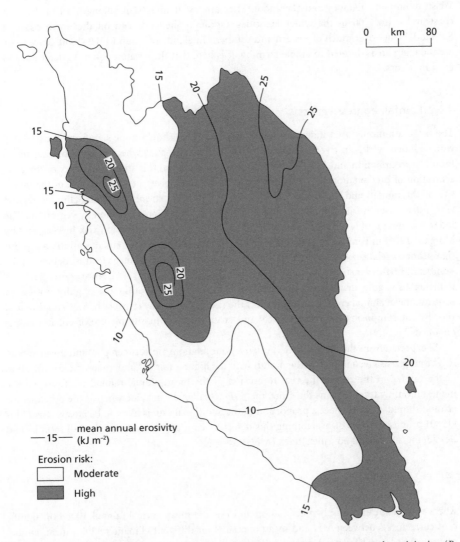

Fig. 4.6 Mean annual erosivity (*EV*) in Peninsular Malaysia estimated from mean annual precipitation (*P*; mm) using the relationship $EV = 9.28\,P - 8838.15$ (after Morgan 1974).

Erosion hazard assessment

relationship with mean annual rainfall. It should be stressed that extrapolating such relationships beyond the data base from which they have been derived in order to apply them elsewhere is dangerous. For example, applying the equation developed for Malaysia to mean annual rainfall totals below 900 mm yields estimates of erosivity that are obviously nonsense. It should also be noted that the results of different researchers are not always comparable because of assumptions made when calculating the R value. The relationship developed between mean annual R and mean annual precipitation by Bollinne et al. (1979) for Belgium is based on rainstorms greater than 1.27 mm, whereas that proposed by Rogler and Schwertmann (1981) for Bavaria, Germany, considers storms only with rainfall greater than 10 mm and I_{30} values greater than 10 mm h^{-1}.

Erosion risk in Great Britain was assessed using the $KE > 10$ index. The annual values are rather low (Morgan 1980b; Fig. 4.7), rising above 1400 J m^{-2} only in parts of the Pennines, the Welsh mountains, Exmoor and Dartmoor. They are less than 900 J m^{-2} along most of the west coast and below 700 in the Outer Hebrides, Orkneys, Shetlands and on the north coast of Scotland. Values over much of eastern and southern England are around 1100–1300. Since these are the main areas devoted to arable farming, it is here that the greatest risk of agricultural soil erosion occurs.

4.1.2 Rainfall aggressiveness

The most commonly used index of rainfall aggressiveness, shown to be significantly correlated with sediment yields in rivers (Fournier 1960), is the ratio p^2/P, where p is the highest mean monthly precipitation and P is the mean annual precipitation. It is strictly an index of the concentration of precipitation into a single month and thereby gives a crude measure of the intensity of the rainfall and, in so far as a high value denotes a strongly seasonal climatic regime with a dry season during which the plant cover decays, of erosion protection by vegetation. The index was used by Low (1967) to investigate regional variations in erosion risk in Peru and by Morgan (1976) in Peninsular Malaysia (Fig. 4.8). Using data from 680 rainfall stations, a low but significant correlation was obtained in Malaysia between p^2/P and drainage texture, defined as the number of first-order streams per unit area ($r = 0.38$; $n = 39$; Morgan 1976). Since drainage texture is analogous to gully density, p^2/P may be regarded as an indicator of the risk of gully erosion. In contrast, mean annual erosivity values reflect the risk of erosion by rainsplash, overland flow and rills. By superimposing the maps of p^2/P and erosivity, a composite picture of erosion risk is obtained (Fig. 4.9).

As expected from the above, there is often a poor relationship between p^2/P and mean annual R. The emphasis given in p^2/P to the month with the highest rainfall underplays the contribution of the rainfall in the rest of the year to erosion. If the mean annual rainfall increases but the highest monthly total remains the same, the p^2/P actually falls in value whereas the potential for erosion should increase, since a proportion of the additional rain is likely to be erosive. Arnoldus (1980) proposed a way of overcoming this defect by considering the rainfall of all months and developing a modified Fournier Index (MFI):

$$MFI = \sum_{1}^{12} \frac{p^2}{P} \qquad (4.1)$$

where p is the mean monthly precipitation and P is the mean annual rainfall. Based on significant correlations between MFI and mean annual R for different climatic regions, mean annual erosivity maps have been produced for the Middle East and Africa north of the equator (Arnoldus 1980) and for 16 countries of the European Union (Gabriels 2002; Fig. 4.10).

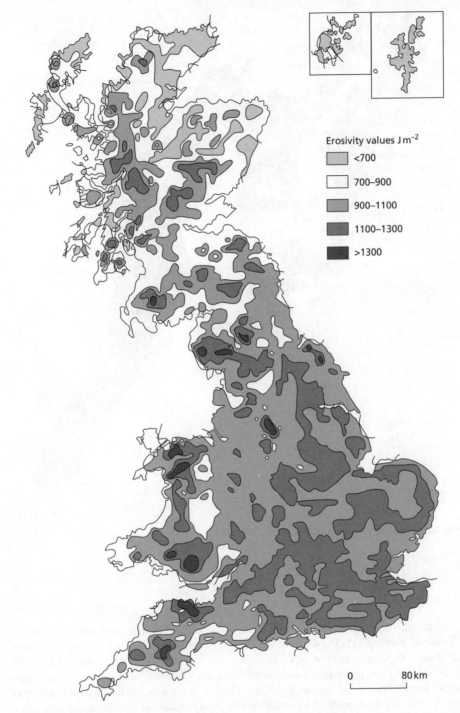

Fig. 4.7 Mean annual erosivity (*KE* > 10) in Great Britain (after Morgan 1980b).

Erosivity values J m^{-2}

- <700
- 700–900
- 900–1100
- 1100–1300
- >1300

0   80 km

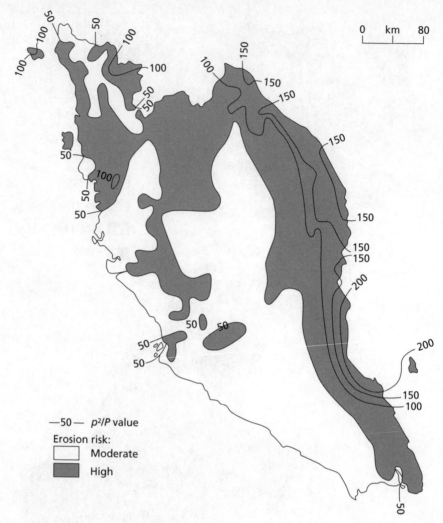

Fig. 4.8 Values of p^2/P in Peninsular Malaysia.

4.1.3 Factorial scoring

A simple scoring system for rating erosion risk was devised by Stocking and Elwell (1973b) for Zimbabwe. Taking a 1:1,000,000 base map, the country was divided on a grid system into units of 184 km². Each unit was rated on a scale from 1 (low risk) to 5 (high risk) in respect of erosivity, erodibility, slope, ground cover and human occupation, the latter taking account of the density of the population and the type of settlement. The five factor scores were summed to give total score, which was then compared with an arbitrarily chosen classification system to categorize areas of low, moderate and high erosion risk. The scores were mapped and areas of similar risk delineated (Fig. 4.11).

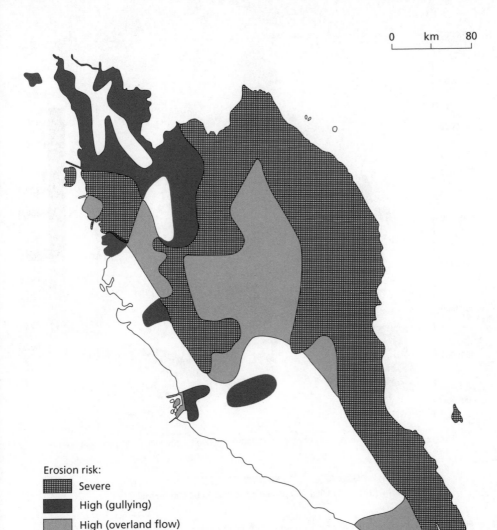

Fig. 4.9 Reconnaissance survey of soil erosion risk in Peninsular Malaysia.

Several problems are associated with the technique. First, the classification may be sensitive to different scoring systems. For example, the use of different slope groups may yield different assessments of the degree of erosion risk. Second, each factor is treated independently, whereas there is often interaction between the factors. Slope steepness may be much more important in areas of high than in areas of low erosivity. Third, the factors are combined by addition. There is no reason why this should be a more appropriate method of combining them than multiplication, although multiplication often results in the score for one factor dominating the total score and,

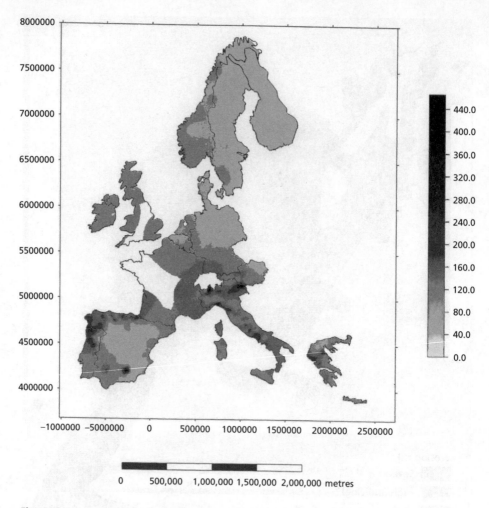

Fig. 4.10 Mean annual erosivity for 16 countries of the European Union based on the modified Fournier index for rainfall stations with ten years or more of monthly rainfall records (after Gabriels 2002). Values under 100 are considered low and values over 300 high. No data were available for much of France and central England.

for that reason, is difficult to use with zero values in the scoring system. Fourth, each factor is given equal weight. Despite these difficulties, the technique is easy to use and has the advantage that factors which cannot be easily quantified in any other way can be readily included. When used carefully, factorial scoring can provide a general appreciation of erosion risk and indicate vulnerable areas where more detailed assessments should be made. A system based on the susceptibility of soils to crusting (four classes), the shear strength of the soil (three classes), land cover (nine classes) and rainfall erosivity (four classes) gives good correlations for the cultivated areas of France between erosion risk and the spatial frequency of muddy floods (Le Bissonnais et al. 2002).

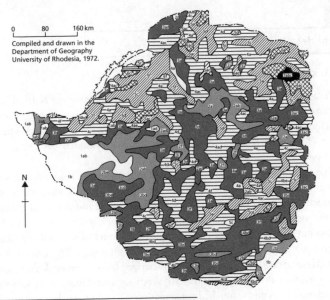

EROSION RISK

Major groups		Factor score	Subgroups according to dominant factors
1	Very low	9–10	
2	Low	11–12	a Erosivity
3	Below average	13–14	b Cover
4	Average	15–16	c Slope
	Above average	17–18	d Erodibility
	High	19–20	e Human occupation
7	Very high	21	f No dominant factor

Categories		Erosivity (J mm m^{-2} h^{-1})	Cover (mm of rainfall) and basal cover est. (%)	Slope (degrees)	Erodibility	Human occupation*
Low	I	below 5000	above 1000 7–10	0–2	orthoferralitic regosoils	Extensive large scale commercial ranching National Parks or Unreserved
Below average	II	5000–7000	800–1000 5–8	2–4	paraferralitic	Large scale commercial farms
Average	III	7000–9000	600–800 3–6	4–6	fersiallitic	Low density CLs (below 5 p.p.km^2) and SCCF
Above average	IV	9000– 11 000	400–600 1–4	6–8	siallitic vertisoils lithosoils	Moderately settled CLs (5–30 p.p.km^2)
High	V	above 11 000	below 400 0–2	above 8	non-calcic hydromorphic sodic	Densely settled CLs (above 30 p.p.km^2)

(*Notes:* Cover, Erodibility and Human occupation are only tentative and cannot as yet be expressed on a firm quantitative basis)
*p.p.km^2 = persons per square kilometre   CL = Communal Lands   SCCF = Small Scale Commercial Farms

Fig. 4.11 Erosion survey of Zimbabwe (after Stocking & Elwell 1973b).

4.2.1 Land capability classification

Land capability classification was developed by the United States Natural Resources Conservation Service as a method of assessing the extent to which limitations such as erosion, soil depth, wetness and climate hinder the agricultural use that can be made of the land. The United States classification (Klingebiel & Montgomery 1966) has been adapted for use in many other countries (Hudson 1981).

The objective of the classification is to divide an area of land into units with similar kinds and degrees of limitation. The basic unit is the capability unit. This consists of a group of soil types of sufficiently similar conditions of profile form, slope and degree of erosion to make them suitable for similar crops and warrant the use of similar conservation measures. The capability units are combined into sub-classes according to the nature of the limiting factor and these, in turn, are grouped into classes based on the degree of limitation. The United States system recognizes eight classes arranged from Class I, characterized by no or very slight risk of damage to the land when used for cultivation, to Class VIII, very rough land that can be safely used only for wildlife, limited recreation and watershed conservation. The first four classes are designated as suitable for arable farming (Table 4.1). Assigning a tract of land to its appropriate class is aided by the use of a flow diagram (Fig. 4.12). The value of land capability assessment lies in identifying the risks attached to cultivating the land and in indicating the soil conservation measures that are required. The classification can be improved by making the conservation recommendations more specific, as is the case with the treatment-oriented scheme developed in Taiwan and tested on hilly land in Jamaica (Sheng 1972a; Table 4.2, Fig. 4.13).

Although the inclusion of many soil properties in the classification may seem to render it useful for land planning generally, it must be appreciated that, as befits a classification evolved in the wake of the erosion scare in the United States in the 1930s, its bias is towards soil conservation. Attempts to use the classification in a wider sphere have only drawn attention to its limitations. The classification does not specify the suitability of land for particular crops; a separate land suitability classification has been devised to do this (FAO 1976). The assigning of a capability class is not an indicator of land value, which may reflect the scarcity of a certain type of land, nor is it a measure of whether a farmer can make a profit, which is much influenced by the market prices of the crops grown and the farmer's skill. The stress laid by the classification on arable farming can also be a disadvantage. Insufficient attention is given to the recreational use of land. The capability classification implies that land is set aside for recreation only when it is too marginal for arable or pastoral farming, but such land is often marginal for recreational use too. This fact highlights the difficulty of incorporating agricultural and non-agricultural activities in a single classification. One approach to this problem is provided by the Canada Land Inventory (McCormack 1971), which employs four separate classifications covering agriculture, forestry, recreation and wildlife. In the UK, separate classifications exist for evaluating the suitability of land for picnic sites and camping sites (George & Jarvis 1979).

Information contained in land capability surveys can be combined with that on erosivity to give a more detailed assessment of erosion risk. A map was produced for England and Wales (Fig. 4.14) by combining data on land capability class (Soil Survey of England and Wales 1979), rainfall erosivity and wind velocity with knowledge of the susceptibility of soils to erosion (Soil Survey of England and Wales 1983). The inclusion of land capability in the assessment procedure

Table 4.1 Land capability classes (United States system)

Class	Characteristics and recommended land use
I	Deep, productive soils easily worked, on nearly level land; not subject to overland flow; no or slight risk of damage when cultivated; use of fertilizers and lime, cover crops, crop rotations required to maintain soil fertility and soil structure.
II	Productive soils on gentle slopes; moderate depth; subject to occasional overland flow; may require drainage; moderate risk of damage when cultivated; use of crop rotations, water-control systems or special tillage practices to control erosion.
III	Soils of moderate fertility on moderately steep slopes, subject to more severe erosion; subject to severe risk of damage but can be used for crops provided plant cover is maintained; hay or other sod crops should be grown instead of row crops.
IV	Good soils on steep slopes, subject to severe erosion; very severe risk of damage but may be cultivated if handled with great care; keep in hay or pasture but a grain crop may be grown once in five or six years.
V	Land is too wet or stony for cultivation but of nearly level slope; subject to only slight erosion if properly managed; should be used for pasture or forestry but grazing should be regulated to prevent plant cover being destroyed.
VI	Shallow soils on steep slopes; use for grazing and forestry; grazing should be regulated to preserve plant cover; if plant cover is destroyed, use should be restricted until land cover is re-established.
VII	Steep, rough, eroded land with shallow soils; also includes droughty or swampy land; severe risk of damage even when used for pasture or forestry; strict grazing or forest management must be applied.
VIII	Very rough land; not suitable even for woodland or grazing; reserve for wildlife, recreation or watershed conservation.

Classes I–IV denote soils suitable for cultivation.
Classes V–VIII denote soils unsuitable for cultivation.

Table 4.2 Treatment-oriented land capability classification

Group	Class	Characteristics and recommended treatments
Suitable for tillage	C1	Up to 7° slope; soil depth normally over 10 cm; contour cultivation; strip cropping; broad-based terraces.
	C2	Slopes 7–15°; soil depth over 20 cm; bench terracing (construction by bulldozers); use of four-wheel tractors.
	C3	Slopes 15–20°; soil depth over 20 cm; bench terracing on deep soil (construction by small machines); silt-pits on shallower soils; use of small tractors or walking tractors.
	C4	Slopes 20–25°; soil depth over 50 cm; bench terracing and farming operations by hand labour.
	P	Slopes 0–25°; soil depth too shallow for cultivation; use for improved pasture or rotational grazing system; zero grazing where land is wet.
	FT	Slopes 25–30°; soil depth over 50 cm; use for tree crops with bench terracing; inter-terraced areas in permanent grass; use contour planting; diversion ditches; mulching.
	F	Slopes over 30° or over 25° where soil is too shallow for tree crops; maintain as forest land.
Wet land, liable to flood; also stony land	P	Slopes 0–25°; use as pasture.
	F	Slopes over 25°; use as forest.
Gullied land	F	Maintain as forest land.

The scheme is most suitable for hilly lands in the tropics.
Source: after Sheng (1972a).

Land Capability Class						
I	**II**		**III**		**IV**	
Permissible slope						
0°–1°	0°–1°	1°–2.5°	0°–2.5°	2.5°–4.5°	0°–4.5°	4.5°–7°
Minimum effective depth (Texture here refers to average textures)						
1 m of Cl or heavier	50 cm of Sal or heavier	50 cm of Sacl or heavier	50 cm of S or LS 25 cm of Sal or heavier	(a) 50 cm of Sacl (b) 25 cm of Cl or heavier	25 cm of any texture	25 cm of Sacl or heavier
Texture of surface soil						
Cl or heavier	Sal or heavier S, or Ls if upper subsoil is Sal or heavier	Sal or heavier	No direct limitations	(a) Sal or heavier (b) Cl or heavier	No direct limitations	Sal or heavier
Permeability 5 or 4 to at least:						
1 m	50 cm	50 cm	No direct limitations	No direct limitations	No direct limitations	
Not worse than 3 to:						
	1 m	1 m	1 m or 50 cm if average texture is Cl is heavier	1 m		
Physical characteristics of the surface soil: permissible symbols						
Not permitted	t1	t1	t1 and t2	t1 and t2	t1 and t2	t1 and t2
Erosion: permissible symbols						
1	1 and 2	1 and 2	1,2 and 3	1,2 and 3	1,2 and 3	1,2 and 3
Wetness criteria: permissible symbols						
Not permitted	w1	w1	w1	w1	w1 and w2	w1 and w2

S   = Sand
Sal = Sandy loam
Ls = Loamy sand
Cl = Clay loam
Sacl = Sandy clay loam

Fig. 4.12 Criteria and flow chart for dermining land capability class according to the Department of Conservation and Extension, Zimbabwe (after Hudson 1981).

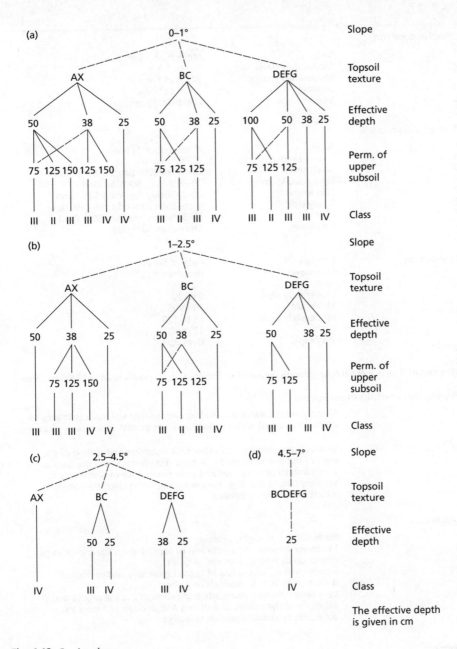

Fig. 4.12 *Continued*

Effective depth (m)

1	Deep	More than 1.5 m
2	Moderately deep	1 m to 1.5 m
3	Moderately shallow	50 cm to 1 m
4	Shallow	25 cm to 50 cm
5	Very shallow	Less than 25 cm

Texture of surface soil

A	Sand	More than 85% sand
X	Loamy sand	80–85% sand
B	Sandy loam	Less than 20% clay; 50–80% sand
C	Sandy clay loam	20–30% clay; 50–80% sand
D	Clay loam	20–30% clay; less than 50% sand
E	Sandy clay	More than 30% clay; 50–70% sand
F	Clay	30–50% clay; less than 50% sand
G	Heavy clay	More than 50% clay

Permeability	Description	Rate of flow*
	Very slow	Less than 1.25
	Slow	1.25 to 5
	Moderately slow	5 to 20
	Moderate	20 to 65
	Moderately rapid	65 to 125
	Rapid	125 to 250
	Very rapid	Over 250

*The rate of flow in mm per hour through saturated undisturbed cores under a head of 12.5 mm of water.

Physical characteristics of surface soil

t1   Slightly unfavourable physical conditions. The soil has a tendency to compact and seal at the surface and a good tilth is not easily obtained.

t2   Unfavourable physical conditions. Compaction and sealing of the surface soil are more severe. A hard crust forms when the bare soil is exposed to rain and sun and poor emergence of seedlings can severely reduce the crop. On ploughing, large clods are turned up which are not easily broken.

Erosion

1   No apparent, or slight, erosion.

2   Moderate erosion: moderate loss of topsoil generally and/or some dissection by runoff channels or gullies.

3   Severe erosion, severe loss of topsoil generally and/or marked dissection by runoff channels or gullies.

4   Very severe erosion: complete truncation of the soil profile and exposure of the subsoil (B horizon) and/or deep and intricate dissection by runoff channels or gullies.

Wetness

w1   Wet for relatively short and infrequent periods.

w2   Frequently wet for considerable periods.

w3   Very wet for most of the season.

Fig. 4.12 *Continued*

Additional Requirements

*Factors affecting cultivation

g		Gravelly or stony
b	Downgrade Class I to II	Very gravelly or stony
o		Bouldery
s		Very bouldery
v	Class VI	Outcrops
r		Extensive outcrops

*Permeability

75 to 500;
otherwise Class IV
Not applicable to
basalts or norites

*Erosion

Class: I: 1
        II: 1; 2
        III: 1; 2; 3

*'t' Factors

Class: II: t1
        III: t1; t2
        IV: t1; t2

*Wetness

Class: II: $w1$
        III: $w1$
        IV: $w2$
        V: $w3$

$w2$ downgrades Class II and III to IVw unless the land is already Class IV on code, in which case it remains as Class IV.

*Note

Any land not meeting the minimum requirements shown on this sheet is Class VI.

Fig. 4.12 *Continued*

Slope Soil Depth	1 Gentle sloping <7°	2 Moderate sloping 7–15°	3 Strongly sloping 15–20°	4 Very strongly sloping 20–25°	5 Steep 25–30°	6 Very steep >30°
Deep (D) >36 inches (>90 cm)	C1	C2	C3	C4	FT	F
Moderately deep (MD) 20–36 inches (50–90 cm)	C1	C2	C3	C4 / P	FT / F	F
Shallow (S) 8–20 inches (20–50 cm)	C1	C2 / P	C3 / P	P	F	F
Very shallow (VS) <8 inches (<20 cm)	C1 / P	P	P	P	F	F

Fig. 4.13 Chart for determining land capability class according to the treatment-oriented scheme of Sheng (1972a; Table 4.2).

means that the map indicates areas with a risk of erosion when land is used in accordance with its capability rating. Erosion in these areas can then be attributed to mismanagement of the land, whereas that which takes place on land being used for an activity not in accordance with the land capability classification can be attributed to misuse of the land.

A system of land capability classification was devised by the Soil Conservation Service of New South Wales, Australia, for the planning of urban land use with particular reference to erosion control (Hannam & Hicks 1980). An example of its application to a small area at Guerilla Bay on the south coast of New South Wales is presented in Fig. 4.15. A map of landform regions is produced by combining information on slope, divided into seven classes, and topographic position, here called terrain, divided into six classes. The potential hazards related to urban land use are tabulated for each region. This information is combined with data on soils, paying attention to erosion hazard and limitations to urban development, to produce a map showing urban capability using five classes. A suitable form of land use is then determined for each area consistent with its capability class, its physical limitations and the degree of disturbance it can sustain without causing excessive erosion and sedimentation.

4.2.2 Land systems classification

Land systems analysis is used to compile information on the physical environment for the purpose of resource evaluation (Cooke & Doornkamp 1974; Young 1976). The land is classified into areal units, termed land systems, which are made up of smaller units, land facets, arranged in a clearly recurring pattern. Since land facets are defined by their uniformity of landform, especially slope, soils and plant community, land systems comprise an assemblage of landform, soil and vegetation types. In a study of 2000 km[2] of the Middle Veld in Swaziland, Morgan et al. (1997b) defined erosion classes based on gully densities mapped from 1:30,000 scale panchromatic aerial pho-

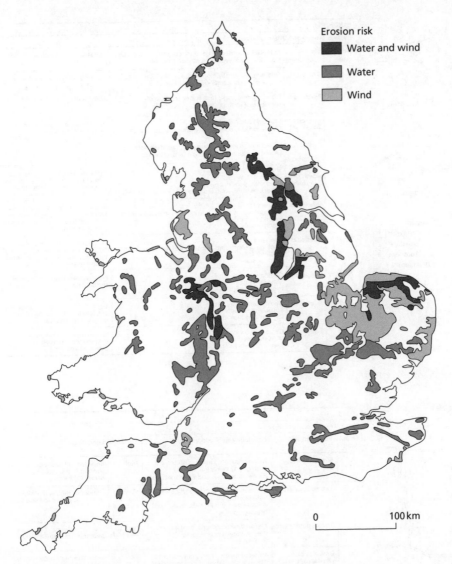

Erosion risk
- Water and wind
- Water
- Wind

0   100 km

Fig. 4.14 Areas susceptible to soil erosion in England and Wales (after Morgan 1985a).

tography. They found that the land systems previously identified by Murdoch et al. (1971) were significantly different ($p < 0.05$) from each other in terms of their frequency distributions of erosion classes for 1972 and 1990 and in the rate of change in severity between the two dates. In a survey of erosion along a pipeline corridor in Georgia, 90 sites were analysed in the field and classified according to erosion severity using a modified version of the Universal Soil Loss Equation and the descriptors of erosion classes given in Table 4.3. When these were compared to the land systems along the right-of-way, it was found that five out of the six comparisons were

LANDFORM

Slope
0–1% 1
1–5% 2
5–10% 3
10–15% 4
15–20% 5
20–25% 6
25–30% 7

Terrain
Hillcrest 1
Sideslope 2
Footslope 3
Drainage plan 5
Swamp 6

LANDFORM

Physical Criteria and Urban Landuse

Slope class	Terrain component	Potential hazards related to topographic location and slope and which will affect urban landuse	Suitable urban landuse
0–5%	Drainage plain	Flooding, seasonally high water-tables, high shrink-swell soils, high erosion hazard	Drainage reserves/stormwater disposal
	Floodplain	Flooding, seasonally high water-tables, high shrink-swell soils, saline soils, gravelly soils	Open space areas, playing fields
	Hillcrests Sideslopes	Shallow soils, stony/gravelly soils Overland flow, poor surface drainage and profile damage	Residential: all types of recreation; large-scale industrial, commercial and institutional development
	Footslopes	Impedence in lower terrain positions, deep soils Others: swelling soils, erodible soils, dispersible soils	
5–10%	Hillcrests Sideslopes Footslopes	Shallow soils Overland flow Deep soils, poor drainage Others: swelling soils, erodible soils, dispersible soils	Residential subdivisions, detached housing, medium-density housing/ unit complexes, modular industrial, active recreational pursuits
10–15%	Sideslopes	Overland flow Geological constraints, possibility of mass movement Swelling soils Erodible soils	Residential subdivisions, detached housing, medium-density housing/ unit complexes, modular industrial, passive recreational
15–20%	Sideslopes	Overland flow Geological constraints, possibility of mass movement Swelling soils Erodible soils	Residential subdivisions, detached housing, medium-density housing/ unit complexes, modular industrial, passive recreational
20–25%	Sideslopes	Geological constraints Mass movement High to very high erosion hazard	Residential subdivision, passive recreational
25–30%	Sideslopes	Geological constraints Possible mass movement High to very high erosion hazard	Upper limit for selective residential use, low-density housing on lots greater than 1 ha, passive recreation
>30%	Sideslopes	Geological constraints Mass movement Severe erosion hazard	Recommend against any disturbance for urban development

SOILS

Summary of Properties of Soils

Map unit	Dominant soils	Lithology and physiography	Erosion hazard	Limitations
A	Shallow gravelly soils	Metasediments; cherts, phyllites, etc. with quartz veination Ridges, sideslopes and some footslopes	High	Impeded soil drainage, shallow soil depth, high stone and gravel contents
B	Swamp alluvial soils	Metasediments; cherts, phyllites, etc. with quartz veination Alluvial parent materials, swamp and drainage plains	Very high to extreme	Seasonally high water-tables, poor to impeded soil drainage
C	Yellow duplex soils	Metasediments; cherts, phyllites, etc. Crests, sideslopes and some footslopes	High to very high	Low to moderate shrink-swell potential, poor soil drainage
D	Drainage plain alluvial soils	Metasediments; cherts, phyllites, etc. Alluvial/colluvial parent materials and surface materials	Very high	Seasonally high water-tables, poor to impeded soil drainage

Fig. 4.15 Urban capability classification for soil erosion control (after Hannam & Hicks 1980).

significantly different in their frequency distributions of erosion classes (Fig. 4.16; Morgan et al. 2004). The land systems differed from each other in either their soil erodibility or slope steepness or, in some cases, both. These results demonstrate that land systems represent dynamic erosion-response units that reflect both the extent of erosion at any one time and its evolution over time.

Chapter 4

86

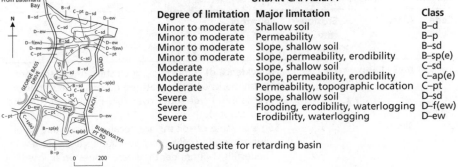

URBAN CAPABILITY

Degree of limitation	Major limitation	Class
Minor to moderate	Shallow soil	B–d
Minor to moderate	Permeability	B–p
Minor to moderate	Slope, shallow soil	B–sd
Minor to moderate	Slope, permeability, erodibility	B–sp(e)
Moderate	Slope, shallow soil	C–sd
Moderate	Slope, permeability, erodibility	C–ap(e)
Moderate	Permeability, topographic location	C–pt
Severe	Slope, shallow soil	D–sd
Severe	Flooding, erodibility, waterlogging	D–f(ew)
Severe	Erodibility, waterlogging	D–ew

〰️ Suggested site for retarding basin

Urban capability
Definitions of the classes used in urban capability assessment are:

Class A. Areas with little or no physical limitations to urban development.

Class B. Areas with minor to moderate physical limitations to urban development. These limitations may influence design and impose certain management requirements on development to ensure a stable land surface is maintained both during and after development.

Class C. Areas with moderate physical limitations to urban development. These limitations can be overcome by careful design and by adoption of site management techniques to ensure the maintenance of a stable land surface.

Class D. Areas with severe physical limitations to urban development which will be difficult to overcome, requiring detailed site investigation and engineering design.

Class E. Areas where no form of urban development is recommended because of very severe physical limitations, to such development, that are very difficult to overcome.

Fig. 4.15 *Continued*

4.2.3 Soil erosion survey

Three types of erosion survey can be distinguished: static, sequential and dynamic. Static surveys consist of mapping, often from aerial photographs, the sheet wash, rills and gullies occurring in an area (Jones & Keech 1966). Erosion hazard is estimated by calculating simple indices such as gully density. Sequential surveys evaluate change by comparing the results of static surveys undertaken on two or more different dates (Keech 1969). Dynamic surveys map both the erosion features and the factors influencing them and seek to establish relationships between the two. A geomorphological mapping system for this purpose was developed by Williams and Morgan (1976). The map portrays information on the distribution and type of erosion, erosivity, runoff, slope length, slope steepness, slope curvature in profile and plan, relief, soil type and land use. As much detail as possible is shown on a single map but, to avoid clutter, it is recommended that overlays are used for erosivity, soils and slope steepness. The legend is shown in Fig. 4.17 and an example, taken from an erosion survey of central Pahang, Malaysia, is shown in Fig. 4.18. Each component of the geomorphological map can be digitized and stored as a separate layer in a geographical information system. Each layer can then be updated as changes occur and also used as a basis for applying a factorial scoring system to assess erosion class or to link to an erosion prediction model.

The disadvantage of using aerial photographs for sequential erosion survey is that they have to be specially commissioned and flown, whereas satellite imagery from LANDSAT TM (Thematic Mapper), SPOT Multispectral (XS) or SPOT PAN (panchromatic) is available on a repetitive basis relatively cheaply. Although their respective ground resolutions of 10, 20 and 30 m are

Erosion hazard assessment

Table 4.3 Coding system for soil erosion appraisal in the field

Code	Class	Erosion rate (t ha^{-1})	Indicators
1	Very slight	<2	No evidence of compaction or crusting of the soil; no wash marks or scour features; no splash pedestals or exposure of tree roots; over 70% plant cover (ground and canopy).
2	Slight	2–5	Some crusting of soil surface; localized wash but no or minor scouring; rills every 50–100 m; small splash pedestals, 1–5 mm depth, where stones or exposed trees protect underlying soil, occupying not more than 10% of the area; soil level slightly higher on upslope or windward sides of plants and boulders; 30–70% plant cover.
3	Moderate	5–10	Wash marks; discontinuous rills spaced every 20–50 m; splash pedestals and exposed tree roots mark level of former surface, soil mounds protected by vegetation, all to depths of 5–10 mm and occupying not more than 10% of the area; slight to moderate surface crusting; 30–70% plant cover; slight risk of pollution problems downstream if slopes discharge straight into water courses.
4	High	10–50	Connected and continuous network of rills every 5–10 m or gullies spaced every 50–100 m; tree root exposure, splash pedestals and soil mounds to depths of 10–50 mm occupying not more than 10% of the area; crusting of the surface over large areas; less than 30% plant cover; danger of pollution and sedimentation problems downstream.
5	Severe	50–100	Continuous network of rills every 2–5 m or gullies every 20 m; tree root exposure, splash pedestals and soil mounds to depths of 50–100 mm covering more than 10% of the area; splays of coarse material; bare soil; siltation of water bodies; damage to roads by erosion and sedimentation.
6	Very severe	100–500	Continuous network of channels with gullies every 5–10 m; surrounding soil heavily crusted; severe siltation, pollution and eutrophication problems; bare soil.
7	Catastrophic	>500	Extensive network of rills and gullies; large gullies (>100 m^2) every 20 m; most of original soil surface removed; severe damage from erosion and sedimentation on-site and downstream.

Source: after Morgan et al. (2004).

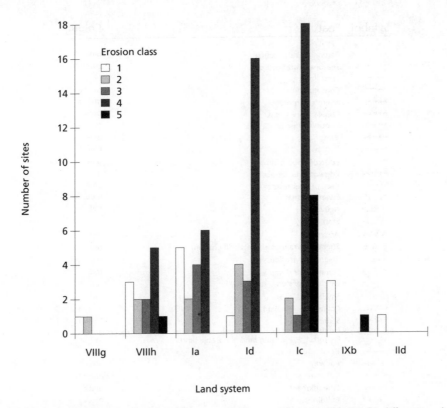

Fig. 4.16 Frequency distributions of erosion severity classes, as defined in Table 4.3, for different terrain units along a pipeline corridor in Georgia. VIII, Mtkvari Basin (g, uplifted and dissected hills; h, uplifted and dissected ridge); I, Caucasus Mountains (a, steeply dissected mountains; c, dissected mountains; d, dissected plateau); II, Kaukaz Mountains (d, low foothills); IX, Rioni Plain (b, piedmont fans and terraces) (after Morgan et al. 2004).

too coarse to map erosion features in detail, they can provide data on areas of bare soil to accuracies in excess of 85 per cent when compared to field data (Buttner & Csillag 1989; Verbyla & Richardson 1996). Satellite imagery is also a source of information on percentage vegetation cover, which can be related with an acceptable degree of accuracy to the Normalized Difference Vegetation Index (*NDVI*) (Mathieu et al. 1997), defined for TM data as:

$$NDVI = \left(\frac{TM4 - TM3}{TM4 + TM3} \right)$$
(4.2)

where *TM4* and *TM3* represent the percentage reflectance values recorded on the near infra-red (0.76–0.90 μm) and red (0.63–0.69 μm) bands of the electromagnetic spectrum respectively. The relationships between *NDVI* and vegetation cover vary depending on the nature of the vegetation. Further, they apply only to green or living plant material and cannot be used to estimate the cover of dead or decaying vegetation. These studies of temporal changes in vegetation and soil conditions indicate that with further research it should be feasible to use remote sensing for continuous monitoring to identify in advance when there is a high risk of erosion so that appropriate protection measures can be implemented.

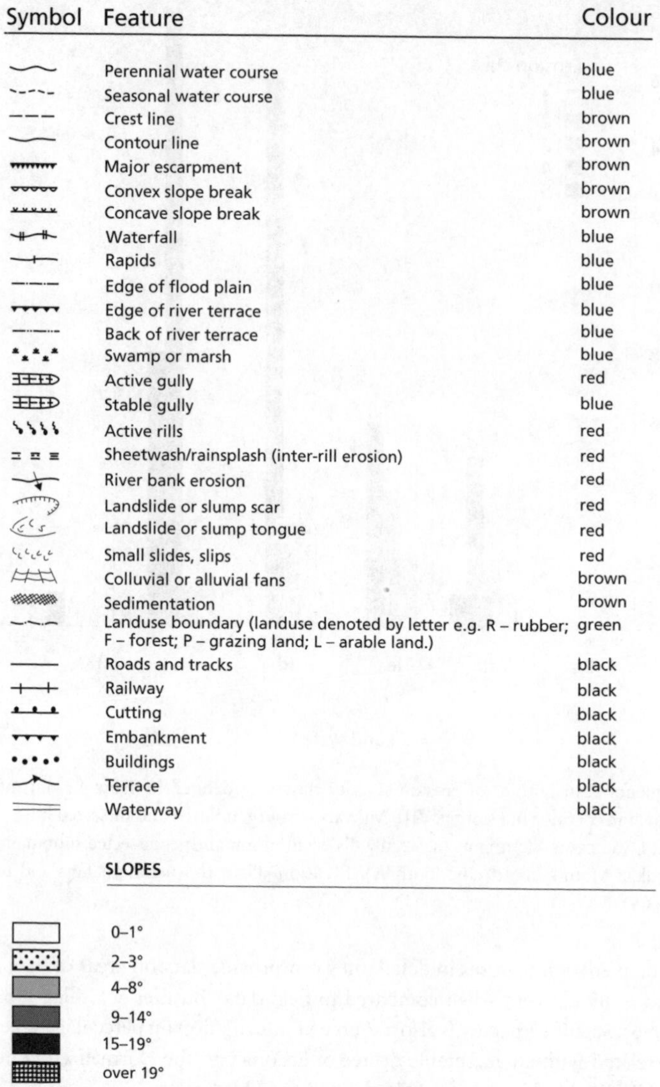

Symbol	Feature	Colour
	Perennial water course	blue
	Seasonal water course	blue
	Crest line	brown
	Contour line	brown
	Major escarpment	brown
	Convex slope break	brown
	Concave slope break	brown
	Waterfall	blue
	Rapids	blue
	Edge of flood plain	blue
	Edge of river terrace	blue
	Back of river terrace	blue
	Swamp or marsh	blue
	Active gully	red
	Stable gully	blue
	Active rills	red
	Sheetwash/rainsplash (inter-rill erosion)	red
	River bank erosion	red
	Landslide or slump scar	red
	Landslide or slump tongue	red
	Small slides, slips	red
	Colluvial or alluvial fans	brown
	Sedimentation	brown
	Landuse boundary (landuse denoted by letter e.g. R – rubber; F – forest; P – grazing land; L – arable land.)	green
	Roads and tracks	black
	Railway	black
	Cutting	black
	Embankment	black
	Buildings	black
	Terrace	black
	Waterway	black

SLOPES

	0–1°
	2–3°
	4–8°
	9–14°
	15–19°
	over 19°

Fig. 4.17 Legend for mapping soil erosion.

Detailed surveys

Detailed surveys of erosion are usually carried out in the field at pre-selected points to give information on the extent and severity of erosion. They are also used as checks on semi-detailed surveys carried out from aerial photographs and satellite imagery. The severity of erosion is rated by a

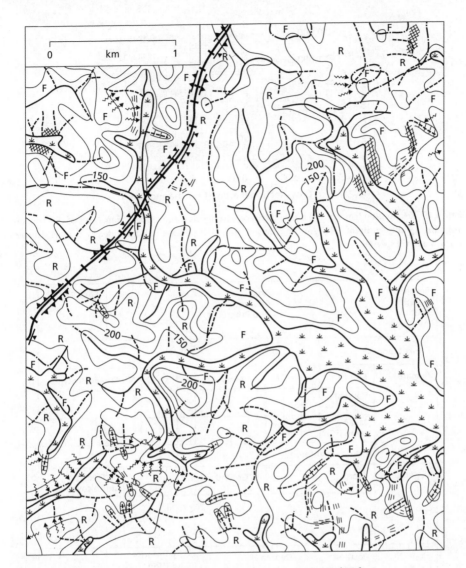

Fig. 4.18 Extract of soil erosion survey map, central Pahang. Contour interval 50 ft.

simple scoring system (Table 4.3) taking account of features that are easily visible, such as the exposure of tree roots, crusting of the soil surface, formation of splash pedestals, the size of rills and gullies and the type and structure of the plant cover. A nomograph can be used to translate these descriptors into an amount of soil loss (Fig. 4.19). Observations are made using quadrat sampling over areas of $1\,m^2$ for ground cover, crusting and depth of ground lowering, $10\,m^2$ for shrub cover and $100\,m^2$ for tree cover, and the density of rills and gullies. A pro-forma for recording data in the field is shown in Fig. 4.20. In interpreting the results of field surveys, it is important to place the data in their time perspective, particularly with respect to likely seasonal variations in the vegetation cover and soil erodibility.

Erosion hazard assessment

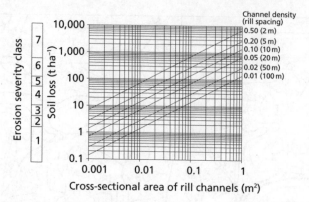

Fig. 4.19 Relationship between soil loss, soil erosion severity class and cross-sectional area of rill channels, assuming a bulk density of $1.4\,\mathrm{Mg\,m^{-3}}$ for the soil material. The lines indicate different channel densities $(\mathrm{m\,m^{-2}})$ and rill spacings (m).

Recorder: Date: Altitude:							Area: Air photo no: Grid Reference:								FACET NO.
Present landuse															
Climate	Month	J	F	M	A	M	J	J	A	S	O	N	D	Erosivity	
	Rainfail (mm)														
	Mean temp (°C)														
	Maximum intensity														
Vegetation	Type				% Ground cover						% Tree and shrub cover				
Slope	Position		Degree					Distance from crest			Shape				
Soil	Depth			Surface texture							Erodibility				
				Permeability			Clay fraction								
Erosion															
REMARKS															
EROSION CODE		0	½		1		2		3		4		5		

Fig. 4.20 Proforma for recording soil erosion in the field (devised by Baker personal communication).

Box 4

Upscaling detailed field surveys to national surveys

If detailed field surveys are carried out at a sufficient number of places and the results are shown to be representative of particular soil or landscape conditions, it is possible to extrapolate the data and produce national scale erosion surveys. The State Soil Conservation Service of Iceland, in collaboration with the Icelandic Agricultural Research Institute, applied a scoring system on a scale of 0 (no erosion) to 5 (very severe erosion) to each of the major types of erosion found in the country (Table B4.1). Colour photographs were produced showing examples of each type of erosion at different levels of severity for use by surveyors in the field when assigning erosion classes. For the national survey, satellite images at 1:10,000 scale were used to identify homogeneous landscape areas. Each area was then visited in the field to determine its constituent erosion types and severity grad-ings. The information was entered into the Arc/Info geographical information system and maps produced of the distribution of erosion classes and erosion types for the whole country (Arnalds et al. 2001; Fig. B4.1).

A distinctive feature of erosion in Iceland is the rofabarð or erosion scarp, usually between 0.2 and 3 m high, formed where a thick mantle of aeolian deposits overlying glacial till or basaltic lava is removed by wind erosion. The root mat from vegetation on the top of the scarp is more resistant than the underlying aeolian material and forms an overhang. The scarp is eroded by wind, raindrop impacts, runoff, needle-ice formation, freeze-thaw and slumping (Arnalds 2000). Actively eroding scarps can retreat at rates of 1 to 8 cm yr^{-1} which, when extrapolated by their length, which can be 14–21 km, give erosion rates of 2000–3000 t ha^{-1} per year. Rofabarð erosion is today the most important form of severe

Table B4.1 Distribution of erosion type and erosion severity in Iceland

Erosion type	Percentage erosion by erosion class				
	1 Little erosion	2 Slight erosion	3 Considerable erosion	4 Severe erosion	5 Very severe erosion
Vegetated land					
Rofabarð	7.6	7.5	7.0	10.3	5.5
Encroaching sand	0.0	0.0	0.0	0.3	0.4
Erosion spots	30.4	39.5	9.6	0.9	0.0
Solifluction	4.1	22.9	21.0	0.9	0.0
Gullies	3.2	5.4	4.3	0.9	0.6
Landslides	1.7	0.4	0.3	0.1	0.0
Deserts/barren land					
Melur (gravel)	43.6	18.3	23.1	0.0	0.0
Lava	8.0	0.5	0.1	0.0	0.0
Sand	0.9	0.7	1.1	9.1	42.9
Sandy gravel	0.0	1.6	19.0	51.9	19.5
Sandy lava	0.0	0.2	4.8	14.7	24.6
Soil remnants	0.1	1.1	1.2	0.5	0.6
Scree	0.3	2.0	8.4	10.5	5.9

Source: after Arnalds et al. (2001).

Continued

Erosion hazard assessment

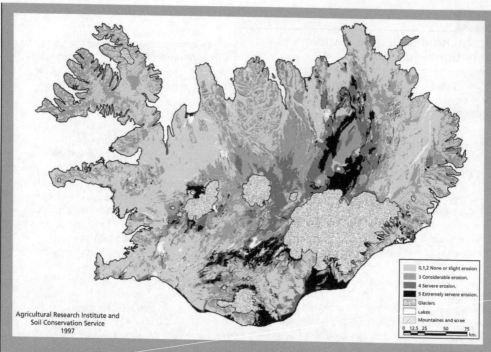

Fig. B4.1 Soil erosion status in Iceland (after Arnalds et al. 2001).

erosion (classes 3, 4 and 5) on vegetated land in the country and therefore a threat to the grazing land resource; yet, overall, rofabarðs occupy only 8.6 per cent of the area of Iceland so their importance in the national survey is relatively small. The most serious erosion is to be found in sandy desert areas where those with erosion gradings of 3, 4 and 5 account for about 22,000 km².

A similar technique was used by McHugh et al. (2002) to survey the extent of erosion in upland England and Wales. Based on field measurements of the dimensions of rills, gullies, hagged peat, sheep scars, areas of poaching, wheel ruts and footpaths, the areal extent and volume of erosion were determined for sample areas, 50 m² in size, distributed at regular intervals on a 1 × 1 km grid. By assuming that the proportion of eroded to non-eroded field sites was

representative of the uplands as a whole, it was concluded that, in 1999, some 2.5 per cent of the uplands, or almost 25,000 ha, was degraded as a result of erosion, although about half of this was revegetated and therefore no longer subject to accelerated soil loss.

The disadvantage of surveys based on single field visits is that they give no information on when or over how long a period the erosion occurred. Unless, as in the case of the study of rofabarðs, repeated measurements are made, it is not possible to determine present-day rates of soil loss. It should also be stressed that the national maps relate to erosion as determined at the scale used in the field to collect the data. The survey of upland England and Wales thus relates to erosion at a plot scale; that for Iceland is at a field scale.

CHAPTER 5

Measurement of soil erosion

Data on soil erosion and its controlling factors can be collected in the field or, for simulated conditions, in the laboratory. Whether field or laboratory studies are used depends on the objective. For realistic data on soil loss, field measurements are the most reliable, but because conditions vary in both time and space, it is often difficult to determine the chief causes of erosion or to understand the processes at work. Experiments designed to lead to explanation are best undertaken in the laboratory, where the effects of many factors can be controlled. Because of the artificiality of laboratory experiments, however, some confirmation of their results in the field is desirable.

5.1

Experimental design

Experiments are usually carried out to assess the influence of one or more factors on the rate of erosion. In a simple experiment to study the effect of slope steepness, it is assumed that all other factors likely to influence erosion are held constant. The study is carried out by repeating an experiment for different slope steepnesses. Decisions are necessary on the range of steepnesses, the specific steepnesses within the range and whether these should reflect some regular progression, e.g. 2°, 4°, 6°, rising in intervals of 2°, or 5°48′, 11°36′, 17°30′, rising in intervals of 0.1 sine values. The range can be selected to cover the most common slope steepnesses or extended to include extreme conditions. A decision is also needed on what levels to set for the factors being held constant, e.g. whether slope length should be 5 or 50 m, or the rainfall intensity 20 or 200 mm h^{-1}.

Measurements are subject to error. Since no single measurement of soil loss can be considered as the absolutely correct value, it is virtually impossible to quantify errors. However, they can be assessed in respect of variability. This requires replicating the experiment several times to determine the mean value of soil loss – for example, for a given slope steepness in the above experiment – and the coefficient of variation of the data. This is frequently rather high. In a review of field and laboratory studies of soil erosion by rainsplash and overland flow, Luk (1981) found typical values of 13–40 per cent for the coefficient of variation. Extreme values of soil loss ranged from ±39 to ±120 per cent. To achieve a measurement of soil loss to an accuracy of ±10 per cent with 95 per cent confidence for three different Canadian soils required six, 25 and 29 replications of the experiment respectively. A decision therefore needs to be made between carrying out between five and ten replications and accepting data with a coefficient of variation of generally between 20 and 30 per cent or completing 30 or more replications to obtain a more accurate measurement. Generally, the variability in field measurement is higher than that in laboratory studies. Roels and Jonker (1985) found coefficients of variation of 20–68 per cent in measurement of interrill erosion using unbounded runoff plots. The higher variability reflects the often highly localized variations in soil conditions, particularly infiltration and cohesion, that occur in the field, whereas in the laboratory, the soils are often processed by drying and sieving to give greater control over experimental conditions.

Systematic errors can be built into an experiment – for example, by starting the study of slope steepness effects on the gentlest slope, carrying out the rest of the study in a sequence of increasing slope steepness and following the same procedure for all the replications. In this way, the soil loss measured at a particular steepness is influenced consistently by the amount of erosion that has taken place on the next lowest slope steepness in the sequence. Remaking the soil surface between the separate runs of the experiment may not entirely eliminate this effect. Strictly, the order of the runs should be randomized but there is often a need to balance randomization with expediency and conduct some runs in sequence where it is very time-consuming to keep switching operating conditions. For instance, in a study involving simulated rainfall of several intensities, it is expedient to complete all the runs at one intensity before proceeding to the next because of the problems encountered in continually altering the settings of the rainfall simulator. It must be appreciated, however, that failure to randomize an experiment may limit the use that can be made of statistical techniques in processing the data. Certain techniques, such as two-way analysis of variance, assume randomness in the data.

With the simple slope steepness experiment outlined above, the experimental design can be described as examining the effect of one variable (slope steepness) with, say, five treatments (2, 4, 6, 8 and 10°) and three replications, giving a total of 15 separate runs. These can be arranged in a random sequence, either in time if the experiment is carried out in the laboratory, or in space if the experiment is set up in the field. Many experiments, however, are designed to investigate the effects of more than one factor – for example, the effect of slope steepness at two different rainfall intensities. The design then becomes one of two variables, slope with five treatments and rainfall intensity with two treatments, giving a total of ten different treatments. With three replications, this gives a total of 30 separate runs. These can be organized in a completely random sequence or, for the reasons of expediency indicated above, in two blocks, one for each rainfall intensity, with the slope treatments randomized within each block. The resulting data can then be analysed to compare treatment means across two blocks for all ten treatments and also to compare block means. Block randomized experiments are particularly useful in the field when, for example, an investigation of the effects of four different land management systems might take up so much land that the results are affected by variability in soil conditions (Fig. 5.1). In this case, if three replications are used, they could be set up as three separate blocks, with the treatments randomized in their position within each block. In many field situations, variations may occur in more than one factor across an experimental site – for example, soil may change across the slope and slope steepness may change down the slope. In this situation, the four land management treatments could be organized on a Latin square design with each treatment located randomly in both across-slope and downslope directions. The results would be analysed to compare across-slope means and downslope means as well as treatment means (Fig. 5.1; Mead & Curnow 1983). In theory, no limits exist on the number of factors that may be incorporated in an experiment. In practice, the size and design of the experiment are usually limited by the amount of time and money available.

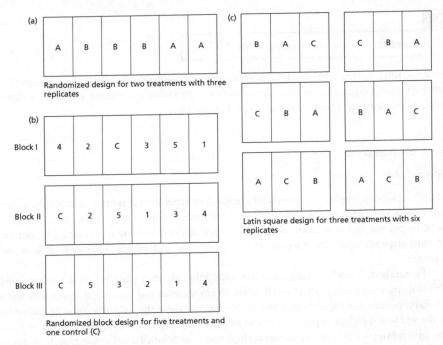

Fig. 5.1 Layout of field experiments: (a) simple randomized design with replicates; (b) randomized block design; (c) Latin square design (after Mead & Curnow 1983).

A problem faced in many experiments is how much data to collect. For instance, if the objective of a study is to measure total soil loss over a period of time, this can be achieved by collecting in bulk all the soil washed or blown from an area or by collecting the soil at regular shorter time intervals so as to obtain a total by summation of the loss in these time periods but also to learn about the pattern of loss over time. Restricting the measurement system to its bare essentials is usually cheaper and the data are generally easier to interpret. However, potentially useful information on whether most of the soil is eroded early or late in a storm is lost. This type of conflict is more apparent where the data are collected automatically. A decision then needs to be made on whether or not to analyse the shorter-period data.

There are no easy solutions to the problems of experimental design. Experiments should have clearly conceived objectives, define what needs to be measured and at what level of accuracy, and be set up in such a way that they can be easily repeated by others. Errors may arise due to different operators or the use of different equipment. In much soil erosion and conservation work, equipment is built by individual researchers and is not commercially available. As a result there is a multiplicity of methods and equipment and little standardization. The techniques described in this chapter are restricted to those in common use.

Field measurements and experiments

Field measurements may be classified into two groups: those designed to determine soil loss from relatively small sample areas or erosion plots, often as part of an experiment, and those designed to assess erosion over a larger area, such as a drainage basin.

5.2.1 Erosion plots

Bounded plots

Bounded plots are employed at permanent research or experimental stations to study the factors affecting erosion. Each plot is a physically isolated piece of land of known size, slope steepness, slope length and soil type from which both runoff and soil loss are monitored. The number of plots depends upon the purpose of the experiment but usually allows for at least two replicates.

The standard plot is 22 m long and 1.8 m wide, although other plot sizes are sometimes used. The plot edges are made of sheet metal, wood or any material that is stable, does not leak and is not liable to rust. The edges should extend 150–200 mm above the soil surface and be embedded in the soil to a sufficient depth so as not to be shifted by alternate wetting-and-drying or freezing-and-thawing of the soil. At the downslope end is positioned a collecting trough or gutter, covered with a lid to prevent the direct entry of rainfall, from which sediment and runoff are channelled into collecting tanks. For large plots or where runoff volumes are very high, the overflow from a first collecting tank is passed through a divisor, which splits the flow into equal parts and passes one part, as a sample, into a second collecting tank (Fig. 5.2). Examples are the Geib multislot divisor and the Coshocton wheel. On some plots, prior to passing into the first collecting tank, the runoff is channelled through a flume where the discharge is automatically monitored. Normally an H-flume is chosen because it is non-silting and unlikely to become blocked with debris. A further automation is to install an automatic sediment sampler to extract samples of the runoff at regular time intervals during the storm for later analysis of its sediment concentration; the time each sample is taken is also recorded. Rainfall is measured with both standard and autographic gauges adjacent to the plots.

A flocculating agent is added to the mixture of water and sediment collected in each tank. The soil settles to the bottom of the tank, and the clear water is then drawn off and measured. The volume of soil remaining in the tank is determined and samples of known volume are taken for drying and weighing. The mean sample weight multiplied by the total volume gives the total weight of soil in the tank. If all the soil has been collected in the tank, this gives the total soil loss from the plot. For tanks below a divisor, the weight of the soil in the tank needs to be adjusted in accordance with the proportion of the total runoff and sediment passing into the tank. Thus for the layout shown in Fig. 5.2, the total soil loss from the plot is the weight of the soil in the first tank plus, assuming one-fifth of the overflow passes through the divisor into the second tank, five times the weight of soil in the second tank. Where automatic sediment sampling occurs, the sediment concentration is determined for each sample. Since the time when each sample was taken during the storm is known, the data can be integrated over time to give a sediment graph. Full details of the equipment, manufacturing instructions for the divisors and operation of erosion plots are found in Agricultural Handbook No. 224 of the United States Department of Agriculture (1979) and in Hudson (1993a).

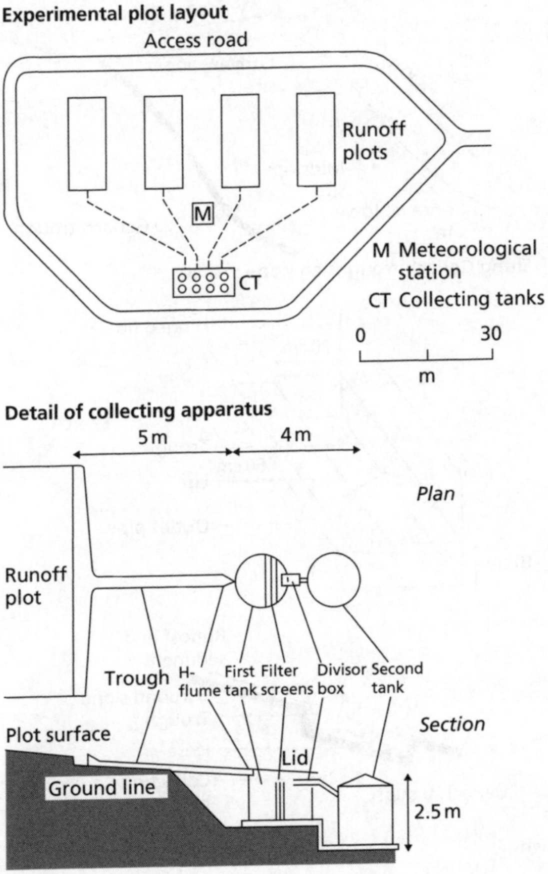

Experimental plot layout

Access road

Runoff plots

M Meteorological station
CT Collecting tanks

0   30
m

Detail of collecting apparatus

5 m   4 m

Plan

Runoff plot

Trough H-flume First Filter tank screens Divisor box Second tank

Plot surface

Ground line

Lid

2.5 m

Section

Fig. 5.2 Typical layout of erosion plots at a soil erosion and conservation research station (after Hudson 1965).

Although the bounded runoff plot gives probably the most reliable data on soil loss per unit area, there are several sources of error involved with its use (Hudson 1957). These include silting of the collecting trough and pipes leading to the tanks, inadequate covering of the troughs against rainfall and the maintenance of a constant level between the soil surface and the sill or lip of the trough. Considerable care is required if adequate data are to be obtained from runoff plots. Hudson (1993a) refers to a catalogue of disasters that have occurred where plots have not been properly designed and managed. These range from the collecting tanks overflowing during extreme events, the tanks floating out of saturated ground, runoff entering the top of the plots, the taps in the collecting tanks being left open, to damage from termites, spiders and baboons! Other problems are that runoff may collect along the boundaries of the plot and form rills which would not otherwise develop, and that the plot is a partially closed system, being cut off from the input of sediment and water from upslope. The data obtained give a measure of soil loss from the entire plot that may be reasonably realistic of losses from fields under similar conditions. The data do not give any indication of the redistribution of soil within a field or along a slope.

Measurement of soil erosion

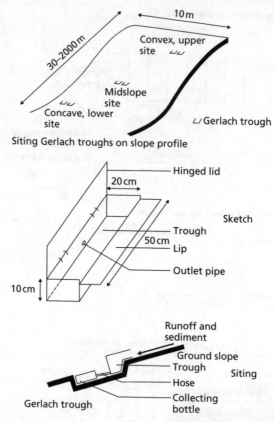

Fig. 5.3 Gerlach troughs.

The size of the plot is important. Plots of only 1 m² in size will allow investigations into infiltration and the effects of rainsplash but are too small for studies of overland flow except as a transporting agent for splashed particles. Thus Govers and Poesen (1988) used plots ranging from 0.5 to 0.66 m² in size to measure interrill erosion. Field studies of soil erodibility in Bedfordshire, England, using small plots showed that erodibility increased with increasing fine sand content in the soils. This reflected the selectivity of fine sands to detachment by rainsplash but did not accord with erodibility assessments at the hillslope scale which incorporated runoff effects (Rickson 1987). Plots must be at least 10 m long for studies of rill erosion. Much larger plots are required for evaluating farming practices such as strip cropping and terracing.

Gerlach troughs

An alternative method of measuring sediment loss and runoff was developed by Gerlach (1966) using simple metal gutters, 0.5 m long and 0.1 m broad, closed at the sides and fitted with a movable lid (Fig. 5.3). An outlet pipe runs from the base of the gutter to a collecting bottle. In a typical layout, two or three gutters are placed side-by-side across the slope and groups of gutters are installed at different slope lengths, arranged *en echelon* in plan to ensure a clear run to each gutter from the slope crest. Because no plot boundaries are used, edge effects are avoided. It is normal to express soil loss per unit width but if an areal assessment is required, it is necessary to assume a catchment area equal to the width of the gutter times the length of the slope. A further assumption is that any loss of water and sediment from this area during its passage downslope is balanced by inputs from adjacent areas. This assumption is reasonable if the slope is straight in plan. On slopes curved in plan, the catchment area must be surveyed in the field. This disadvantage is offset by the flexibility of monitoring soil loss at different slope lengths and steepnesses within an open system. Because of their cheapness and simplicity, Gerlach troughs can be employed for sample measurements of soil loss at a large number of selected sites over a large area.

Changes in ground level

The simplest way of measuring changes in ground level over time is to use what is technically known as an erosion pin but, in reality, is a 250–300 mm long nail, 5 mm in diameter, driven through a washer into the soil (Emmett 1965). The head of the nail should be some 20–30 mm above the soil surface, the washer flush with the surface and the base of the nail sufficiently far into the ground not to be disturbed by changes in soil volume due to wetting-and-drying or freeze–thaw. Periodic measurements of the gap between the head of the nail and the washer indicate the extent to which the surface has been lowered; where the washer has become buried, the depth of the material above the washer indicates the depth of deposition. Measurements need to be taken over many years before a consistent pattern of ground lowering can be separated from shorter-term fluctuations in level due to changes in soil volume. A large number of pins, usually installed on a grid system, is needed to obtain representative data over a large area.

A disadvantage of erosion pins is that they can be easily disturbed by livestock and wildlife or stolen by the local population, who can find supposedly better uses for iron or steel nails. They can also be difficult to relocate in subsequent surveys. An alternative approach is to establish a network of metal pegs, set unobtrusively in concrete at ground level so that their position remains stable over time. A portable aluminium girder is placed across any two adjacent pegs from which vertical readings of the depth to the soil surface can be taken at regular intervals. Between readings the girder is removed. The method, pioneered by Hudson (1964) on rangeland in Zimbabwe, is known as the erosion bridge. Again long-period measurements are needed to isolate trends in erosion or deposition from short-term fluctuations in soil level.

Splash erosion measurements

Erosion plots, Gerlach troughs, erosion pins and erosion bridges provide information on erosion by rainsplash, overland flow and rills combined. Attempts to assess the relative contributions of each are based on separate measurements of splash erosion and rill erosion with the balance being attributed to overland flow. Splash erosion has been measured in the field by means of splash boards (Ellison 1944) or small funnels or bottles (Sreenivas et al. 1947; Bollinne 1975). These are inserted in the soil to protrude 1–2 mm above the ground surface, thereby eliminating the entry of overland flow, and the material splashed into them is collected and weighed. An alternative approach is the field splash cup (Morgan 1981; Fig. 5.4), where a block of soil is isolated by enclosing it in a central cylinder and the material splashed out is collected in a surrounding catching

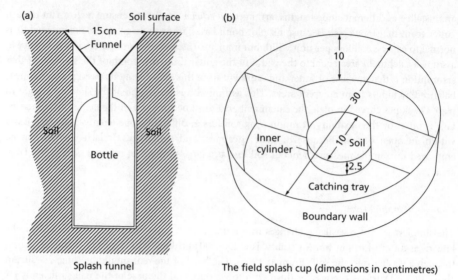

Fig. 5.4 Methods of measuring splash erosion in the field, after (a) Bollinne (1975) and (b) Morgan (1981).

tray. Because the quantity of splashed material measured per unit area depends upon the diameter of the funnels and cups, the following correction has to be applied to determine the real mass of particles detached by splash:

$$MSR = MSe^{0.054D} \tag{5.1}$$

where MSR is the real mass of splashed material per unit area ($g\,cm^{-2}$), MS is the measured splash per unit area ($g\,cm^{-2}$) and D is the diameter of the cup or funnel (cm) (Poesen and Torri 1988).

Rill erosion measurement

The simplest method of assessing rill erosion is to establish a series of transects, 20–100 m long, across the slope and positioned one above the other. The cross-sectional area of the rills is determined along two successive transects. The average of the two areas multiplied by the distance between the transects gives the volume of material removed. By knowing the bulk density of the soil, the volume is converted into the weight of soil loss and this, in turn, is related to an area defined by the length and distance apart of the transects. Since this method ignores the contribution of interrill erosion to the sediment carried in the rills and also depends upon being able to identify distinctly the edge of the rills, it is likely to underestimate rill erosion by 10–30 per cent.

Gully erosion measurement

Erosion from relatively small gullies can be assessed using the same profiling technique as described above for rills. For larger gullies, sequential surveys using aerial photography are more suitable. Keech (1992) studied panchromatic stereo-photography at 1:25,000 scale of St Michael's Gully in the Mhondoro Communal Area of Zimbabwe, for 1946, 1956, 1968, 1971, 1976 and 1984.

Using a stereoplotter, a stereomodel of the gully was set up for each year of survey using horizontal and vertical controls taken from 1:50,000 scale topographical maps. Difficulty was experienced in deciding where the up-slope boundary of the gully should be drawn and then in interpreting that boundary consistently along the periphery of the gully on the different sets of photography. Once a working procedure was established, the position of the edge of the gully was plotted for the different dates. Some 126 points or nodes were then identified along the defined margin on the 1946 photography, against which changes in position could be measured. Using the plotting machine to read off heights, cross-sections of the gully were determined at intervals down the gully. From this information, the extent of retreat of the gully headwall and side walls and the volume of material removed from the gully between different dates were calculated. More recent surveys of gullies use large-scale (1:10,000) aerial photography to construct high-resolution digital elevation models (DEMs). Rates of gully erosion are calculated from differences in elevation between DEMs of different dates (De Rose et al. 1998; Betts et al. 2003). Digitally generated DEMs can match the accuracy of traditional analytical photogrammetric techniques and are much faster to produce (Baily et al. 2003).

5.2.2 Catchments

Sediment yield

The sediment yield of a catchment is obtained from measurements of the quantity of sediment leaving a catchment along the river over time. Recording stations are established at the exit point for the automatic measurement of discharge, using weirs and depth recorders, and suspended sediment concentrations in the river water. Water samples are taken at set times with specially designed integrated sediment samplers, or the sediment concentration is monitored continuously by recording the turbidity of the water. Bucket samples are not recommended because they contain only the surface water and cannot provide information on the sediment being transported throughout the depth and width of the river channel. With measurements made at set times, there is a need to extrapolate the data to cover the period between samples. The standard approach is to establish a sediment discharge rating curve in which the sediment concentration (C) is related to the water discharge (Q) by the equation:

$$C = aQ^b \tag{5.2}$$

The accuracy of this method is highly dependent on the frequency of sampling. Walling et al. (1992) compared results from regular 7-day, 14-day and 28-day sampling programmes of suspended sediment load in the River Exe, Devon, England over a two-year period, with results from continuous sampling using a turbidity recorder. Estimates of the sediment load were found to vary between 10 and 300 per cent of that obtained from continuous monitoring. The likelihood of underestimation increased as the sampling frequency decreased, since, with longer sampling intervals, the record is likely to include fewer flood events. Monthly sampling for estimating annual sediment yield of larger catchments, 1000–3000 km^2 in size, can result in underestimations of 60–80 per cent (Phillips et al. 1999). Of course, it should be recognized that the turbidity meter does not give a true record because the measurements are subject to errors associated with the influence of the particle size of the sediment load, the magnitude of the sediment concentration, the presence of organic matter and the need to keep the sensors clear of algae. Despite these problems, the method is currently the best available to provide estimates of suspended sediment yield, especially if high frequency data are needed. It should be recognized, however,

Measurement of soil erosion

that, because of the need for regular calibration and maintenance, such data come at greater cost compared to using rating curves.

Reservoir surveys

Sedimentation rates in lakes and reservoirs can indicate how much erosion has taken place in a catchment upstream, provided the efficiency of the reservoir as a sediment trap is known. Rapp et al. (1972b) used repeated surveys of designated transects across four reservoirs in the Dodoma region of Tanzania in relation to a set level or benchmark. Using manual soundings of depth from a boat, a contour map was made of the bottom of the reservoir. The volume of the reservoir was obtained by adding up the partial volumes (V):

$$V = h(A + B)/2 \tag{5.3}$$

where V is the volume of sediment (m^3), h is the upper contour interval (m), A is the area of the upper contour (m^2) and B is the area of the lower contour interval (m^2). To this total, the volume below the lowest contour was added, calculated as the product of the area of the lowest contour and the mean depth to that contour from the bottom. The reservoir volume was then compared with the initial volume. The reduction in volume represents the volume of sediment accumulated in the reservoir. Assuming an average dry bulk density of $1.5\,Mg\,m^{-3}$, the volume was converted into a mass and divided by the area of the catchment upstream to give an erosion value in $t\,ha^{-1}$. This was further adjusted by assuming that the sediment in the reservoir represented about half that eroded from the catchment. Dividing the adjusted total by the number of years since the reservoir was built gave a mean annual erosion rate. More rapid reservoir surveys can be made using an echo-sounder to obtain depth readings, an electro-distance measuring theodolite or laser theodolite to fix the position of the sounding, and a digital elevation model (*DEM*) to produce the contour map (Brabben et al. 1988). Sedimentation rates in farm ponds can be analysed using a similar methodology (Verstraeten & Poesen 2001).

Potential sources of error are quite high with reservoir surveys, the most important being the estimate of the trap efficiency of the reservoir, which requires knowledge of the frequency and sediment concentration of flows carried over the spillway in times of flood and errors arising in calculating the capacity of the reservoir. Even small errors in the latter may give rise to errors up to two orders of magnitude in determining the volume of deposited material. Care is also required to ensure that the source of sediment is from erosion in the contributing catchment. For many small reservoirs in south-east Australia, up to 85 per cent of the sediment comes from trampling of the shoreline by cattle (Lloyd et al. 1998). Compared with data from suspended sediment concentrations, the data from reservoir surveys incorporate material transported as bed load as well as suspended sediment.

Tracers

The most commonly used tracer in soil erosion measurement is the radioactive isotope, caesium-137 (^{137}Cs). Caesium-137 was produced in the fall-out of atmospheric testing of nuclear weapons from the 1950s to 1970s. It was distributed globally in the stratosphere and deposited on the earth's surface in the rainfall. Regionally, the amount deposited varies with the amount of rain but, within small areas, the deposition is reasonably uniform. The isotope is strongly and quickly adsorbed to clay particles within the soil. By analysing the isotope content of soil cores collected on a grid system varying in density from 10×10 to $20 \times 20\,m$, the spatial pattern of isotope loading is

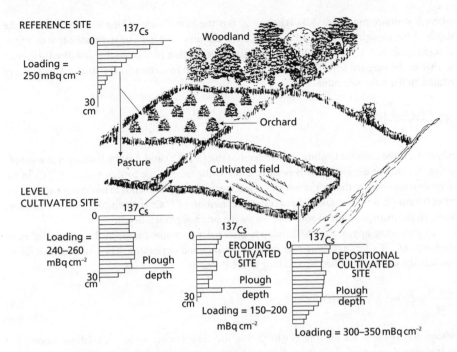

Fig. 5.5 Schematic representation of the effect of erosion and deposition upon the loading and profile distribution of caesium-137 (after Walling & Quine 1990).

established. Figure 5.5 (Walling & Quine 1990) shows a typical situation. In the pasture land at the top of the slope, the isotope is concentrated at the surface; its presence in small amounts at depth is the result of earthworm activity in the soil. On the arable site, the isotope is more uniformly distributed with depth as a result of disturbance of the soil by ploughing. The decline in isotope loading by about 40 per cent on the steeper slope is a result of erosion. At the bottom of the slope, there is an increase in loading due to deposition of material. Since the deposition has been active for some years, some of the isotope lies below the present plough depth. Spatial variations in isotope loading in comparison with those at a reference site, usually in either woodland or grassland, have been interpreted successfully, in many parts of the world, as indicating the patterns of erosion and deposition (Ritchie & Ritchie 2001). When comparing the results with those from erosion plots, it should be noted that they reflect the sum of all the processes by which soil can be redistributed over a field or a hillside and not just the outcome of interrill and rill erosion.

The method provides qualitative information on the patterns of soil erosion and deposition in the landscape over a period of 30–50 years. A conversion model is required to turn this information into estimates of erosion rates (Walling et al. 2002a). The simplest and most widely used model is the proportional approach (Mitchell et al. 1980; De Jong et al. 1983):

$$Y = 10 \times \frac{BdX}{100TP} \tag{5.4}$$

where Y is mean annual soil loss $(t\,ha^{-1}\,yr^{-1})$, B is the bulk density of the soil $(kg\,m^{-3})$, d is the depth of the plough or cultivated layer (m), T is the time that has elapsed since the start of the accumulation of ^{137}Cs and P is the ratio of the concentration of ^{137}Cs in the mobilized sediment to that in the original soil. The term X is the percentage reduction in the total ^{137}Cs inventory relative to the reference value, i.e.

$$X = \left(\frac{A_{ref} - A}{A_{ref}}\right) \times 100 \tag{5.5}$$

where A_{ref} is the caesium loading at the reference site $(Bq\,m^{-2})$ and A is the loading at the sample point. For sites of deposition, X is positive and P is the ratio of the concentration of ^{137}Cs in the deposited sediment to that in the mobilized sediment. The method tends to underestimate erosion rates because it does not take into account the dilution of ^{137}Cs concentration on erosion sites through the incorporation of soil from below the original plough depth.

An alternative approach is to develop a mass balance model based on the accumulation or depletion of ^{137}Cs through time as a result of erosion and deposition. The mass balance for an eroding site can be described by (Walling & He 1999a):

$$\frac{dA(t)}{dt} = I(t) - \left(\lambda + \frac{R}{d_m}\right)A(t) \tag{5.6}$$

where $A(t)$ is the ^{137}Cs loading $(Bq\,m^{-2})$, t is the time (years) since ^{137}Cs fallout began, $I(t)$ is the annual input of ^{137}Cs from fallout at time t $(Bq\,m^{-2}\,yr^{-1})$, λ is the decay constant for ^{137}Cs $(= 0.023\,y^{-1})$, R is the erosion rate $(kg\,m^{-2}\,y^{-1})$ and d_m is the average plough depth represented as a cumulative mass depth $(kg\,m^{-2})$. For an eroding soil,

$$Y = \frac{10dB}{P}\left[1 - \left(1 - \frac{X}{100}\right)^{1/(t-1963)}\right] \tag{5.7}$$

and for a depositional site,

$$R' = \frac{A(t) - A_{ref}}{\int_{1963}^{t} C_d(t')e^{-\lambda(t-t')}dt'} \tag{5.8}$$

where R' is the deposition rate $(kg\,m^{-2}\,yr^{-1})$ and $C_d(t')$ is the concentration of ^{137}Cs in the deposited sediment in year t' $(Bq\,kg^{-1})$.

If, for the situation shown in Fig. 5.5, it is assumed that the data relate to 1990, the ^{137}Cs loading on the eroding cultivated site is $200\,mBq\,cm^{-2}$, the bulk density of the soil is $1.4\,Mg\,m^{-3}$ and P is equal to 2.0, a typical value (He & Walling 1996), the erosion rate is calculated at $22.9\,t\,ha^{-1}\,yr^{-1}$. The proportional method yields $19.4\,t\,ha^{-1}\,yr^{-1}$. If it is assumed for the depositional site that the loading is $300\,mBq\,cm^{-2}$, the loading is distributed uniformly throughout the 25 cm plough depth to give a concentration of $2.5\,Bq\,kg^{-1}$ and the conditions are unvarying from year to year, the deposition rate is calculated as $7.4\,t\,ha^{-1}\,yr^{-1}$. Generally, soil erosion rates obtained using caesium-137 compare well with measured rates from erosion plots and instrumented catchments (Theocharopoulos et al. 2003; Zhang et al. 2003).

Recent developments of the approach allow for the likelihood that some or all of the ^{137}Cs fallout will be removed by erosion before it can be adsorbed to the soil (Walling & He 1999a), as

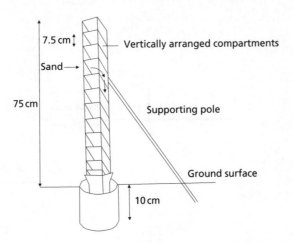

7.5 cm

Vertically arranged compartments

Sand

75 cm

Supporting pole

Ground surface

10 cm

Fig. 5.6 Bagnold sand catcher.

well as the additional effects of the redistribution of soil by tillage, the loss of soil through harvesting of root crops and the movement of soil in three dimensions over the landscape (van Oost et al. 2003). Where it is possible to identify separately the fallout from the accident at the Chernobyl nuclear power station in 1986, the method can be used to determine erosion rates over the past 15 years (Golosov 2003). The potential of tracers with shorter half-lives than the 30.12 years of ^{137}Cs is being investigated to see if they can be used to determine erosion rates over shorter periods. The most promising are unsupported lead-210 (^{210}Pb) (Walling & He 1999b) and beryllium-7 (^{7}Be) (Walling et al. 2000).

5.2.3 Sand traps

The techniques for measuring wind erosion are less well established than those for monitoring water erosion. The problem is to design an aerodynamically sound trap to catch soil particles while allowing the air to pass freely through the device. The build-up of back pressure causes resistance to the wind that is deflected from the traps. By careful adjustment of the ratios between the sizes of inlet, outlet and collecting basins, a satisfactory trap can be produced. An example is the Bagnold catcher (Fig. 5.6), consisting of a series of boxes placed one above the other so as to catch all the particles moving through a unit width of air flow at different heights. The disadvantage of the Bagnold catcher is that it cannot be reoriented as wind direction changes. Devices such as the Big Spring Number Eight sampler (Fryrear 1986; Fig. 5.7) and the Wilson and Cooke bottle sampler (Sterk & Raats 1996; Fig. 5.8) overcome this by being mounted at different heights on a mast to which a wind vane is attached, allowing the whole apparatus to rotate so that the capture tubes always face the wind. These catchers have efficiencies between 75 and 100 per cent, depending on the size of the particles being carried in the wind (Goosens et al. 2000). The material caught in the traps is collected and weighed after each period of observation. A trap was developed by Janssen and Tetzlaff (1991) where the wind-blown material falls from the collector on to a tray mounted on top of a balance. The weight of the material is recorded automatically, thereby giving a continuous record of particle movement throughout a storm. An alternative

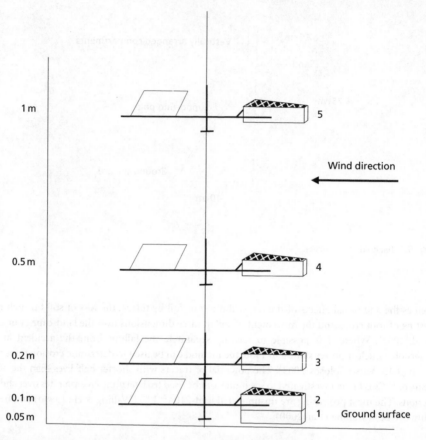

Fig. 5.7 Big Spring Number Eight sampler. Sizes of inlets for the different catchers are: (1) 10 × 20 mm; (2) 20 × 20 mm; (3) 30 × 20 mm; (4) 50 × 20 mm; and (5) 50 × 20 mm.

automatically recording sensor, the saltiphone (Spaan & van den Abeele 1991) uses a microphone to record the impacts of saltating particles. The main disadvantage of the saltiphone is that it does not collect the material, so its particle-size distribution cannot be determined.

A particular problem of wind erosion measurements is to determine the number of samplers required and their best location. Fryrear et al. (1991) recommended placing sampling masts, each with a cluster of traps at different heights, in a radial pattern from the centre of a field, whereas Sterk and Stein (1997) set up the masts on a grid system. Recent studies have shown that grid and random sampling tend respectively to overestimate and underestimate sediment transport. Better results are obtained with a nested sampling scheme based on placing a parallelogram over the area and locating masts at regular 500-m intervals along each side. Then from each of these masts, a compass bearing is selected randomly between 0 and 360° and further masts are located along this bearing at 200, 500 and 1000 m intervals (Chappell et al. 2003). It is clear from these studies that a large number of sample points is needed to obtain reliable data.

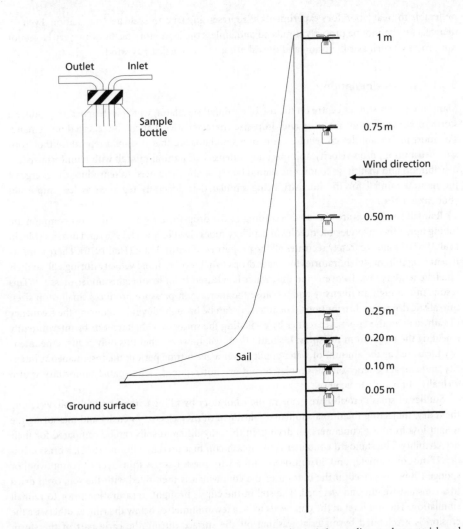

Fig. 5.8 Wilson and Cooke bottle sampler. The dimensions can be varied according to the material used to build the sampler. The sampler used by the National Soil Resources Institute had an inlet pipe of 10 mm diameter and an outlet pipe of 15 mm diameter.

Laboratory experiments

The key questions arising with laboratory studies concern the scale of the experiment, the greater influence of boundary effects and the extent to which field conditions are simulated. It is not usually possible to construct a scaled-down version of field conditions – for example, by using a small plot to represent a large hillslope – because scale-equivalence cannot be maintained in raindrops and soil particles without affecting their basic properties or behaviour. It is therefore preferable to treat laboratory experiments as representing true-to-scale field simulation. Even so, many factors cannot be properly simulated and, unless the laboratory facilities are very large, nor can processes such as rill erosion and the saltation of soil particles by wind.

5.3.1 Rainfall simulation

Many laboratory studies centre on the use of a rainfall simulator, which is designed to produce a storm of known energy, intensity and drop-size characteristics that can be repeated on demand. The most important design requirements of a simulator are that it should reproduce the drop-size distribution, drop velocity at impact and intensity of natural rainfall with a uniform spatial distribution and that these conditions should be repeatable. The need to reproduce the energy of the natural rainfall for the intensity being simulated is generally regarded as less important (Bubenzer 1979).

Rainfall simulators are classified according to the drop-formers used. The most common are tubing tips, either hypodermic needles or capillary tubes (De Ploey et al. 1976) and nozzles (Morin et al. 1967). None accurately recreates all the properties of natural rain (Hall 1970). There is insufficient height in most laboratories for water drops to achieve terminal velocity during fall, so their kinetic energy is low. To overcome this, water is released from low heights under pressure. This results in too high an intensity and, because the increase in pressure produces small drop sizes, unrealistic drop-size distributions. The intensity can be brought down by reducing the frequency of rain striking the target area, either by oscillating the spray over the target or by intermittently shielding the target from the spray. It should be noted, however, that this only reduces the intensity measured as the amount of water applied over a given time period; the instantaneous intensity and impact energy of the simulated rain are not reduced. True spatial uniformity is also virtually impossible to achieve.

Studies of splash erosion are made in the laboratory by filling containers with soil, weighing them dry, subjecting them to a simulated rainstorm of pre-selected intensity and measuring the weight loss from the containers on drying. In this way different soils can be compared for their detachability. The standard container is the splash cup first used by Ellison (1944), a brass cylinder 77 mm in diameter and 50 mm deep with a wire mesh base. A thin layer of cotton wool or sponge rubber is placed in the bottom of the cup, which is then filled with the soil, oven dried to a constant weight and weighed. The soil in the cup is brought to saturation prior to rainfall simulation. The soil level in the cup needs to be a few millimetres below the rim, as otherwise the erosion is accelerated by material washing off the surface during the early part of the storm (Mazurak & Mosher 1968); this effect can be enhanced if the initial raindrop impacts create a pitted or rough surface or if the surface is too loosely packed (Kinnell 1974). Towards the end of the storm, the splash loss will be reduced because the level of the soil within the cup becomes too low and splashed particles are intercepted by the rim of the cup (Bisal 1950).

Where several soils are being compared it is important to treat them all the same. Moldenhauer (1965) recommends collecting them from the field under uniform moisture conditions, achieved by flooding the surrounding area with 100 mm of water and covering it with polythene sheet for 48 hours before taking the sample. This gives a condition close to field capacity. When the samples are brought back to the laboratory they should be dried to a constant weight and split into 200 g portions. These are poured into the splash cups uniformly, reversing the direction of pouring regularly so that all the large aggregates do not accumulate at one end. Differences in the moisture content and surface distribution of stones and aggregates are the main reasons for the high level of variability in measurements of splash erosion in laboratory experiments (Luk 1981). Although the soils are free draining, their condition does not represent the field situation because they are uncompacted and not subject to suction. Attempts to simulate soil suction on such a small scale have generally not been successful and it is difficult to achieve repeatable compaction. Fortunately, a soil in a loose saturated state is at its most erodible, so laboratory experiments reproduce the worst conditions for erosion. Further processing of the soil – for example, passing it through 4 mm or finer sieves to remove stones or break up aggregates – may reduce the variability in the splash measurements but can produce misleading results compared with small plot studies in the field (Rickson 1987).

5.3.2 Runoff simulation

Where the target is in the form of a small soil plot, the rainfall simulator may be supplemented by a device to supply a known quantity of runoff at the top of the plot, instead of relying solely on runoff resulting from the rainfall. The set-up at the Experimental Geomorphology Laboratory, University of Leuven, Belgium, consists of an aluminium and plexiglass flume, 4 m long and 0.4 m wide, which is fed by water at its upper end through a cylinder with ten openings. The discharge is controlled by a tap through which water is pumped to the cylinder from a container; a constant water level is maintained in the container (Savat 1975). The slope of the flume can be adjusted to different steepnesses and rainfall simulators positioned over the top. The main problem with a small flume of this type is that edge effects are difficult to eliminate. They can be reduced, however, by collecting the sediment and runoff washing off a narrow strip down the centre of the plot instead of collecting for the whole plot width. The short length of the flume makes it difficult to simulate rill erosion although the supply of runoff at the top of the slope can be adjusted to simulate the effect of different slope lengths. Sediment can also be added to the runoff upslope of the test soil.

A much longer combined rainfall and runoff simulator with adjustable slope has been built at the Centre for Geomorphology in Caen, France, to enable rill erosion to be investigated under controlled conditions (Govers et al. 1987). The flume is 20.4 m long, 1.4 m wide and 0.7 m deep. The base of the flume is covered with a 15-cm thick layer of gravel overlain by a perforated plastic sheet on top of which 40 cm thickness of test soil is placed. A rainfall simulator, comprising 644 flexible capillary tubes, is mounted 1.7 m above the top of the flume. A fine wire-mesh is placed 70 cm below the capillaries to break up the 2.9 mm diameter drops into a drop-size distribution ranging from 0.6 to 4.6 mm, with a median volume drop diameter of 2.8 mm. The long flume in the Soil Erosion Laboratory, University of Toronto, Canada, is constructed from ten flexible modules, each 2.45 m long, 0.85 m wide and 0.31 m deep, which can be tilted and combined to produce slopes of varying profile shape (Bryan & Poesen 1989). A rainfall simulator comprising nine cone-jet spray nozzles is placed 8 m above the flume. The 6 m long and 2 m wide flume at the National Soil Resources Institute, Cranfield University, England, is made up of three modules, which can be arranged to produce straight, convex, concave or convexo-concave slope shapes of varying steepnesses or to form two straight side slopes and a valley floor. Nine cone-jet nozzles provide rainfall simulation.

5.3.3 Wind tunnels

Almost all the basic studies on wind erosion have been carried out in the laboratory in wind tunnels. Wind is supplied by a fan that either sucks or blows air through the tunnel. The tunnel is shaped so that air enters through a honeycomb shield, serving as a flow straightener, into a convergence zone, where flow is constricted, passes through the test section and leaves through a divergence zone, in which flow is diffused. A mesh screen at the outlet traps most of the sand

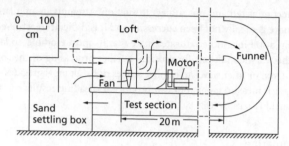

Fig. 5.9 Details of a closed-circuit wind tunnel.

particles, while allowing the air to blow through without the build-up of back pressure. Realistic wind profiles are produced only in the test section and the value of the tunnel depends on the length of this. Small tunnels, such as the one described by De Ploey and Gabriels (1980) with a test section only 1.5 m long, do not permit a satisfactory sand flow to be attained. A test section at least 15 m long is required for this, though Bagnold (1937) was able to simulate sand flow in a 10-m long tunnel by feeding a stream of sand into its mouth.

The principles of wind tunnel design are described by Pope and Harper (1966). Most of the tunnels used for soil erosion research are of the open-circuit type where air is drawn in from outside. These are cheaper than closed-circuit tunnels and afford easier control over the air flow in the test section because there is less upstream contact between the air and the boundary walls and therefore less turbulence. In a closed-circuit tunnel (Fig. 5.9), the fan is located in a loft above the test section and dry centrally heated air is forced downwards and into the test section through a funnel where unwanted turbulence is likely to be induced. Closed-circuit tunnels are less noisy and give good control over air humidity. The latter is important because it influences the critical wind velocity for particle movement. Soil trays are placed on the floor of the tunnel in the test section and these can be removed for weighing. Thus the amount of erosion can be determined by weight loss.

Wind and water erosion have traditionally been studied separately. In order to enhance understanding of their interactions, a combined wind tunnel and rainfall simulation facility has been built at the International Centre for Eremology at the University of Ghent, Belgium (Gabriels et al. 1997). The closed-circuit tunnel has a 12 m long and 1.2 m wide test section with an adjustable ceiling height of 1.8–3.2 m; a rainfall simulator, comprising adjustable spray nozzles, is installed in the ceiling.

5.4

Composite experiments

In recent years, considerable use has been made of rainfall simulators in the field. They have now virtually replaced the natural runoff plot as the major research tool in the USA for studies of soil erodibility and vegetation cover effects. Field rainfall simulators provide the advantages of field conditions for soils, slope and plant cover, all of which are difficult to reproduce in the laboratory, with the benefits of a repeatable storm.

For studies with small plots, the rainfall simulators used in the laboratory, like the one designed by Meyer and Harmon (1979), can be transferred directly to the field. Larger simulators can be built by joining together modules of the smaller ones. Modules of a simulator built by Bazzoffi et al. (1980) have been combined to form a field simulator capable of covering a standard runoff plot (Zanchi et al. 1981). Alternatively, large-area simulators can be purpose-designed and built. The best example is the rotating-boom simulator (Swanson 1965) in which nozzles are mounted on a series of spray booms positioned radially from a central stem. The nozzles spray water from a height of 3 m and the booms rotate at four revolutions per minute to give a uniform distribution of rain over a circular area large enough to accommodate two plots, 10×4.3 m.

A major difficulty with field rainfall simulators is their limited portability, especially in difficult terrain. Several designs for simple, portable simulators have been produced (Imeson 1977; Kamphorst 1987; Cerdà et al. 1997) with the ability to generate rainfall at intensities between 40 and $120 \, mm \, h^{-1}$ with reasonably realistic drop-size distributions. However, their limited water storage and low height means that they produce only short duration storms of low energy. They also cover very small plot sizes, often less than $1 \, m^2$.

Less use has been made of field wind tunnels partly because, as noted above, small ones, which would be the most portable, are of limited value. Since experimental field plots are not used in wind erosion research, there is little demand for large permanently sited tunnels. However, some large wind tunnels have been developed for field use (Zingg 1951; Chepil et al. 1955; Bocharov 1984; Fryrear 1984).

Box 5

Sediment budgets

If a full picture of soil erosion by water over a landscape is to be established, it is necessary to examine the various sources of sediment, the pathways along which it is moved to the catchment outlet and the opportunities for deposition on the way. If separate measurements are made of each of these components, they can be used to establish a sediment budget.

The selection of measurement sites to establish the pattern of sediment movement poses a problem of sampling. Since it is only possible to take measurements at specific points in the landscape, it is important that these are representative of the catchment as a whole and not biased towards known places of high erosion, as otherwise an unrealistic assessment of the sediment budget will be obtained. One approach (Fig. B5.1) is to divide a catchment into sub-basins and set up recording stations to measure the sediment yield at the mouth of each. Within each sub-basin, slope profiles are chosen at regular intervals or at random along the mid-slope line, which lies half-way between the main river and the divide. Bounded erosion plots or Gerlach troughs are installed along each profile either randomly or in set positions. Amphlett (1988) recommends that within such a nested arrangement, the respective sample catchments should differ in size by an order of magnitude. Thus, within a $100 \, km^2$ catchment, the first level of sub-catchment should be $10 \, km^2$ in size, the next level $1 \, km^2$, the next $0.1 \, km^2$ and so on, down to $10 \, m^2$ or even $1 \, m^2$ erosion plots. To this basic structure, there is a need to establish measurement points along known pathways of sediment movement such as tractor wheelings and trackways, to set up networks of erosion pins or sediment traps to collect material in depositional sites, and to monitor erosion of river banks using either erosion pins inserted horizontally into the banks or repeated surveys. Data may also be obtained from sediment surveys in ponds and reservoirs. The information obtained is then used to estimate the contribution of the different sources to the sediment yield of the catchment.

Continued

Measurement of soil erosion

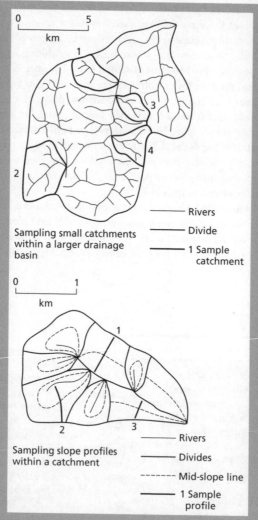

0   5
km

1

3

4

2

Sampling small catchments
within a larger drainage
basin

———— Rivers

———— Divide

━━━━ 1 Sample
catchment

0   1
km

1

2   3

Sampling slope profiles
within a catchment

———— Rivers

———— Divides

------ Mid-slope line

━━━━ 1 Sample
profile

Fig. B5.1 Sampling scheme based on nested catchments.

The value of sediment budgets in understanding the behaviour of catchments over time was demonstrated by Trimble (1983) for the 360 km² catchment of Coon Creek, Wisconsin. He compared budgets for two time periods, 1853–1938 and 1938–75. In the first period, the catchment suffered severe soil erosion because farming was practised with poor management; most of the eroded sediment was deposited as colluvium on the lower slopes. In the second period, soil conservation measures were introduced. Although these reduced hillslope erosion by 26 per cent, they seemed to have no impact on the sediment yield from the catchment. With erosion control in place, the runoff reaching the lower slopes was no longer carrying material at transport capacity and had excess energy available for erosion. As a result, some 20 per cent of the colluvial material was remobilized, the tributary valleys and part of the upper main valley became sediment sources rather than sinks and the sediment yield of the catchment remained unchanged.

Table B5.1 gives the sediment budgets for four small catchments in England (Walling et al. 2002). Suspended sediment yield was measured at the catchment outlets and from field drains. Measurements were made of erosion on river banks and sediment storage in the river channels and wetlands. Erosion and deposition rates in fields within the catchments were estimated from caesium-137 loadings in the soil. Fingerprinting techniques (Collins & Walling 2002) were used to discriminate between sediments derived from the catchment surface, subsurface tile drains and channel banks. The results show that while surface erosion is the most important source of sediment, field drains and other subsurface pathways are as important as surface runoff in transporting this material to the river. Field drains account for between 30 and 60 per cent of the sediment output. A substantial amount of sediment goes into storage within the catchment so that sediment delivery ratios are between 14 and 27 per cent. River banks account for only between 6 and 10 per cent of the sediment.

Table B5.1 Sediment budgets (t) for Rosemaund and Smisby catchments, England, 1997–1998

Soils Slopes Sub-catchment		Rosemaund Silty clay loams <5° Belmont	Jubilee	Smisby Clay Mean of 2.2° Lower Smisby	New Cliftonthorpe
Input	Surface erosion	699.9	143.2	1041.2	425.8
	Subsurface erosion	6.8	2.1	6.0	1.7
	Bank erosion	14.1	4.7	14.4	3.4
Storage	In-field	194.2	61.7	292.2	153.2
	Field-to-channel	401.4	47.4	558.2	213.4
	In-channel	2.3	0.8	2.3	0.5
	Wetlands	0.0	0.0	0.0	2.5
Transport	Surface runoff	43.6	15.6	136.4	43.9
	Subsurface pathways	60.7	18.5	54.4	15.3
	Field drains	67.5	20.6	60.4	17.0
Sediment delivery ratio (%)		17.1	26.7	19.7	14.2
Output	Suspended sediment load	122.9	40.1	208.9	61.3

Source: after Walling et al. (2002b).

CHAPTER 6

Modelling soil erosion

The techniques described in Chapter 5 allow rates of erosion to be determined at different positions in the landscape over various spatial and time scales. However, it is not possible to take measurements at every point in the landscape. It also takes time to build up a sufficient data base to ensure that the measurements are not biased by an extreme event or a few years of abnormally high rainfall. Long-period measurements are required to study how erosion rates respond to changes in land use and climate or the use of erosion-control measures. In order to overcome these deficiencies, models can be used to predict erosion under a wide range of conditions. The results of the predictions can then be compared with the measurements to ensure their validity. If the predictions are sufficiently accurate, the method may be used to estimate erosion in other areas of similar conditions.

Most of the models used in soil erosion studies are of the empirical grey-box type (Table 6.1). They are based on defining the most important factors and, through the use of observation, measurement, experiment and statistical techniques, relating them to soil loss. However, as our understanding of the mechanics of erosion processes grows, the opportunity exists for developing white-box and physically based models.

6.1

Defining objectives

Models are of necessity simplifications of reality. Decisions need to be made on a suitable level of complexity or simplicity depending on the objective. The starting point for all modelling must therefore be a clear statement of the objective. This may be prediction or explanation. Managers, planners and policy-makers require relatively simple predictive tools to aid decision-making, albeit about rather complex systems. Researchers seek models that describe how the system functions in order to enlighten understanding of the system and how it responds to change. This chapter is concerned only with predictive models since these are most useful for practical applications but, increasingly, many of these models are developed by researchers and have a sound physical base reflecting our knowledge of how the erosion system works.

Decisions need to be made on whether prediction should be for a year, a day, a storm or short periods within a storm; and whether it should be for a field, a hillslope or a drainage basin. These differences in temporal and spatial perspectives will influence the processes that need to be

Table 6.1 Types of models

Type	Description
Physical	Scaled-down hardware models usually built in the laboratory; need to assume dynamic similitude between model and real world.
Analogue	Use of mechanical or electrical systems analogous to system under investigation, e.g. flow of electricity used to simulate flow of water.
Digital	Based on use of digital computers to process vast quantities of data.
(a) Physically based	Based on mathematical equations to describe the processes involved in the model, taking account of the laws of conservation of mass and energy.
(b) Stochastic	Based on generating synthetic sequences of data from the statistical characteristics of existing sample data; useful for generating input sequences to physically based and empirical models where data are only available for short periods of observation.
(c) Empirical	Based on identifying statistically significant relationships between assumed important variables where a reasonable data base exists. Three types of analysis are recognized: (i) black-box, where only main inputs and outputs are studied; (ii) grey-box, where some detail of how the system works is known; (iii) white-box, where all details of how the system operates are known.

Source: after Gregory and Walling (1973).

included in the model, the way they are described and the type of data required for model validation and operation. Good scientific practice should force all modellers to specify the design requirements of their models before they are developed. A user should be able to expect that a model has been tested for the conditions for which it was designed and, from this information, should then be able to select the most appropriate model to meet a specific set of objectives. Any attempt to use a model for conditions outside those specified should be viewed as bad practice and, at best, speculative.

The objective has implications for how well a model must perform. Screening models are simple in concept and designed to identify problem areas. Usually it is sufficient if predictions are of the right order of magnitude. Assessment models need to predict with greater accuracy because they are used for evaluating the severity of erosion under different management systems and may be intended as design tools for the selection of conservation practices. For examining the on-site consequences of erosion, such as the loss of soil depth and declines in productivity, predictions of erosion are required for periods of 20–30 years either on an annual basis or as an annual average over the time span. More detailed, event models are needed for assessing the off-site effects – for example, pesticide use, which may create a problem for only a few days following application, and comparing pollutant loads in rivers from different agricultural management systems.

Our understanding of erosion processes is greatest over very short time periods of only a few minutes. While it might be feasible to apply this understanding to slightly longer periods, continuous extrapolation is not possible. A single event or a storm is probably the upper limit to

Modelling soil erosion

which relationships established for instantaneous conditions can be applied. Thus longer-term modelling can only be achieved by summing the predictions for individual storms. The alternative is to develop models empirically using data collected on an annual or mean annual basis. The scale of operation must also be considered. The detailed requirements for modelling erosion over a large drainage basin differ from those demanded by models of soil loss from a short length of hillslope or at the point of impact of a single raindrop. Scale influences the number of factors that need to be incorporated in the model, those that can be held constant and those that can be designated primary factors around which the model must be constructed.

In addition to the area over which the model should operate being defined, the behaviour of the model at the boundary of that area needs to be considered. The simplest assumption is that a model applies to a single area of land with a well defined boundary across which there is no transfer of water or sediment. Such an assumption is clearly unrealistic for most hillslope applications, since water and sediment pass downslope from one segment to another and, where the slope is variable in plan, may also move across-slope to concentrate in hollows or areas of flow convergence. Most erosion models allow for transfer of material across the lower boundary but few deal with more complicated boundaries, such as the junction between the hillslope and the river channel. Most soil erosion models also consider the boundaries as fixed in their location over time, ignoring the possible movement of the lower hillslope boundary upslope as a result of bank erosion and the gradual lowering of the summit by erosion.

The model should be formulated conceptually, representing it by a flow chart such as that shown in Fig. 6.1. Viewing the model in this way enables the structure of the system, the logical order of transfers of energy and matter through the system, the system variables and the interactions between the variables to be defined. It is also a good test of the level of scientific understanding of the system and the degree to which the model might need to be simplified because of insufficient knowledge. A model can be simplified by concentrating on those processes which have the greatest influence over the output and ignoring those which have very little effect. Once the processes operating within the model have been identified, they need to be described mathematically. The methods used can range from simple equations, often expressing a statistical relationship, to complex expressions related to the fundamental physics or mechanics of the process. The former are more common in empirical models, whereas the latter provide the foundation for physically based models.

It is generally recognized that a good model should satisfy the requirements of reliability, universal applicability, easy use with a minimum of data, comprehensiveness in terms of the factors and erosion processes included and the ability to take account of changes in climate, land use and conservation practice. Unfortunately, many of the ideal characteristics of models conflict with each other. Ease of operation can mean that input data are readily available – for example from tables in an accompanying user's manual. Such data, however, are at best only guide values and there may be uncertainty in how well they describe actual conditions.

6.2 Empirical models

The simplest model is a black-box type relating sediment loss to either rainfall or runoff. A typical relationship is:

$$Qs = aQ^b \tag{6.1}$$

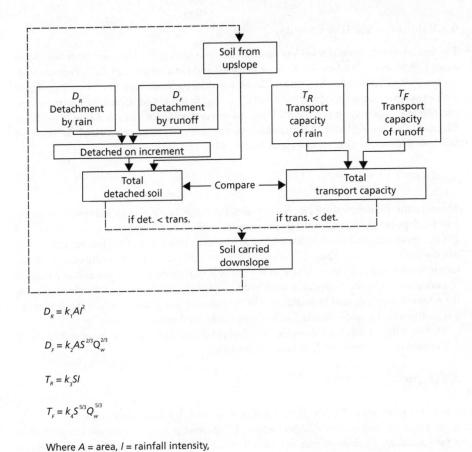

$$D_R = k_1 A I^2$$

$$D_F = k_2 A S^{2/3} Q_w^{2/3}$$

$$T_R = k_3 S I$$

$$T_F = k_4 S^{5/3} Q_w^{5/3}$$

Where A = area, I = rainfall intensity,

S = ground slope ($\sin \theta$),

Q_w = runoff

Fig. 6.1 Flow chart of the model of the processes of soil erosion by water (after Meyer & Wischmeier 1969).

where Qs is the sediment discharge and Q is the water discharge. It is not always possible to establish the values of a and b with confidence. The relationship between sediment and water discharge may vary with the volume of runoff and therefore change seasonally. During a single storm the value of b often differs on the rise of the flood wave from that on the recession. The main disadvantage of this type of model is that it gives no indication of why erosion takes place.

Greater understanding of the causes of erosion is achieved by grey-box models. These frequently culminate in expressing the relationship between sediment loss and a large number of variables with a regression equation. A problem with these empirical models is that the equations cannot be extrapolated beyond their data range, either to more extreme events or to other geographical areas.

Modelling soil erosion

6.2.1 Universal Soil Loss Equation

The first attempt to develop a soil loss equation for areas such as hillslopes and fields was that of Zingg (1940), who related erosion to slope steepness and slope length (eqn 3.11). Further developments led to the addition of a climatic factor based on the maximum 30-minute rainfall total with a two-year return period (Musgrave 1947), a crop factor, to take account of the protection-effectiveness of different crops (Smith 1958), a conservation factor and a soil erodibility factor. Changing the climatic factor to the rainfall erosivity index (R) ultimately yielded the Universal Soil Loss Equation (USLE; Wischmeier & Smith 1978). The equation is:

$$E = R \times K \times L \times S \times C \times P \tag{6.2}$$

where E is the mean annual soil loss, R is the rainfall erosivity factor, K is the soil erodibility factor, L is the slope length factor, S is the slope steepness factor, C is the crop management factor and P is the erosion-control practice factor. Since K represents mean annual soil loss per unit of R, E has the same units as K. Thus, if K is in $t\,ha^{-1}$ for one unit of metric R, multiplication by the metric R value will give the value of E in $t\,ha^{-1}$. If K is in US tons per acre per unit of American R, multiplication by the American R value yields a value of E in US tons per acre. In both cases, the E value is then adjusted by multiplying by the values of L, S, C and P, which are dimensionless coefficients. The individual factors in the equation are derived as follows.

R. This is the rainfall erosivity index, described in section 4.1.2, based on mean annual EI_{30}. If E is in foot-tons per acre and I_{30} is in inches per hour,

$$R = EI_{30}/100 \tag{6.3}$$

R is in American units. If E is in $MJ\,ha^{-1}$ and I_{30} is in $mm\,h^{-1}$, R is in metric units.

K. This is the soil erodibility index (section 3.2) defined as mean annual soil loss per unit of R for a standard condition of bare soil, recently tilled up-and-down slope with no conservation practice and on a slope of 5° and 22 m length. K can be defined for a unit of both American and metric R. Wherever possible, K should be based on measured values. If the value is obtained from the nomograph (Fig. 3.1), it should be remembered that this is an estimated value and will be subject to error.

LS. The factors of slope length (L) and slope steepness (S) can be combined in a single index, which expresses the ratio of soil loss under a given slope steepness and slope length to the soil loss from the standard condition of a 5° slope, 22 m long, for which $LS = 1.0$. The appropriate value can be obtained from nomographs (Wischmeier & Smith 1978) or from the equation:

$$LS = \left(\frac{x}{22.13}\right)^{n} (0.065 + 0.045s + 0.0065s^{2}) \tag{6.4}$$

where x is the slope length (m) and s is the slope gradient in per cent. As seen in section 3.3, the value of n should be varied according to the slope steepness.

C. The crop management factor represents the ratio of soil loss under a given crop to that from bare soil. Since soil loss varies with the erosivity and the morphology of the plant cover (section 3.4), it is necessary to take account of changes in these during the year in arriving at an annual value. For arable farming, the year is divided into periods corresponding to different stages of crop growth. These are defined as: (i) fallow, inversion ploughing to seed-bed establishment; (ii) seed-bed, secondary tillage for formation of the seed-bed to 10 per cent crop cover; (iii) establishment, 10–50 per cent crop cover; (iv) development, 50–75 per cent crop cover; (v) maturity, 75 per cent crop cover to harvest; and (vi) residue or stubble, harvest to ploughing or new seeding. For the last period, three options are considered: leaving the residue in the field without seeding; leaving the residue and seeding an off-season crop; and removing the residue and leaving the ground bare. For a given crop, separate ratio values are obtained for each period from tables summarizing data collected over many years by the United States Natural Resources Conservation Service at their experimental stations (Wischmeier & Smith 1978). The values vary not only with the crop but also, for a single crop, with yield, plant density and the nature of the previous crop. The individual values for each period are weighted according to the percentage of the mean annual R value falling in that period. The weighted values are then summed to give the annual C value.

The method just described allows C factor values to be determined for the crop rotations and management practices found in the USA. For other countries, detailed information for calculating the C factor in this way does not always exist and it may be more appropriate to use average annual values. Table 6.2 gives typical ranges of values for different crops and management systems.

P. Values for the erosion-control practice factor are obtained from tables of the ratio of soil loss where the practice is applied to the soil loss where it is not. With no erosion-control practice, $P = 1.0$. Values cover contouring and contour strip-cropping and vary with the slope steepness (Table 6.3). Where diversion terracing is adopted, the value for contouring is used for the P factor and the LS factor is adjusted for the slope length represented by the horizontal spacing between the terraces. This method is not appropriate for bench terracing, for which values of P for different types have been established from research in Taiwan (Chan 1981).

As the example in Table 6.4 shows, the equation is normally used to predict mean annual loss. Since it was derived from and tested on data from experimental stations in the USA, which, when combined, represent over 10,000 years of record, it is widely accepted as reliable. When tested against a data set of 2300 plot years, it gave predictions of mean annual soil loss within ±5 t ha^{-1} 84 per cent of the time (Wischmeier & Smith 1978). The USLE has become the standard technique of soil conservation workers. The equation can be rearranged so that, if an acceptable value of E is chosen, the slope length (L) required to reduce soil loss to that value can be calculated, all other conditions being unchanged. In this way, appropriate terrace spacings can be determined. Alternatively, the required C value may be predicted and the tables searched to find the most suitable cropping practice that will give that value.

Although the equation is described as universal, its data base, though extensive, is restricted to the USA east of the Rocky Mountains, to slopes where cultivation is permissible, normally 0–7°, and to soils with a low content of montmorillonite; it is also deficient in information on the erodibility of sandy soils. Several attempts have been made to apply the equation more widely. Extensive data have been collected on R, K, C and P factor values so that it can be used in West Africa (Roose 1975) and India (Singh et al. 1981). Modifications have been made to the R value and data collected on C factor values so that it can be applied to Bavaria, Germany (Schwertmann et al. 1987), where it has been used in association with erosion surveys to quantify the risk of erosion (Auerswald & Schmidt 1986).

In addition to the limitations of its data base, there are theoretical problems with the equation. Soil erosion cannot be adequately described merely by multiplying together six factor values. There is considerable interdependence between the variables. Some of these are considered. For instance, rainfall influences the R and C factors and terracing the L and P factors. Other interactions, however, such as the greater significance of slope steepness in areas of intense rainfall (section 3.3) are ignored. The rainfall erosion index is based on studies of drop-size distributions

Table 6.2 *C*-factor values for the Universal Soil Loss Equation

Practice	Average annual C-factor
Bare soil	1.00
Forest or dense shrub, high mulch crops	0.001
Savanna or prairie grass in good condition	0.01
Overgrazed savanna or prairie grass	0.10
Maize, sorghum or millet; high productivity; conventional tillage	0.20–0.55
Maize, sorghum or millet; low productivity; conventional tillage	0.50–0.90
Maize, sorghum or millet; high productivity; chisel ploughing into residue	0.12–0.20
Maize, sorghum or millet; low productivity; chisel ploughing into residue	0.30–0.45
Maize, sorghum or millet; high productivity; no or minimum tillage	0.02–0.10
Cotton	0.40–0.70
Meadow grass	0.01–0.025
Soya beans	0.20–0.50
Wheat	0.10–0.40
Rice	0.10–0.20
Groundnuts	0.30–0.80
Palm trees, coffee, cocoa with cover crops	0.10–0.30
Pineapple on contour; residue removed	0.10–0.40
Pineapple on contour; with surface residue	0.01
Potatoes; rows downslope	0.20–0.50
Potatoes; rows across slope	0.10–0.40
Cowpeas	0.30–0.40
Strawberries; with weed cover	0.27
Pomegranate; with weed cover	0.08
Pomegranate; clean-weeded	0.56
Ethiopian tef	0.25
Sugar cane	0.13–0.40
Yams	0.40–0.50
Pigeon peas	0.60–0.70
Mungbean	0.04
Chilli	0.33
Coffee: after first harvest	0.05
Plantains: after establishment	0.05–0.10
Papaya	0.21

Source: after Wischmeier and Smith (1978), Roose (1977), Singh et al. (1981), El-Swaify et al. (1982), Hurni (1987), Hashim and Wong (1988).

of rains, which, as seen in section 3.1.2, may have limited applicability. One important factor to which soil loss is closely related, namely runoff, is not dealt with explicitly but is incorporated within the *R* factor.

During the 1970s, the Universal Soil Loss Equation was widely used for estimating sheet and rill erosion in national assessments of soil erosion in the USA. By necessity, the equation was applied to conditions beyond its data base – for example, to rangeland and forest land in the western USA. As a result of the experience, a number of changes were made that are now incorporated in the Revised Universal Soil Loss Equation (RUSLE) (Renard et al. 1991). These include:

Table 6.3 P-factor values for the Universal Soil Loss Equation

Erosion-control practice	P-factor value
Contouring: 0–1° slope	0.60*
Contouring: 2–5° slope	0.50*
Contouring: 6–7° slope	0.60*
Contouring: 8–9° slope	0.70*
Contouring: 10–11° slope	0.80*
Contouring: 12–14° slope	0.90*
Level bench terrace	0.14
Reverse-slope bench terrace	0.05
Outward-sloping bench terrace	0.35
Level retention bench terrace	0.01
Tied ridging	0.10–0.20

* Use 50% of the value for contour bunds or if contour strip cropping is practised.
Source: after Wischmeier and Smith (1978), Roose (1977), Chan (1981a).

Table 6.4 Prediction of soil loss using the Universal Soil Loss Equation

Problem
Calculation of mean annual soil loss on a 100-m long slope of 7° on soils of the Rengam Series under maize cultivation with contour bunds spaced at 20 m intervals, near Kuala Lumpur, Malaysia.

Equation
Mean annual soil loss = $R \times K \times LS \times C \times P$

Estimating R (Rainfall erosion index)
In the absence of locally published maps of R values, three different procedures for estimating R from mean annual precipitation (P) are used.

Method 1

Mean annual precipitation (P)	= 2695 mm
From Roose (1975), mean annual rainfall erosion index (R) in American units	= 0.5 P
	= 0.5 × 2695
	= 1347.5
Conversion to $Mg\,mm\,ha^{-1}h^{-1}$	= 1347.5 × 17.3
	= 23,311.8

Method 2

From Morgan (1974), mean annual erosivity ($KE > 25$)	= 9.28 P − 8,838
	= (9.28 × 2695) − 8838
	= 16,171.6 $J\,m^{-2}$
	= 161.716 $Mg\,ha^{-1}$

Table 6.4 *Continued*

Multiply by I_{30} (use $75\,\mathrm{mm\,h^{-1}}$; maximum value $= 161.716 \times 75$ recommended by Wischmeier and Smith (1978)

$$= 12,128.7$$

Method 3
From Foster et al. (1981), mean annual $EI_{30} = 0.276\,P \times I_{30}$
$(\mathrm{kg.m.mm})(\mathrm{m^2\,h^{-1}})$

$$= 0.276 \times 2695 \times 75$$
$$= 55,786.5$$

Divide by 100 to convert to $\mathrm{Mg\,mm\,ha^{-1}\,h^{-1}} = 55.79$

Best estimate: discard result from Method 3, which is rather low.
Take average value of Methods 1 and 2: $= 17,720$

Estimating K (Soil erodibility index)
From Whitmore and Burnham (1969), the soils have a 43% clay, 8% silt, 9% fine sand and 40% coarse sand content; organic content is about 3%.
Using the nomograph (Fig. 3.1), gives a first approximation K value $= 0.005$

Estimating LS (Slope factor)
For slope length (l) and slope steepness (s) in metres and percent respectively,

$$LS = (l/22)^{0.5}(0.065 + 0.045s + 0.0065s^2)$$

With contour bunds at $20\,\mathrm{m}$ spacing, $l = 20\,\mathrm{m}$ and $s = 12\%$ (approximation of 7°)

$$LS = (20/22)^{0.5}(0.065 + (0.045 \times 12) + (0.0065 \times 12^2))$$
$$LS = 0.95 \times 1.54$$
$$LS = 1.46$$

Estimating C (Crop management factor)
According to Table 6.2, the C value for maize ranges between 0.2 and 0.9, depending on the productivity. For many tropical farming conditions, C for maize lies between 0.4 and 0.9 (Roose 1975), depending on the cover.

During the three-month period from seeding to harvest, the cover is likely to vary from 9 to 45 per cent in the first month, from 55 to 93 per cent in the second month, and from 45 to 57 per cent in the third month. Therefore, we might assume C values of 0.9, 0.4 and 0.7 for the three respective months.

Maize can be planted at any time of year in Malaysia but assume planting after the April rains, allowing growth, ripening and harvesting in June and July, which are the driest months. Land is under dense secondary growth prior to planting (assume $C = 0.001$) and allowed to revert to the same after harvest ($C = 0.1$).

Of the mean annual precipitation, 32 per cent falls between January and April inclusive, 10 per cent in May, 6 per cent in June, 7 per cent in July and 45 per cent between August and December. Assuming that erosivity is directly related to precipitation amount, these values can be used to describe the distribution of the R factor throughout the year.

From the above information, the following table is constructed.

Table 6.4 *Continued*

Months	C value	Adjustment factor (% R value)	Weighted C value (col 2 × col 3)
January–April	0.001	0.32	0.00032
May	0.9	0.10	0.09
June	0.4	0.06	0.024
July	0.7	0.07	0.049
August–December	0.1	0.45	0.045
Total			0.20832

C-factor for the year = 0.208

Estimating *P* (Erosion-control practice factor)
From Table 6.3, *P* value for contour bunds = 0.3

Soil loss estimation
Mean annual soil loss = 17,720 × 0.005 × 1.46 × 0.208 × 0.3
$$= 8.07 \, t \, ha^{-1}$$

revisions to the *R* factor values in the USA; the development of a seasonally variable *K* factor, obtained by weighting instantaneous estimates of *K* by the proportion of the annual *R* for successive 15-day periods; modifications to the *LS* factor to take account of the susceptibility of soils to rill erosion; and a new procedure for computing the *C*-factor value through the multiplication of various sub-factor values. The sub-factors included in the *C* factor take account of prior land use, crop canopy, surface cover (mulches and ground vegetation) and surface roughness.

The Universal Soil Loss Equation was developed as a design tool for soil conservation planning but, because of its simplicity, attempts have been made to use it as a research technique. Applying the equation to purposes for which it was not intended, however, cannot be recommended (Wischmeier 1978). Since it was designed for interrill and rill erosion, it should not be used to estimate sediment yield from drainage basins or to predict gully or stream-bank erosion. Care should be taken in using it to estimate the contribution of hillslope erosion to basin sediment yield because it does not estimate deposition of material or incorporate a sediment delivery ratio. Since the equation was developed to estimate long-term mean annual soil loss, it cannot be used to predict erosion from an individual storm. If applied in this way, it provides an estimate of the average soil loss expected from a number of such storms and this may be quite different from the actual soil loss in any single storm. The equation should not be applied to conditions for which factor values have not been determined and therefore need to be estimated by extrapolation. These qualifications apply equally to the Revised Universal Soil Loss Equation.

6.2.2 SLEMSA

The Soil Loss Estimator for Southern Africa (SLEMSA) was developed largely from data from the Zimbabwe highveld to evaluate the erosion resulting from different farming systems so that appropriate conservation measures could be recommended. The technique has since been adopted throughout the countries of Southern Africa. The equation is (Elwell 1978):

$$Z = K \times X \times C \qquad (6.5)$$

where Z is mean annual soil loss $(t\,ha^{-1})$, K is mean annual soil loss $(t\,ha^{-1})$ from a standard field plot, 30 m long, 10 m wide, at 2.5° slope for a soil of known erodibility (F) under a weed-free bare fallow, X is a dimensionless combined slope length and steepness factor and C is a dimensionless crop management factor.

K. The value of K is determined by relating mean annual soil loss to mean annual rainfall energy (E) using the exponential relationship:

$$\ln K = b\ln E + a \tag{6.6}$$

where E is in $J\,m^2$, and the values of a and b are functions of the soil erodibility factor (F):

$$a = 2.884 - 8.2109F \tag{6.7}$$

$$b = 0.4681 + 0.7663F \tag{6.8}$$

X. The topographic factor (X) adjusts the value of soil loss calculated for the standard condition to that for the actual conditions of slope steepness and slope length. The value of X is obtained from:

$$X = (L)^{0.5}[(0.76 + 0.53s + 0.076s^2)/25.65] \tag{6.9}$$

where L is the slope length (m) and s is the percentage slope.

The procedure is modified if contour ridges are used as a conservation measure. In this case, instead of L and s being defined in the downslope direction, L becomes the maximum length of the ridges and s is the across-slope grade (in per cent) of the ridge. If the across-slope grade is less than 1 per cent, the value of X is first determined for a 1 per cent grade and then multiplied by the term Y, defined by:

$$Y = s/(0.572s + 0.428) \tag{6.10}$$

C. The crop management factor (C) adjusts the value of soil loss for the standard bare soil condition to that from a cropped field. The value is dependent upon the percentage of the rainfall energy intercepted by the crop (i). For crops and natural grassland with $i < 50$ per cent and for dense pastures and mulches with $i \geq 50$ per cent, the following relationship is used:

$$C = e^{(-0.06i)} \tag{6.11}$$

For crops and natural grasslands when $i \geq 50$ per cent, the relationship is:

$$C = (2.3 - 0.01i)/30 \tag{6.12}$$

The value of i is obtained by weighting the percentage crop cover in each ten-day period by the percentage of the mean annual rainfall energy (E) occurring in that period and summing the values.

Guide values for soil erodibility (F) and average percentage crop covers for use with SLEMSA are given in Table 6.5. An example of its use is shown in Table 6.6.

Table 6.5 Input values for soil erodibility and crop cover for use in SLEMSA

Soil erodibility (F factor)

Soil texture	Soil type	F value
Light	sands loamy sands sandy loams	4
Medium	sandy clay loam clay loam sandy clay	5
Heavy	clay heavy clay	6

Subtract the following from the F value:

1	for light-textured soils consisting mainly of sands and silts
1	for restricted vertical permeability within one metre of the surface or for severe soil crusting
1	for ridging up-and-down the slope
1	for deterioration in soil structre due to excessive soil loss in the previous year (>20 t ha^{-1}) or for poor soil management
0.5	for slight to moderate surface crusting or for soil losses of 10–20 t ha^{-1} in the previous year.

Add the following to the F value:

2	for deep (>2 m) well drained, light-textured soils
1	for tillage techniques which encourage maximum retention of water on the surface, e.g. ridging on the contour
1	for tillage techniques which encourage high surface infiltration and maximum water storage in the profile, e.g. ripping, wheel-track planting
1	for first season of no tillage
2	for subsequent seasons of no tillage

Crop cover ratings (C)

Crop	Average percentage cover
Cotton	40–65
Cowpeas	40–55
Tobacco	11–54
Sorghum	50–70
Sunflower	20–59
Groundnuts	55–65
Velvet beans	46–70
Coffee	60–80
Maize	42–80
Rotational grass	80–98
Soya beans	40–65
Rice	70–80
Weed fallow	100

After Elwell and Wendelaar (1977), Elwell (1978).

Table 6.6 Prediction of soil loss using SLEMSA

Problem

Calculation of mean annual soil loss on a 100-m long slope of 7° on soils of the Rengam Series under maize cultivation with contour bunds at 20 m spacing, near Kuala Lumpur, Malaysia (compare with Table 6.4).

Basic equation

Soil loss $(Z) = K \times X \times C$

Estimating K

Mean annual precipitation (P)	$= 2695\,mm$
Mean annual rainfall energy (E) (Morgan 1974)	$= 9.28P - 8838$
	$= (9.28 \times 2695) - 8838$
	$= 16,172\,J\,m^{-2}$

Soil erodibility (F)

From Table 6.5, for a sandy clay, $F = 5$

Calculation of K

$$\ln K = b\ln E + a$$

where $a = 2.884 - 8.1209F$

$b = 0.4681 + 0.7663F$

Therefore, a	$= 2.884 - (8.1209 \times 5)$
	$= -37.7205$
and b	$= 0.4681 + (0.7663 \times 5)$
	$= 4.2996$
$\ln K$	$= (4.2996 \times 9.6910) - 37.7205$
	$= 3.9469$

which gives $K = 51.78$

Estimating X

For unridged land

$X = \sqrt{l}\,(0.76 + 0.53s + 0.0765s^2)/25.65$

where l is slope length (m) and s is slope steepness (%)

For contour bunds every 20 m, $l = 20$ and $s = 7°$ or 12%

$X = \sqrt{20}\,(0.76 + (0.53 \times 0.12) + (0.0765 \times 12^2)/25.65$

$X = 4.47 \times 18.064/25.65$

$X = 3.15$

Estimating C

Calculation of the proportion of rainfall intercepted by the plant cover (i) by weighting the percentage crop cover for each cropping season by the proportion of the mean annual rainfall occurring in that season.

For the conditions described in Table 6.4 and taking values from Table 6.5

Months	% rainfall	% cover	Value of i (col 2 × col 3)
January–April	0.32	100	32.00
May–July	0.23	45	10.35
August–December	0.45	90	40.50
Total			82.85

$C = (2.3 - 0.01i)/30$

$= (2.3 - (0.01 \times 82.85))/30$

$= 0.049$

Soil loss estimation

Mean annual soil loss $= 51.78 \times 3.15 \times 0.049$

$= 7.99\,t\,ha^{-1}$

6.2.3 The Morgan, Morgan and Finney method

Morgan et al. (1984) developed a model to predict annual soil loss from field-sized areas on hillslopes, which, while endeavouring to retain the simplicity of the Universal Soil Loss Equation, encompassed some of the recent advances in understanding of erosion processes. The approach was revised by Morgan (2001). The model separates the soil erosion process into a water phase and a sediment phase (Fig. 6.2). The sediment phase is a simplification of the scheme described by Meyer and Wischmeier (1969; Fig. 6.1). It considers soil erosion to result from the detachment of soil particles by raindrop impact and runoff and the transport of those particles by overland flow. The process of transport by rainsplash is ignored. Thus, the sediment phase comprises three predictive equations, one for the rate of particle detachment by rainsplash, one for the rate of particle detachment by runoff and one for the transport capacity of overland flow. The inputs to these equations of rainfall energy and runoff volume respectively are obtained from the water phase. The model uses 12 operating functions (Table 6.7) for which 19 input parameters are required (Table 6.8). Typical values for soil parameters are given in Table 6.9, for effective hydrological depth of the soil in Table 6.10 and for the E_t/E_0 ratio, interception and the crop management factors in Table 6.11.

The effects of soil conservation practices can be allowed for within the separate phases of the model. For example, agronomic measures will bring about changes in evapotranspiration, interception and crop management, which will affect respectively the volume of runoff, the rate of detachment and the transport capacity.

Table 6.7 Operating functions for the Morgan–Morgan–Finney method of predicting soil loss

Water phase

$ER = R \times (1 - A)$
$LD = ER \times CC$
$DT = ER - LD$
$KE(DT) = DT(11.9 + 8.7 \log I)$
$KE(LD) = LD ((15.8 - PH^{0.5}) - 5.87)$
$KE = KE(DT) + KE(LD)$
$Q = R \exp(-R_c/R_o)$
$R_c = 1000 \, MS \times BD \times EHD(E_t/E_o)^{0.5}$
$R_o = R/R_n$

Sediment phase

$F = K \times KE \times 10^{-3}$
$H = ZQ^{1.5} \sin S(1 - GC) \times 10^{-3}$
$Z = 1/(0.5COH)$
$J = F + Z$
$G = CQ^2 \sin S \times 10^{-3}$

ER = effective rainfall (mm)
LD = leaf drainage (mm)
DT = direct throughfall (mm)
KE = kinetic energy of the rainfall ($J \, m^{-2}$)
Q = volume of overland flow (mm)
F = annual rate of soil particle detachment by raindrop impact ($kg \, m^{-2}$)
H = annual rate of soil particle detachment by runoff ($kg \, m^{-2}$)
J = annual rate of total soil particle detachment ($kg \, m^{-2}$)
G = annual transport capacity of overland flow ($kg \, m^{-2}$)

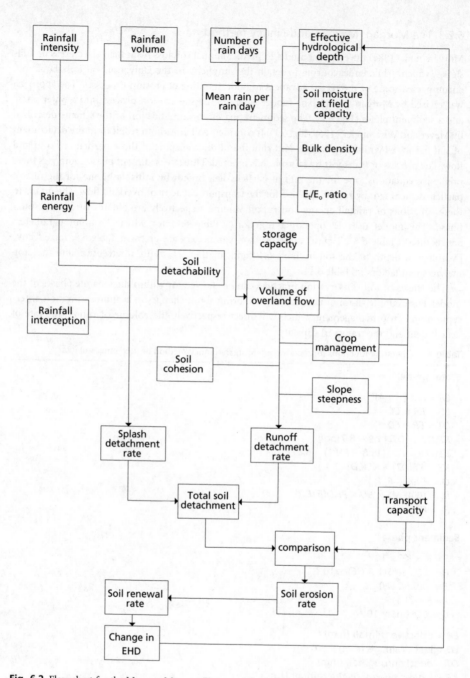

Fig. 6.2 Flow chart for the Morgan–Morgan–Finney method of predicting soil loss.

Table 6.8 Input parameters to the Morgan–Morgan–Finney method of predicting soil loss

Factor	Parameter	Definition
Rainfall	R	Annual or mean annual rainfall (mm)
	R_n	Number of rain days per year
	I	Typical value for intensity of erosive rain (mm h^{-1}); use 10 for temperate climates, 25 for tropical climates and 30 for strongly seasonal climates (e.g. Mediterranean type or monsoon)
Soil	MS	Soil moisture content at field capacity or 1/3 bar tension (wt %)
	BD	Bulk density of the top soil layer (Mg m^{-3})
	EHD	Effective hydrological depth of soil (m); value depends on vegetation/crop cover, presence or absence of surface crust, presence of impermeable layer within 0.15 m of the surface
	K	Soil detachability index (g J^{-1}) defined as the weight of soil detached from the soil mass per unit of rainfall energy
	COH	Cohesion of the surface soil (kPa) as measured with a torvane under saturated conditions
	SD	Total soil depth (m) defined as the depth of soil surface to bedrock
	W	Rate of increase in soil depth by weathering at the rock–soil interface (mm yr^{-1})
	V	Rate of increase in effective hydrological layer (mm yr^{-1}) as a result of crop management practices and the natural breakdown of vegetative matter into humus
Landform	S	Slope steepness (°)
Land cover	A	Proportion (between 0 and 1) of the rainfall intercepted by the vegetation or crop cover
	E_t/E_o	Ratio of actual (E_t) to potential (E_o) evapotranspiration
	C	Crop cover management factor; combines the C and P factors of the Universal Soil Loss Equation
	CC	Percentage canopy cover, expressed as a proportion between 0 and 1
	GC	Percentage ground cover, expressed as a proportion between 0 and 1
	PH	Plant height (m), representing the height from which raindrops fall from the crop or vegetation cover to the ground surface
Time	N	Number of consecutive years for which the model is to operate

The model compares the predictions of total detachment by rainsplash and runoff with the transport capacity of the runoff and assigns the lower of the two values as the annual rate of soil loss, thereby denoting whether detachment or transport is the limiting factor. The predictions obtained by the model are most sensitive to changes in annual rainfall and soil parameters when erosion is transport-limited and to changes in rainfall interception and annual rainfall when erosion is detachment-limited. Thus, good information on rainfall and soils is required for successful prediction. An example of the use of the model is presented in Table 6.12.

Like the Universal Soil Loss Equation, the model cannot be used to predict soil loss from individual storms or from gully erosion. The model has been used to predict soil loss in Indonesia

Modelling soil erosion

Table 6.9 Typical values for soil parameters used in the Morgan–Morgan–Finney method of predicting soil loss

Soil type	MS	BD	K	COH
Sand	0.08	1.5	1.2	2
Loamy sand	–	–	0.3	2
Sandy loam	0.28	1.2	0.7	2
Loam	0.20	1.3	0.8	3
Silt	–	–	1.0	–
Silt loam	0.25	1.3	0.9	3
Sandy clay loam	–	–	0.1	3
Clay loam	0.40	1.3	0.7	10
Silty clay loam	–	–	0.8	9
Sandy clay	–	–	0.3	–
Silty clay	0.30	–	0.5	10
Clay	0.45	1.1	0.05	12

Source: summarized in Morgan et al. (1982), Morgan (2001).

Table 6.10 Recommended values for the Effective Hydrological Depth (m) in the Morgan–Morgan–Finney method of predicting soil loss

Condition	EHD
Bare soil without surface crust; no impermeable barrier in top 0.2 m	0.09
Bare shallow soils on steep slopes; crusted soils	0.05
Row crops (e.g. wheat, barley, maize, beans, rice)	0.12
Row crops intercropped with legumes/grasses	0.15
Mature forest, dense secondary forest	0.20
Rubber, oil palm	0.15
Cocoa, coffee	0.12
Banana	0.18
Savanna/prairie grass	0.14
Cultivated grass	0.12
Cotton	0.10
Groundnut	0.12

Where terracing is used, add 0.01 to *EHD* to take account of the resulting increase in water storage.
Source: after Morgan (2001).

(Besler 1987) and Nepal (Shrestha 1997) and incorporated within a geographical information system for predicting erosion in Mediterranean Europe (De Jong 1994; Paracchini et al. 1997; De Jong et al. 1999). The model can be used to predict soil loss from small catchments by dividing the catchment into land units of similar soils, slopes and land cover (section 4.2.2) and routing the annual runoff and sediment production over the land surface from one unit to another. The runoff for each unit is then calculated as the summation of the runoff generated on that unit and the runoff flowing into it from upslope; the transport capacity is calculated for the combined runoff. Similarly, the total sediment available for transport is the summation of that detached on

Table 6.11 Typical values for plant parameters used in the Morgan–Morgan–Finney method of predicting soil loss

Plant/crop	A	E_t/E_o	C
Wet rice		1.35	0.1–0.2
Wheat	0.43	0.59–0.61	0.1–0.2 (winter sown)
			0.2–0.4 (spring sown)
Maize	0.25	0.67–0.70	0.2
Barley	0.30	0.56–0.60	0.1–0.2
Millet/sorghum		0.62	0.4–0.9
Cassava/yam			0.2–0.8
Potato	0.12	0.70–0.80	0.2–0.3
Beans	0.20–0.25	0.62–0.69	0.2–0.4
Groundnuts	0.25	0.50–0.87	0.2–0.8
Cabbage/Brussels sprouts	0.17	0.45–0.70	
Banana		0.70–0.77	
Tea		0.85–1.00	0.1–0.3
Coffee		0.50–1.00	0.1–0.3
Cocoa		1.00	0.1–0.3
Sugar cane		0.68–0.80	
Sugar beet	0.12–0.22	0.73–0.75	0.2–0.3
Rubber	0.20–0.30	0.90	0.2
Oil palm	0.30	1.20	0.1–0.3
Cotton		0.63–0.69	0.3–0.7
Cultivated grass		0.85–0.87	0.004–0.01
Prairie/savanna grass	0.25–0.40	0.80–0.95	0.01–0.10
Forest/woodland	0.25–0.35 (coniferous and tropical)	0.90–1.00	0.001–0.002 (with undergrowth)
	0.15–0.25 (temperate broad-leaved)		0.001–0.004 (no undergrowth)
Bare soil	0	0.05	1.00

C values should be adjusted by the following P ratios if mechanical soil conservation measures are practised: contouring, multiply by 0.6; contour strip-cropping, multiply by 0.35; terracing, multiply by 0.15.
Source: summarized in Morgan et al. (1982).

that unit and the material transported into the unit from upslope, and this is compared to the transport capacity. When operated in this way, the model is able to identify the major source areas of sediment and the locations of deposition within a catchment and to provide the information for determining the sediment delivery ratio (Morgan 2001).

6.2.4 Wind Erosion Prediction Equation

A similar technique to the Universal Soil Loss Equation has been developed for predicting wind erosion (Woodruff & Siddoway 1965; Skidmore & Williams 1991; Fryrear et al. 1998), taking account of soil erodibility (I), wind energy, expressed by a climatic factor (C), surface roughness (K), length of open wind blow (L) and the vegetation cover (V). The equation is:

$$WE = f(I, C, K, L, V) \tag{6.13}$$

Table 6.12 Prediction of soil loss using the Morgan–Morgan–Finney method

Problem

Calculation of mean annual soil loss on a 100 m long slope of 7° on soils of the Rengam Series under maize cultivation with contour bunds at 20 m spacing, near Kuala Lumpur, Malaysia (compare with Tables 6.4 and 6.6)

Estimating input parameter values

MS	Estimate for a sandy clay; assume similar to a sandy loam. From Table 6.9	= 0.28
BD	Estimate for a sandy clay; assume similar to a clay loam. From Table 6.9	= 1.3
EHD	Guide value from Table 6.10	= 0.12
K	Estimate for a sandy clay; since clay content will resist detachment but sand content will not, assume similar to a sandy loam. From Table 6.9	= 0.3
COH	Estimate for a sandy clay; assume similar to a sandy loam. From Table 6.9	= 2
S	Slope angle of 7° gives sine value	= 0.1219
R	Annual rainfall	= 2695
R_n	Number of rain days	= 185
I	Typical intensity of erosive rain; guide value from Table 6.8	= 25
A	Proportion of the rainfall intercepted by the land cover. Calculations for conditions given in Table 6.4, taking data from Table 6.11	

Months	Number of months	Land use	A value	Weighted A value (col 2 × col 4)
January–April	4	forest	0.25	1.00
May–July	3	maize	0.25	0.75
August–December	5	forest regrowth	0.25	1.25
Total				3.00

Average weighted A value = 3.00/12   = 0.25

E_t/E_o Calculation of time-weighted average value, taking data from Table 6.11

Months	Number of months	Land use	E_t/E_o value	Weighted E_t/E_o value (col 2 × col 4)
January–April	4	forest	0.90	3.6
May–July	3	maize	0.70	2.1
August–December	5	forest regrowth	0.90	4.5
Total				10.2

Average weighted E_t/E_o value = 10.2/12   = 0.85

CC   Calculation of time-weighted average value based on assumed percentage canopy covers under the different land uses

Table 6.12 *Continued*

Months	Number of months	Land use	CC value	Weighted CC value (col 2 × col 4)
January–April	4	forest	0.90	3.60
May	1	maize	0.10	0.10
June	1	maize	0.70	0.70
July	1	maize	0.55	0.55
August–December	5	forest regrowth	0.25	1.25
Total				6.20

Average weighted CC value = 6.20/12   = 0.52

GC   Calculation of time-weighted average value based on assumed percentage ground cover under the different land uses

Months	Number of months	Land use	GC value	Weighted GC value (col 2 × col 4)
January–April	4	forest	0.90	3.6
May–July	3	maize	0.10	0.3
August–December	5	forest regrowth	0.20	1.0
Total				4.9

Average weighted GC value = 4.9/12   = 0.41

PH   Calculation of time-weighted average value based on height of fall (m) of rainfall from lowest component of the vegetation or crop cover

Months	Number of months	Land use	PH value	Weighted PH value (col 2 × col 4)
January–April	4	forest (with understorey and litter layer)	0.01	0.04
May	1	maize	0.15	0.15
June	1	maize	0.80	0.80
July	1	maize	1.50	1.50
August–December	5	forest regrowth	0.25	1.25
Total				3.74

Average weighted PH value = 3.74/12   = 0.31

C   Calculation of time-weighted average value, taking data from Table 6.11

Months	Number of months	Land use	C value	Weighted C value (col 2 × col 4)
January–April	4	forest	0.001	0.004
May–July	3	maize	0.2	0.6
August–December	5	forest regrowth	0.1	0.5
Total				1.104

Average weighted C value = 1.104/12   = 0.09

Table 6.12 *Continued*

Estimating effective rainfall (ER)
From Table 6.7

ER = 2695 × (1 − 0.25)
= 2021.25 mm

Estimating volumes of leaf drainage (LD) and direct throughfall (DT)
From Table 6.7

LD = 2021.25 × 0.52
= 1051.05 mm
DT = 2021.25 − 1051.05
= 970.20 mm

Estimating kinetic energy of the effective rainfall
From Table 6.7

$KE(DT)$ = 970.20 (11.9 + 8.7 log 25)
= 23,343 J m^{-2}
$KE(LD)$ = 1051.05 ((15.8 × 0.31$^{0.5}$) − 5.87)
= 3077 J m^{-2}
KE = 23,343 + 3077
= 26,421 J m^{-2}

Estimating volume of overland flow
From Table 6.7
R_c = 1000 × 0.28 × 0.12 × 0.85$^{0.5}$
= 30.98 mm
R_n = 2695/185
= 14.57 mm
Q = 2695 e$^{-30.98/14.97}$
= 340.1 mm

Estimating rate of soil particle detachment
From Table 6.7, annual rate of detachment by raindrop impact

F = 0.3 × 26,421 × 10^{-3}
= 7.93 kg m^{-2}

From Table 6.7, annual rate of detachment by runoff

Z = 1/(0.5 × 2)
= 1
H = 1 × 340.1$^{1.5}$ × 0.1219 × 0.59 × 10^{-3}
= 0.45 kg m^{-2}

Total annual rate of detachment

J = 7.93 + 0.45
= 8.38 kg m^{-2}

Estimating annual transport capacity of overland flow

G = 0.09 × 340.1^2 × 0.1219 × 10^{-3}
= 1.27 kg m^{-2}

Soil loss estimation
Mean annual soil loss equals the lower of the values of J and G = 1.27 kg m^{-2}
Convert to t ha^{-1} = 10 × 1.27   = 12.7 t ha^{-1}

where WE is the mean annual wind erosion ($t\,ha^{-1}$), I is the mean annual wind erosion for a given soil ($t\,ha^{-1}$) in a level dry bare field at Garden City, Kansas (Table 3.5), C is a dimensionless number to adjust the value of I for the actual climatic conditions compared with those at Garden City, K is a dimensionless number to adjust I for roughness produced by tillage, L is in m and V is the quantity of vegetative cover expressed as flattened wheat straw equivalent in $kg\,DM\,ha^{-1}$. The equation allows for interactions between the factors and so it cannot be solved by multiplying the values of the various factors together. Factor relationships are rather complex and predictions can only be obtained by using complex nomographs or specially developed equations applied in a set sequence (Table 6.13).

Table 6.13 Prediction of wind erosion

Basic equation

$WE = f(I, C, K, L, V)$

Estimating I (soil erodibility index)
Values represent the potential annual soil loss ($t\,ha^{-1}$) for a level bare dry field near Garden City, Kansas. They have been determined from wind tunnel studies (section 3.2; Table 3.7) and may be estimated by knowing the percentage of dry stable aggregates larger than 0.84 mm in the soil. The values are adjusted by a term Is to take account of the steepness of the windward slope (Woodruff & Siddoway 1965).

Estimating C (climatic index)
The index takes account of wind erosion as a function of wind velocity and soil moisture content. The former is expressed by the mean annual wind velocity (v, ms^{-1}) measured at a height of 9 m and the latter by the Thornthwaite precipitation effectiveness index ($P - E$). The index is then expressed as a percentage of its value of 2.9 for Garden City Kansas. Thus:

$$C = \frac{100}{2.9} \cdot \frac{v^3}{(P-E)^2}$$

where

$$P - E = 115 \sum_{i=1}^{12} \left(\frac{P}{T-10} \right)^{10/9}$$

in which P is the mean precipitation for month i (inches) and T is the mean temperature for month i (°F). In application, a minimum value of 18.4 is adopted for mean monthly $T - 10$ and maximum value of 13 mm is used for mean monthly precipitation.

Estimating K (ridge roughness index)
The roughness of ridges produced by tillage and planting equipment is expressed by a roughness factor (R) calculated from

$R = 4H^2/l$

where H is the ridge height (mm) and l is the distance between the ridges. Values of K are expressed by:

$K = 1$	$R < 2.27$
$K = 1.125 - 0.153\ln R$	$2.27 \leq R < 89$
$K = 0.336 \exp (0.0032R)$	$R \geq 89$

Modelling soil erosion

Table 6.13 *Continued*

Estimating L (length of open wind blow)

The value of L is calculated as a function of equivalent field length (D) and the distance sheltered by any trees, shelterbelts, field hedges or windbreaks. The equivalent field length is calculated from measurements of actual field length (l, m), field width (w, m), field orientation expressed as the clockwise angle between field length and north (ϕ, rad), and wind direction clockwise from north (θ, rad):

$$D = \frac{lw}{l\left|\cos\left(\frac{\pi}{2}+\theta-\phi\right)\right| + w\left|\sin\left(\frac{\pi}{2}+\theta-\phi\right)\right|}$$

The value of L is the determined from:

$$L = D - 10H$$

where H is the height of the shelterbelt.

Estimating V (vegetation cover)

The vegetation cover index depends on standing live biomass, standing dead residue and flattened crop residue. The original work on the effects of vegetation was carried out for flattened wheat straw and so all weights of living or dead vegetative matter need to be converted into flattened wheat straw equivalents, defined as 254 mm tall stalks lying flat on the soil in rows perpendicular to the wind direction, with 254 mm row spacing and stalks oriented perpendicular to the wind direction. Values of V for various crops are given in Woodruff and Siddoway (1965) and for perennial rangeland grasses in Lyles and Allison (1981). For a wide range of crops, an estimate of the flattened wheat straw or small grain equivalent weight of the residue (SGe, kg ha^{-1}) can be obtained from:

$$SGe = 0.162Rw/d + 8.708(Rw/d\gamma)^{1/2} - 271$$

where Rw is the weight of the standing residue of the crop (kg ha^{-1}), d is the average stalk diameter (cm) and γ is the average specific weight of the stalks (Mg m^{-3}) (Lyles & Allison 1981).

Applying the equation

Step 1: $E1 = l.ls$

Step 2: $E2 = E1.K$

Step 3: $E3 = E2.C$

Step 4: $E4 = (F^{0.3484} + E3^{0.3484} - E2^{0.3484})^{2.87}$

where $F = E2\{1 - 0.1218(L/Lo)^{-0.3829}\exp(-3.33L/Lo)\}$

  $Lo = 1.56 \times 10^6 (E2)^{-1.26}\exp(-0.00156E2)$

Step 5: $E5 = g E4^h$

where $g = 952.5^{(1-h)}$

  $h = \dfrac{1}{0.0537 + 0.9436\exp(-0.000112V)}$

Step 6: $WE = E5$

6.3

Physically based models

With greater concern over the past few decades about the off-site consequences of erosion and identification of non-point source pollution, efforts have been made to develop models that will predict the spatial distribution of runoff and sediment over the land surface during individual storms in addition to total runoff and soil loss. In meeting these objectives, empirical models possess severe limitations. They cannot be universally applied and most are not able to simulate the movement of water and sediment over the land or be used on scales ranging from individual fields to small catchments. A more physically based approach to modelling is thus required. In reality, physically based models still rely on empirical equations to describe erosion processes and so should strictly be termed process-based models.

Physically based models incorporate the laws of conservation of mass and energy. Most of them use a particular differential equation known as the continuity equation, which is a statement of the conservation of matter as it moves through space over time. The equation can be applied to soil erosion on a small segment of a hillslope as follows. There is an input of material to the segment from the segment upslope and an output of material to the segment downslope. Any difference between input and output relates to either erosion or deposition on the segment. Using the symbol ∂ to denote a change in the value, the continuity equation can be expressed as a mass balance (Bennett 1974; Kirkby 1980b):

$$\frac{\partial(AC)}{\partial t} + \frac{\partial(QC)}{\partial x} - e(x,t) = q_s(x,t) \tag{6.14}$$

where A is the cross-sectional area of the flow, C is the sediment concentration in the flow, t is time, x is the horizontal distance downslope, e is the net pick-up rate or erosion of sediment on the slope segment and q_s is the rate of input or extraction of sediment per unit length of flow from land external to the segment – for example, from the sides of a convergent slope surface. On a plane slope segment, $q_s = 0$.

Almost all physically based erosion models owe their origin to the relatively simple scheme developed by Meyer and Wischmeier (1969; Fig. 6.1) to test whether a mathematical approach to simulating erosion is feasible. By applying the model to consecutive downslope segments in turn, sediment is routed over the land surface and the pattern of erosion evaluated along a complete soil profile. The first range of models developed from the Meyer–Wischmeier approach used mathematical descriptions for the processes of runoff generation, flow of runoff over the land and the detachment and transport of sediment but the K, L, S, C and P factors to account for soil, slope, land cover and land management. Examples include the Agricultural Nonpoint Source Model (AGNPS; Young et al. 1989), the Guelph model for evaluating the effects of Agricultural Management systems on Erosion and Sedimentation (GAMES; Dickinson et al. 1986), the Areal Nonpoint Source Watershed Environment Response Simulation (ANSWERS; Beasley et al. 1980) and CREAMS (Chemicals, Runoff and Erosion from Agricultural Management Systems; Knisel 1980; Foster et al. 1981). Over the past ten years, a new generation of process-based models has been developed. Three examples are presented here.

6.3.1 WEPP

WEPP (Water Erosion Prediction Project) is a process-oriented model designed to replace the Universal Soil Loss Equation for the routine assessment of soil erosion by organizations involved in soil and water conservation and environmental planning (Nearing et al. 1989b). The overall package contains three computer models: a profile version, a watershed version and a grid version. The profile version estimates soil detachment and deposition along a hillslope profile and the net total soil loss from the end of the slope. It can be applied to areas up to about 260 ha in size. The watershed and grid versions allow estimates of net soil loss and deposition over small catchments. The models take account of climate, soils, topography, management and supporting practices. They are designed to run on a continuous simulation but can be operated for a single storm. A separate model, CLIGEN (Nicks et al. 1995), is used to generate the climatic data on rainfall, temperature, solar radiation and wind speed for any location in the USA for input to WEPP.

The erosion model within WEPP applies the continuity equation for sediment transport downslope in the form (Foster & Meyer 1972):

$$dQs/dx = D_i + D_f \tag{6.15}$$

where Qs is the sediment load per unit width per unit time, x is the distance downslope, D_i is the delivery rate of particles detached by interrill erosion to rill flow and D_f is the rate of detachment or deposition by rill flow. The interrill erosion rate (D_i) is given by:

$$D_i = K_i I^2 CeGe(Rs/w) \tag{6.16}$$

where K_i is an expression of interrill erodibility of the soil, I is the effective rainfall intensity, Ce expresses the effect of the plant canopy, Ge expresses the effect of ground cover, Rs is the spacing of rills and w is the width of the rill computed as a function of the flow discharge. The canopy effect is estimated by:

$$Ce = 1 - Fe^{-0.34PH} \tag{6.17}$$

where F is the fraction of the soil protected by the canopy and PH is the canopy height. The ground cover effect is estimated by:

$$Ge = e^{-2.5gi} \tag{6.18}$$

where gi is the fraction of the interrill surface covered by ground vegetation or crop residue.

The rate of detachment of soil particles by rill flow (D_f) is given by:

$$D_f = D_c(1 - Qs/Tc) \tag{6.19}$$

where D_c is the detachment capacity, Qs is the sediment load in the flow and Tc is the sediment load at transport capacity. Detachment capacity is expressed as:

$$D_c = K_r(\tau - \tau_c) \tag{6.20}$$

where K_r is the rill erodibility of the soil, τ is the flow shear stress acting on the soil and τ_c is the critical flow shear stress for detachment to occur. The transport capacity of the flow is obtained from:

$$Tc = k_t \tau^{3/2} \tag{6.21}$$

where k_t is a transport coefficient and τ is the hydraulic shear acting on the soil. Net deposition occurs if the sediment load is greater than the transport capacity, in which case eqn 6.19 becomes:

$$D_f = v_s/q(Tc - Qs) \qquad (6.22)$$

where v_s is the settling velocity of the particles and q is the flow discharge per unit width. The initial particle input to the model is the distribution of sediment-size classes in the original soil. No particle-size selectivity occurs in detachment but deposition is particle-size selective. The particle-size distribution of the sediment load is recalculated once deposition occurs. The model assumes that all the material detached on the interrill areas is delivered to the rills.

WEPP has undergone a major programme of testing and evaluation. When tested without calibration for nearly 4000 storm events across nine experimental stations in the USA, the model gave predictions of mean annual soil loss at the erosion plot scale of similar accuracy to those of the USLE and RUSLE, with a coefficient of determination (r^2) of 0.85 (Zhang et al. 1996). When applied to data from erosion plots under semi-arid pinyon–juniper woodland, New Mexico, the model successfully predicted runoff from disturbed land, where the vegetation, litter cover and stones had been removed, but underpredicted for undisturbed land by a factor of nearly three, possibly due to difficulties in characterizing the hydraulic conductivity of the soil. In tests, again without calibration, on 1194 events on 15 small watersheds the model overpredicted sediment yield for three watersheds at Holly Springs, Mississippi, by a factor of two. This was attributed to specific problems with simulating the effects of cutting maize for silage and a rather high rate of weed growth after harvest. Predictions of runoff and sediment yield for the other watersheds, however, were good, with errors being greatest for the lower values (Liu et al. 1997). These tests show that where the soil, slope, land cover and land management can be adequately described by the choice of input parameter values, the model performs well.

6.3.2 GUESS

GUESS (Griffith University Erosion Sedimentation System) is a mathematical model that simulates the processes of erosion and deposition along a hillside (Rose et al. 1983). The model is designed as a guide to farmers, scientists and extension workers concerned with the evaluation and control of soil erosion by water. Soil erosion at any position on the slope and at any time during the storm is related to a sediment flux that depends on the hydrological conditions and the sediment concentration.

The model differs from WEPP in separating the surface soil into two parts: that which is the original soil and possesses a certain degree of cohesion and that comprising recently detached material with no cohesion. As in WEPP, no selectivity occurs in detachment but deposition is particle-size selective. GUESS, however, describes the soil in terms of 50 particle-size classes instead of eight, all of equal mass, and determined according to their settling velocity. The continuity equation is modified accordingly and takes the form:

$$\left(\frac{\partial Qs_i}{\partial x}\right) + \frac{\partial(C_i h)}{\partial t} = e_i + e_{di} + r_i + r_{di} - d_i \qquad (6.23)$$

where Qs_i is the sediment load of sediment class i, C_i is the concentration of sediment class i in the flow, e_i is the rate of detachment of particles of sediment class i in the original soil by raindrop impact, e_{di} is the rate at which recently detached soil of sediment class i is redetached by raindrop impact, r_i is the rate of detachment of particles of sediment class i by flow, r_{di} is the rate at which recently detached soil of sediment class i is redetached by the flow, and d_i is the rate of deposition.

The rate of detachment of soil particles of sediment class i by raindrop impact is given by:

$$e_i = aC_eI/N \qquad (6.24)$$

where a is the detachability of the soil, C_e is the fraction of the soil surface exposed to raindrops, I is the rainfall intensity and N is the number of particle size classes. The same equation is used to calculate the rate of redetachment of soil particles by raindrop impact (e_{di}) except that a is replaced by a_d, an expression of redetachability of the soil. The detachment rate of soil particles by flow is modelled as a function of stream power (Ω), defined as the product of flow shear stress (τ) and flow velocity (v) above a critical value (Ω_c):

$$r_i = (1-H)F(\Omega - \Omega_c)/IJ \qquad (6.25)$$

where $(1-H)$ is the fraction of the original soil surface exposed to runoff, F is the fraction of excess stream power which is used in detachment, generally assumed to be ≈ 0.1, and J is the amount of unit stream power necessary to detach a unit mass of soil. For redetachment of soil particles, the rate is:

$$r_{di} = \left(\frac{\alpha_i HF}{g}\right)\left(\frac{\sigma}{\sigma - \rho}\right)\left(\frac{\Omega - \Omega_c}{h}\right)\frac{M_i}{M} \qquad (6.26)$$

where α_i is a dimensionless parameter with a value dependent on the depth of flow (usually ≥ 1 but taken as 1 for shallow flow), H is the fraction of the surface soil covered by recently deposited material, σ is the submerged sediment density, ρ is the density of water, h is the depth of flow, M_i is the mass fraction of sediment class i to that of the total mass (M) of material being redetached. The rate of deposition of sediment class i is given by:

$$d_i = \alpha_i v_{si} C_i \qquad (6.27)$$

where v_{si} is the fall velocity of particles of sediment class i.

Unlike in the WEPP model, not all the sediment detached or redetached by raindrop impact is contributed to the flow. A limitation is placed so that:

$$C_{rain} = (HaI)/(\Sigma v_{si}/N) \qquad (6.28)$$

where C_{rain} is the contribution of detachment by raindrop impact to the sediment concentration in the flow, and H approximates to 0.9. The transport capacity of the flow (C_{max}) is represented by (Rose & Hairsine 1988):

$$C_{max} = \left(\frac{F\rho}{\Sigma v_{si}/N}\right)\left(\frac{\sigma}{\sigma - \rho}\right)sv \qquad (6.29)$$

The equation is applied either to the overland flow or, if rills are present, by assigning all the flow to the rills and, by dividing by the number of rills, calculating the water discharge and velocity per rill. The sediment flux in the rill can then be multiplied by the number of rills to give the total sediment flux for the slope.

While it is possible for the actual sediment concentration (C) in the flow to equal C_{max}, it is usually lower than C_{max} by an amount depending on the strength of the soil matrix (β). The erodibility parameter (β) can be defined from the relationship:

$$C = C_{max}^{\beta} \tag{6.30}$$

The model calculates average values of C and C_{max} for an event so that β can be evaluated from:

$$\beta = \frac{\ln \overline{C}}{\ln \overline{C}_{max}} \tag{6.31}$$

With bare tilled soils, β approximates to 1.0 but where the ground surface is protected by a vegetation cover in close contact with the soil, it can fall to between 0.6 and 0.8 in value (Presbitero et al. 1995). On a heavily consolidated soil, β can be as low as 0.1–0.4 (Hashim et al. 1995; Paningbatan et al. 1995). The model needs to be calibrated to determine the values of J, Ω_c and β. An extensive testing programme is under way in Australia, the Philippines, Malaysia and Thailand.

6.3.3 EUROSEM

EUROSEM (European Soil Erosion Model; Morgan et al. 1998; Fig. 6.3) is an event-based model designed to compute the sediment transport, erosion and deposition over the land surface throughout a storm. It can be applied to either individual fields or small catchments. Compared with other models, EUROSEM simulates interrill erosion explicitly, including the transport of water and sediment from interrill areas to rills, thereby allowing for deposition of material *en route*. This is considered more realistic than assigning all or a set proportion of the detached material to the rills. A more physically based approach to simulating the effect of vegetation or crop cover is used, taking account of the influence of leaf drainage. Soil conservation measures can be allowed for by choosing appropriate values of the soil, microtopography and plant cover parameters so as to describe the conditions associated with each practice. Unlike other models, however, EUROSEM does not describe the eroded sediment in terms of its particle size.

Equation 6.14 is used as the continuity equation in which the term (e), the net rate of pick-up of sediment, is defined as:

$$e = D_i + D_f \tag{6.32}$$

The rate of detachment by raindrop impact (D_i) is calculated from:

$$D_i = k.KE^{1.0}.e^{-2h} \tag{6.33}$$

where k is the detachability of the soil by raindrop impact and h is the depth of surface water. The kinetic energy (KE; $J\,m^{-2}\,mm^{-1}$) of the rainfall is determined separately for the direct through-fall (DT) and the leaf drainage (LD) using relationships proposed by Brandt (1990):

$$KE(DT) = 8.95 + 8.44 \log I \tag{6.34}$$

$$KE(LD) = (15.8 \times PH^{0.5}) - 5.87 \tag{6.35}$$

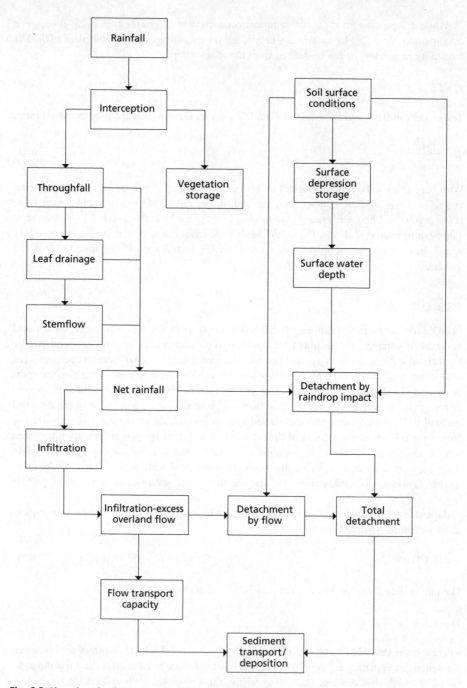

Fig. 6.3 Flow chart for the European Soil Erosion Model (EUROSEM).

where I is the intensity of the rainfall $(\mathrm{mm\,h^{-1}})$ and PH is the height of the plant canopy (m). The rate of detachment of soil particles by flow is modelled as a balance between detachment and deposition:

$$D_f = \eta w v_s (C_{max} - C) \tag{6.36}$$

where η is an expression of the efficiency of the detachment process, its value being dependent upon soil cohesion, and w is the width of flow. When the sediment concentration in the flow exceeds transport capacity, η assumes a value of 1.0 and the term e in eqn 6.14 becomes negative and represents a net deposition rate. Equation 2.27 is used to describe the transport capacity of the flow in terms of unit stream power, expressed as the product of slope and velocity.

The model can be applied to overland flow over a relatively smooth surface or to a slope containing rills or other defined channels such as plough furrows. When rills are present the model places all the runoff into the rills and, taking account of the number of rills, calculates the flow velocity and depth per rill. If the rills overflow, a unified rill model developed by Dr Roger Smith (USDA) is used to calculate the hydraulic conditions of the flow as a function of the wetted perimeter.

EUROSEM has been used to simulate erosion at an erosion plot scale in England (Quinton 1994), Kenya (Mati 1999), Costa Rica, Nicaragua and Mexico (Veihe et al. 2001), Oklahoma, USA (Quinton & Morgan 1998) and the Netherlands (Folly et al. 1999) (Table 6.14). It has also been

Table 6.14 Comparison of observed storm runoff and storm soil loss with values predicted by EUROSEM

Location	Storm runoff (mm)		Storm soil loss $(\mathrm{t\,ha^{-1}})$	
	Observed	Predicted	Observed	Predicted
Coschocton C5, USA				
29 August 1966	5.4	16.3	0.11	0.49
28 November 1968	0.8	0.0	0.00	0.00
22 September 1970	14.9	13.9	0.46	0.45
22 February 1975	18.7	13.1	0.50	0.38
Embori, Kenya				
15 May 1994	1.5	5.5	0.34	0.99
16 May 1994	7.9	12.9	2.85	4.95
18 May 1994	2.9	4.6	0.74	0.69
22 May 1994	14.4	21.0	4.10	3.95
9 June 1994	5.8	5.8	2.69	2.62
23 June 1994	1.9	2.2	0.43	0.68
2 July 1994	0.0	1.7	0.00	0.48
7 July 1994	0.0	1.9	0.00	0.03
Catsop, The Netherlands				
13 May 1987	0.17	0.08	0.02	0.02
15 December 1989	0.08	0.23	no data	1.54
22 January 1993	0.06	0.16	0.07	0.48
30 May 1993	0.09	0.16	0.07	0.48
14 October 1993	0.54	0.22	0.22	0.56

Source: after Quinton and Morgan (1998), Mati (1999), Folly et al. (1999).

used to evaluate the effectiveness of contour grass strips as an erosion-control measure near Cochabamba, Bolivia (Quinton & Rodriguez 1999) and adopted as the erosion model in an investigation of alternative policies for reducing the losses of nitrogen, phosphorus and soil from agricultural land in Norway (Vatn et al. 1996).

6.3.4 Wind erosion models

Although considerable research is taking place to develop a physically based wind erosion model, the work is still in the developmental stage. Particular difficulties arise in simulating the high degree of complexity and randomness of the detachment and transport of soil particles by wind; in representing a three-dimensional sediment transport system as a simpler two-dimensional one; and in describing the physics of the system by mathematical equations that can be solved. Many of the present models apply only to rather limited conditions such as sediment movement at transport capacity or bare soil. The most comprehensive state-of-the-art model that can simulate the effects of soil, land cover and tillage is the Wind Erosion Prediction System (WEPS) (Hagen 1991), which is a daily simulation model applicable at a field scale. Research is under way to link WEPS and WEPP into a single process-based model (Fox et al. 2001). Unfortunately, it is extremely difficult to validate wind erosion models because suitable field measurements of erosion are hard to obtain (see section 5.1.3).

6.4

Sensitivity analysis

Sensitivity analysis indicates by how much the output of a model alters in relation to a unit change in the value of one or more of the inputs. It should be carried out on all models to show that the model behaves rationally and to indicate how accurately values of the inputs need to be measured or estimated. Rational behaviour is generally judged on whether the level of sensitivity of the factors in the model matches what is expected in reality and on whether the relationships between the output and the controlling factors accord with what is observed in the field or in laboratory experiments. For example, EUROSEM gives rational output with respect to relationships between soil loss and slope steepness and the percentage cover of low-growing vegetation but is less satisfactory with tree covers because it underpredicts soil particle detachment by leaf drainage (Morgan 1996).

Sensitivity analysis can be extended to evaluate whether the interaction between factors is correctly simulated and whether a model gives plausible results when operated under extreme conditions. It may also be used to identify which processes could be excluded without significant loss of information should a decision be made to simplify a model.

Despite its simplicity of purpose, sensitivity analysis is not straightforward. A choice has to be made about which sensitivity index to use (Table 6.15). Absolute sensitivity (McCuen 1973) is the simplest. It describes the rate of change in output with respect to a change in the value of input. However, it does not take account of the relative magnitudes of the values. For example, a 10 per cent change in a typical value of saturated hydraulic conductivity might increase the input value by 20–25 mm h^{-1}, whereas a 10 per cent change in the value of Manning's n might increase the the input value by 0.001–0.002. To compare these changes, the relative sensitivity index, which normalizes the input and output around a base value chosen by the user,

Table 6.15 Sensitivity indices

Absolute sensitivity (AS) (McCuen 1973):

$$AS = \frac{(O_2 - O_1)}{(I_2 - I_1)}$$

where O_1 and O_2 are values of model output obtained with values of I_1 and I_2 of input parameter I.

Relative sensitivity (RS) (McCuen 1973):

$$RS = \frac{\left(\dfrac{O_2 - O_1}{O}\right)}{\left(\dfrac{I_2 - I_1}{I}\right)}$$

where I_1 and I_2 are values of input parameter with a chosen range, plus and minus a percentage of a base value I, and O_1, O_2 and O are their respective output values.

Average linear sensitivity (ALS) (Nearing et al. 1989a):

$$ALS = \frac{\left(\dfrac{O_2 - O_1}{\overline{O}}\right)}{\left(\dfrac{I_2 - I_1}{\overline{I}}\right)}$$

where $\overline{O}$ and $\overline{I}$ represent the respective average values of the two input and output values.

should be used. The relative sensitivity index has been modified by Nearing et al. (1989a), by normalizing the input and output in relation to their mean values, to produce an average linear sensitivity index.

For complex models like WEPP and EUROSEM, sensitivity analysis needs to be undertaken separately for the different outputs, such as runoff and soil loss. If sensitivity analysis is carried out for each input parameter individually, the results may be misleading with respect to reality because interactions within the model mean that it is illogical to consider, for example, a change in saturated hydraulic conductivity without also changing related parameters such as capillary drive or porosity. There may also be interactions between parameters that are not obviously inter-related. Soil loss may be sensitive to cohesion or soil erodibility when vegetation cover is less than 30 per cent but not sensitive when vegetation cover exceeds 80 per cent. To cover these situations, a very comprehensive sensitivity analysis is required like that undertaken by Veihe and Quinton (2000), who used 4000 simulations of EUROSEM, selecting input values for each parameter randomly using Monte Carlo analysis.

Clearly if a predictive model is to give satisfactory estimates of soil loss, a high level of accuracy is required in the values of the most sensitive inputs. The critical question is whether this accuracy can be obtained, particularly for inputs that are highly variable spatially, such as saturated hydraulic conductivity, or difficult to measure, such as Manning's n. Knowing the errors involved in the input data is important because they are a potential source of error in any predictions.

Model validation

The success of any model must be judged by how well it meets its objectives or requirements. With a predictive model this means deciding on the time and space scale for which predictions are required and the level of accuracy. When making a judgement on the utility of a model, it is necessary to distinguish between failures due to misuse, those related to inadequate input data and those associated with the structure of the model or its operating functions. In the latter case, failure may result from poor conceptualization of the problem, omission of important factors or inaccurate representation of a particular element in the model by the operating function or equation employed. The solution is to modify or in some instances completely rethink the model.

The accuracy of model predictions is usually tested by comparing predicted with measured values and applying some measure of goodness-of-fit. The data used for validation should, of course, always be different from those used to develop the model. Criteria for validation are by no means clear-cut and need to be set for individual models in relation to their objective. In many cases, a qualitative assessment is all that is required. Thus, Meyer and Wischmeier (1969) judged the success of their model by how well it reproduced the patterns of sediment movement found in the field. For many applications, it may be sufficient to show that a model predicts the correct location of erosion and sedimentation.

Where the objective is to predict amounts of soil loss, it is necessary to compare the predictions with measured values. This can be achieved by dividing the predicted by the measured value to give a ratio. Ideally, the ratio should be equal to 1.0 but, since this is rarely the case, its value has to be related to some guideline in order to judge whether it is acceptable – for example, the ratio might lie between 0.75 and 1.5 in value or, less stringently, between 0.5 and 2.0. The percentage of predictions that are acceptable can be used as a measure of a model's success. Although the use of ratios is a simple and effective method of validation, they cannot be applied where non-linear relationships are involved. In this case, the simplest measure is to examine the absolute differences between predicted and observed values. These types of tests, however, assume that a given level of error is equally important over the full range of values, whereas, for many purposes, higher levels may be acceptable at very low values of soil loss. A predicted value for mean annual soil loss of $0.02\,t\,ha^{-1}$ may be acceptable in relation to a recorded value of $0.002\,t\,ha^{-1}$ for evaluating erosion-control strategies, since both values indicate that there is no on-site erosion problem. However, a predicted value of $200\,t\,ha^{-1}$ against a recorded value of $20\,t\,ha^{-1}$ would not be acceptable, since it would give a very misleading impression of the severity of the problem.

Many workers use correlation and regression analyses in order to test the strength of association between predicted and observed values. Experience with the WEPP model indicates that coefficients of determination (r^2) greater than 0.5 should be deemed acceptable (Zhang et al. 1996) and that the best models are unlikely to give values in excess of 0.76 (Nearing 1998). A high and significant value of the correlation coefficient, however, does not imply that a model is performing well, merely that the model is producing results with a degree of precision and with the right trend. Unless the predicted and observed values have a 1:1 relationship, the model predictions are inaccurate to a certain degree. Ideally, a regression equation for the relationship should have a slope of 1.0 and pass through zero. Statistical tests should be applied to determine whether particular relationships differ significantly from these two conditions. A least-squares regression approach assumes that there will be some error in the predicted value but not in the observed value. This assumption is certainly questionable because, as seen in Chapter 5, all measurement techniques are subject to measurement errors. It is preferable to recognize this and use a reduced major axis regression (Kermack & Haldane 1950; Till 1973), which allows for errors in both sets of values. A model can only be as good as the data that are fed into it. If much of the input data contains errors of 10 per cent or more, it is unrealistic to expect a model to give predictions to accuracies of 5 per cent or less. It is also unreasonable to expect a model to perform better than the natural variability in erosion as indicated by data from replicate erosion plots (Nearing 2000).

The efficiency coefficient (CE), proposed by Nash and Sutcliffe (1970), is now increasingly used as an alternative to the correlation coefficient to express the performance of a model:

$$CE = \frac{\sum (X_{obs} - X_{mean})^2 - \sum (X_{pred} - X_{obs})^2}{\sum (X_{obs} - X_{mean})^2} \qquad (6.37)$$

where X_{obs} is the observed value, X_{mean} is the mean of a set of observed values and X_{pred} is the predicted value. The efficiency parameter is thus a measure of the variance in the predictions from the one-to-one prediction line with the measured values. In a comparative study using 1600 years of plot data for 20 locations in the USA, Tiwari et al. (2000) obtained CE values of 0.71, 0.80 and 0.72 for WEPP, USLE and RUSLE respectively. The better result with the USLE was attributed to better input data. Annual soil loss predictions with the revised Morgan–Morgan–Finney model gave CE values of 0.58 for runoff and 0.65 for soil loss, although tests on a second data set gave respective values of 0.94 and 0.84 (Morgan 2001). Generally a CE value greater than 0.5 is considered satisfactory (Quinton 1997) and one should not expect values to exceed 0.7 (Nearing 1998).

It is also necessary to know whether a prediction has been arrived at in the right way. For example, if soil loss is limited by the detachment rate and the model predicts the right total but as a result of a limitation in transport capacity, it cannot be considered a successful model and it cannot be used for soil conservation, since any strategy based on its predictions will reflect a wrong diagnosis. It is therefore essential that validation of process-based models be extended to cover all of their constituent modules. Given the very limited measured data available for such validation and the associated measurement errors, De Roo (1996) concludes that, at present, it is virtually impossible to validate a process-based model in any meaningful way. Despite the advances made in recent years in developing physically based models, a simple empirical model is often more successful in predicting soil erosion and is usually easier to use.

Box 6

Uncertainty in model predictions

The difficulty of obtaining an exact fit between predicted and observed values is a reflection of uncertainty in model prediction. Uncertainties arise from: errors in the measured values; the high spatial variability of some input parameters, which means that they cannot be properly represented by a single value; the need to estimate some parameter values because they cannot be easily measured; selection of parameter values to characterize field conditions, particularly surface roughness, crusting, stoniness and vegetation cover; and errors in the model structure or the operating equations, particularly where empirical equations are used to describe physical processes.

Continued

Modelling soil erosion

Many model users attempt to overcome problems of uncertainty in input parameter values by splitting a data set of observed values into two parts. The first part is used to calibrate the model by determining the values required of up to four or five input parameters in order to obtain good fits between the predicted and observed values. A check is made to ensure that the calibrated values fall within what is reasonable for field conditions. The calibrated model is then applied to other events. Unfortunately, calibration may only work for very specific conditions. Calibrations based on winter storms may not be applicable to summer storms. Calibrations for bare soil conditions are unlikely to apply to vegetated conditions. Quinton and Veihe (2000) attempted to calibrate EUROSEM for erosion plots in Central America and found that every plot had its own characteristics. It was not possible to obtain a general calibrated model. It is even questionable whether calibration of process-based models is desirable, since it compromises the basis of a physically based approach (De Roo 1996).

For complex models, like WEPP and EUROSEM, it is clear that the same output can be obtained from many different sets of input data (Brazier et al. 2000). One way of dealing with the uncertainty is to use multiple data sets, choosing values of each input parameter randomly within a range of values that is representative of likely field conditions, to apply some criteria for removing those data sets that yield predictions with unacceptable errors and then to use the remaining data sets for predictive modelling with more tightly constrained values of the input parameters. The model outputs from the reduced data sets are expressed as a range of values – for example, the minimum, maximum and median values or the values for the 5th, 50th and 95th quantiles. In tests with EUROSEM, Quinton (1997) used Monte Carlo analysis to generate 625 data sets and found that only 14.4 per cent of these predicted erosion with a *CE* value greater than 0.5. In more comprehensive work with WEPP, Brazier et al. (2000) carried out two different studies using a methodology known as Generalized Likelihood Uncertainty Estimation (GLUE; Beven & Binley 1992). One study involved 2.7 million simulations of annual erosion, of which 37,000 yielded acceptable results, and the other involved 3 million simulations, of which 66,000 were

Table B6.1 Example of uncertainty in prediction of mean annual soil loss using multiple simulations of the WEPP model

Observed soil loss class (kg per m-width)	Difference between predicted and observed value at different percentage points (lower limits) of the statistical distribution of predicted values		
	5%	50%	95%
0.0–0.6	0.16	1.29	2.81
0.6–2.0	0.00	0.07	2.92
2.0–4.2	0.28	1.76	4.62
4.2–8.0	0.00	2.00	4.01
8.0–14.2	−30.16	−5.00	11.55
14.2–35.1	−37.32	−15.85	9.59
35.1–73.3	−41.36	−15.74	8.18
73.3–110.8	−63.13	−40.31	−16.39
110.8–140.5	−125.22	−104.73	−79.37
140.5–166.5	−95.50	−59.28	4.45

Predictions are based on data sets selected from multiple simulations using the GLUE methodology.
Source: after Brazier et al. (2001).

acceptable. Table B6.1 shows the differences between the values of the 5th, 50th and 95th quantiles of the statistical distribution of predicted values from the acceptable data sets and the observed values for different magnitudes of annual soil loss. It is clear from these studies that model predictions have a high level of uncertainty associated with them. The results also confirm the tendency of WEPP to overpredict at low values and underpredict at high values.

Given the large number of simulations required to produce data sets from which predictions can be made with reduced levels of uncertainty, it is unlikely that many model users will attempt this type of analysis, particularly when, as shown by Quinton (1997), it may reduce the level of variability around the prediction but will not necessary result in a better fit between observed and predicted values. Reducing the level of uncertainty associated with model predictions is vital if models are to become widely used. Modellers need to recognize that most users do not want multiple predictions with different levels of uncertainty but wish to operate with a single predicted value in which they can have confidence.

CHAPTER 7

Strategies for erosion control

The aim of soil conservation is to reduce erosion to a level at which the maximum sustainable level of agricultural production, grazing or recreational activity can be obtained from an area of land without unacceptable environmental damage. Since erosion is a natural process, it cannot be prevented. But it can be reduced to a maximum acceptable level or soil loss tolerance. This should be considered as a performance criterion that erosion-control measures are expected to achieve.

7.1

Soil loss tolerance

Theoretically, soil erosion should be maintained at a rate that equals or is below the natural rate at which new soil forms. Unfortunately, it is difficult to recognize when this balance exists. Although rates of soil loss can be measured (see Chapter 5), rates of soil formation are so slow that they cannot be easily determined. According to Buol et al. (1973), rates of soil formation throughout the world range from 0.01 to 7.7 mm yr^{-1}. The fastest rates are exceptional, however, and the average is about 0.1 mm yr^{-1} (Zachar 1982).

An alternative approach that avoids the need to measure the rate of new soil formation directly is to estimate the rate required to match the rate of removal by erosion and solution in areas where an equilibrium condition might be presumed to exist. Using data from small watersheds under forest and grassland, Alexander (1988) found the required rates to be between 0.3 and 2 t ha^{-1} annually with the majority being below 1 t ha^{-1} which, assuming a bulk density for the soil of 1.0 Mg m^{-3}, is equivalent to 0.1 mm yr^{-1}. Such a rate, however, may be a rather conservative indicator for the development of an agriculturally productive soil.

Bennett (1939) and Hall et al. (1979) suggest that in soils of medium to moderately coarse texture on well managed crop land the annual rates of formation of the A horizon can exceed 11 t ha^{-1}. This is because the subsoil can be improved by incorporating it with the top soil during tillage and by adding fertilizers and organic matter. It is against this background that values for soil loss tolerance are set so as to maintain an adequate rooting depth and avoid significant reductions in yield while the surface layer of soil is removed by erosion (McCormack & Young 1981). Soil loss tolerance is then defined as the maximum permissible rate of erosion at which soil fertility can be maintained over 20–25 years. A mean annual soil loss of 11 t ha^{-1} is generally accepted as appropriate but values as low as $2\,t\,ha^{-1}$ are recommended for particularly sensitive areas where soils are thin or highly erodible (Hudson 1981). Where soils are deeper than 2 m, the subsoils are capable of improvement and reductions in crop yield are unlikely to be brought about by erosion over the next 50 years or more, some scientists favour increasing the tolerance to $15-20\,t\,ha^{-1}$ (Schertz 1983).

These recommendations on soil loss tolerance are based solely on agricultural considerations. They ignore problems of pollution and sedimentation that arise as plant nutrients and pesticides leave a field either in solution in the runoff or attached to sediment particles. Particular concern relates to the removal of nitrogen, phosphorus and organic matter but, in the context of erosion, most attention is given to the removal of phosphorus, which can occur in both soluble and particulate forms. Between 45 and 90 per cent of the annual phosphorus contribution to water courses is in particulate form (Kronvang 1990; Hasholt 1991; Sibbesen et al. 1994; Sibbesen & Sharpley 1997), with some 18–49 per cent of this being bioavailable, which means that it is potentially available for uptake by algae (Sharpley & Smith 1990) and therefore a contributor to eutrophication of water bodies. Although no guidelines exist regarding acceptable rates of particulate phosphorus within runoff, some simple calculations can provide a rough indicator. If it is assumed that a mean annual flow-weighted concentration of soluble phosphorus of $10\,\mu g\,l^{-1}$ produces a risk of eutrophication (Sharpley & Smith 1990) and that this represents some 20 per cent of the total annual transport of phosphorus, annual rates in excess of $1\,kg\,l^{-1}$ would be of concern. Since, very broadly, measured data show that values of mean annual phosphorus concentrations in $kg\,l^{-1}$ relate proportionately to mean annual soil loss values in $t\,ha^{-1}$ (Sharpley & Smith 1990; Sibbesen et al. 1994), this would equate to an annual erosion rate of $1\,t\,ha^{-1}$, thereby supporting the recommendation made by Moldenhauer and Onstad (1975) that a soil loss tolerance no greater than $1\,t\,ha^{-1}$ may be required to reduce the effects of non-point source pollution from agricultural land to acceptable levels.

A pragmatic approach to determining the level of soil loss tolerance is to decide what level of environmental damage is acceptable using the criteria described in Table 4.3 as a guide. For most purposes, soil tolerance would be set to equate with the upper limits of damage classes 1, 2 or 3 since the damage is unacceptable at higher levels. This approach provides the opportunity to set tolerance levels in relation to local conditions and locally set objectives rather than seeking some universal value.

7.2

Principles of soil conservation

From the discussion of the mechanics of the detachment and transport of soil particles by rainsplash, runoff and wind (Chapter 2), it follows that the strategies for soil conservation must be based on: covering the soil to protect it from raindrop impact; increasing the infiltration capacity of the soil to reduce runoff; improving the aggregate stability of the soil; and increasing surface roughness to reduce the velocity of runoff and wind. The various conservation techniques can be described under the headings of agronomic measures, soil management and mechanical methods. Agronomic measures utilize the role of vegetation to protect the soil against erosion. Soil management is concerned with ways of preparing the soil to promote plant growth and improve its structure so that it is more resistant to erosion. Mechanical or physical methods, often involving engineering structures, depend on manipulating the surface topography – for example, installing terraces or windbreaks – to control the flow of water and air. Agronomic measures combined with

Table 7.1 Effect of various soil conservation practices on the detachment and transport phases of erosion

Practice	Control over					
	Rainsplash		Runoff		Wind	
	D	T	D	T	D	T
Agronomic measures						
Covering soil surface	*	*	*	*	*	*
Increasing surface roughness	–	–	*	*	*	*
Increasing surface depression storage	+	+	*	*	–	–
Increasing infiltration	–	–	+	*	–	–
Soil management						
Fertilizers, manures	+	+	+	*	+	*
Increasing surface roughness (tillage)	–	–	*	*	*	*
Subsoiling, drainage	–	–	+	*	–	–
Mechanical measures						
Contouring, ridging	–	+	+	*	+	*
Terraces	–	+	+	*	–	–
Shelterbelts	–	–	–	–	*	*
Waterways	–	–	–	*	–	–

D, detachment, T, transport; – no control; + moderate control; * strong control.

good soil management can influence both the detachment and transport phases of erosion, whereas mechanical methods are effective in controlling the transport phase but do little to prevent soil detachment (Table 7.1).

Preference is always given to agronomic measures. These are less expensive and deal directly with reducing raindrop impact, increasing infiltration, reducing runoff volume and decreasing wind and water velocities. They are more easily fitted into existing farming systems and more relevant to maintaining or restoring biodiverse plant communities. Mechanical measures are largely ineffective on their own because they cannot prevent the detachment of soil particles. Their main role is in supplementing agronomic measures, being used to control the flow of any excess water and wind that may arise. Many mechanical works are costly to install and maintain. Some, such as terraces, create difficulties for farmers. Unless the soils are deep, terrace construction exposes the less fertile subsoils and may therefore result in lower crop yields. On irregular slopes, terraces will vary in width, making for inefficient use of farm machinery, and only where slopes are straight in plan can this problem be overcome with parallel terrace layouts. Moreover, there is a risk of terrace failure in severe storms. When this occurs, the sudden release of water ponded up on the hillside can do more damage than if no terraces had been constructed. For all these reasons, terracing is often unpopular with farmers.

7.3

Drivers and constraints

Globally, soil erosion associated with agricultural land is more widespread than that associated with other land uses. Erosion control is therefore strongly influenced by the factors that encourage or discourage farmers from adopting soil conservation. In non-agricultural areas, erosion is important wherever it threatens the integrity of a resource – for example, destruction of roads, trackways and footpaths, sedimentation of water bodies or exposure of buried pipelines. The driver for erosion control is to protect the resource and prevent liability for environmental damage, particularly in societies that adopt the principle of the 'polluter pays'. Moreover, organizations like highways agencies, engineering companies and pipeline companies will install protective measures because they want to retain their reputations for sensitive management of the environment.

7.3.1 Perceptions of erosion

The relevance of conservation measures to a farming system depends in part of how farmers and others perceive the erosion problem and its consequences. Most farmers are aware of the problem and its effects. The notion of the peasant farmer damaging land through ignorance is severely mistaken. The small-scale farmer is as much an experienced and efficient practioner of land husbandry as the large-scale commercial farmer, but with a different objective, namely that of survival rather than profit (Hudson 1981). If a farmer destroys land by overcropping or overgrazing, it is because there is no alternative employment from which to make a living. Farmers who work marginal land because there is no other land available are generally well aware of the damage they cause. Surveys of small-scale farmers in Sierra Leone showed that the majority correctly perceived an erosion problem on their land and associated it with high rainfall, steep slopes and lack of vegetation cover (Millington 1987). Studies of large-scale farmers in erosion-prone areas of Ohio revealed that over 40 per cent of them knew they owned land on which erosion was severe enough to affect productivity (Napier 1990). However, since farmers appreciate erosion mainly through its effects on yield, sediment accumulation on footslopes and formation of gullies, they underestimate its seriousness compared with more scientific assessments based on reductions in the depth of soil (Stocking & Clark 1999; Holden & Shiferaw 1999).

Most farmers are concerned with the effects of erosion on productivity and on increased costs of seeds for replanting a crop destroyed by erosion, fertilizers to maintain soil fertility, and water storage or irrigation facilities to provide additional water for crop growth (Shaxson 1987). Farmers' decisions, however, tend to be a compromise between preventing long-term soil damage by erosion and maximizing short-term income. Generally, the farmer is not unwilling to change practices but will do so only if substantial benefits arise and the investment costs can be recovered. Where a land user does not perceive such benefits, soil conservation is unlikely to be adopted. For example, where land is cheap and readily available, the incentive is to make a rapid profit and then move on to new land when the soil becomes unproductive. For the subsistence farmer, a particular problem is whether or not a new practice increases the risk of crop failure. Taking up higher yielding crop varieties that may also help to reduce erosion by increasing biomass will not be an acceptable gamble if, in one year out of ten, the crop yields less than the minimum required for subsistence, even though much higher yields will be obtained in the other nine years (Hudson 1981). These pressures invariably mean that, regardless of good intentions, the concept of a farmer managing the land according to an ethic of land stewardship is an inappropriate base for soil conservation (Hudson 1988).

7.3.2 Land tenure

The arrangements by which tenure is granted to the land user can influence attitudes towards soil conservation. Where farmers own the land, they are more likely to consider the long-term consequences of their actions and adopt soil-protection measures unless the need for short-term survival dictates otherwise. Tenure systems based on short-term cultivation rights, share-cropping and collectives generally lead to poor management because of uncertainty about whether conservation work carried out on the land will be rewarded. Legal title to the land, however, may be less important than a guaranteed long-term right to use the land, as seen in Vietnam and China, where soil conservation is practised even though private ownership of the land is not allowed (Phien & Tu Siem 1996; Li et al. 1998).

The overall size of a farm does not necessarily influence the frequency or type of soil conservation measures employed. Hallsworth (1987) cites data from the Nyasa, Rift Valley and Western Provinces of Kenya that show that 51 per cent of the farmers with holdings of less than 1 ha use fertilizers to maintain productivity, whereas on farms of over 10 ha the figure is still only 68 per cent. Equivalent figures for the use of terracing are 24 per cent and 21 per cent respectively. For mulching, however, the figures are 5 and 17 per cent respectively, the lower figure reflecting the difficulty of supplying mulch material on very small holdings. More important than farm size is the layout of the farm and, in particular, its degree of fragmentation, since many conservation measures, such as terraces, become impractical when the land is held in small and scattered parcels. Mulching is more difficult because there may be a considerable distance between the strip of land from which the mulch is taken and the strip on which it needs to be applied. Land consolidation programmes may not improve the erosion status, however, because, as seen in northern France, the reorganization of holdings into larger fields means the removal of field boundaries and increases in the length over which surface runoff can flow.

7.3.3 Labour

All soil conservation work implies extra labour. It is needed for the building and maintenance of terraces and for the growing of additional soil-protective crops either in rotation or by inter-cropping (Table 7.2). The likelihood of any soil conservation measure being adopted depends on whether the farmer and associated family can meet the increased labour demands (Stocking & Abel 1992). Surveys in northern Thailand showed that while 76 per cent of the farmers believed that soil conservation would bring about an improvement in yields, 85 per cent said they did not have the labour resource to implement the measures (Harper & El-Swaify 1988). While it may be

Table 7.2 Examples of labour requirements for soil conservation

Conservation practice	Labour (person-days ha^{-1})
Construction of contour terraces	100–300
Construction of bench terraces	200–500
Construction of bench terraces with stone side slopes (Inca type)	1000–1800
Annual maintenance of terraces	40–60
Contour stone bunds	50–60
Contour grass barriers	40–45
Intercropping an additional crop with maize (extra labour over maize monoculture)	150–200
Construction of Katumani-type pits	100–120

Sources: Wenner (1981), Alfaro Moreno (1988), IFAD (1992), Stocking and Abel (1992), Gichangi et al. (1992).

thought that in countries with high population densities and high rates of unemployment, labour-intensive conservation schemes would be beneficial, this view ignores the fact that farmers may give a higher priority in their use of labour to other activities, such as marketing their produce, traditional craftwork, house maintenance and off-farm employment. Moreover, a poor small-scale farmer is often physically unable to put more hours of work into the land because of poor diet and poor health. If the family tries to secure additional sources of income by the young men going to the towns or to the mines, an inadequate workforce is left behind on the farm, comprising grandparents, whose working life is over, young women, whose working time is limited by duties of housekeeping and child bearing, and children. In many countries, AIDS and other diseases are substantially reducing labour supply on family farms.

The organization of labour within the family is frequently highly specialized by age and sex. Men may do the land clearing and ploughing and women the weeding and collection of fuel wood; men may look after the cash crops and women the food crops. Knowing which groups perform which tasks is important, so that extension workers can target the most appropriate sector of the labour force. When soil conservation introduces new tasks, these have to be assigned in acceptable ways without creating stresses elsewhere in the system. Replacing hand-hoeing with ox ploughs may enable more land to be cultivated by the men but will result in much more weeding for the women (Milner & Douglas 1989).

The availability of labour can be improved if farmers have sufficient income to hire additional workers but, more often than not, serious constraints exist, particularly at times of land preparation and harvest when the work has to be done in a very short time. Other ways of improving labour supply are through communal schemes where farmers and their families come together to build terraces, dig the soil or weed the crop for what is, at the same time, a social occasion.

7.3.4 Access to soil conservation

The ability of farmers to adopt soil conservation measures will depend on their access to all appropriate resources, not just labour. These may vary from knowledge of new systems to an ability to afford the necessary inputs of capital to take them up. Whether or not farmers have the cash to purchase the additional seeds, fertilizer or machinery required to support a more conservation-oriented farming system will clearly affect its uptake; many poor farmers have insufficient security to support loans and would consider the risk of borrowing money too high. Most credit agencies would also view small-scale farmers as unacceptable risks and restrict access to credit to larger-scale and wealthier land owners.

It is pointless designing a soil conservation programme that requires levels of input to which the targeted farmers have no access. However, it should be recognized that many farmers use their own initiative, technical skill and labour to develop soil conservation measures where they benefit from so doing. Between 1948 and 1978, farmers in the Machakos District of Kenya (Tiffen et al. 1994) made large investments in terracing, tree planting and hedging as well as improving cultivation techniques in order to grow coffee, cotton, oranges and papaya. Despite a rapidly increasing population density of 3 per cent per year, the cash value of farm output per hectare rose tenfold and soil loss was reduced by one-third. Farmers on land holdings of less than 5 ha in western Santa Catarina, Brazil, adopted intercropping of maize and velvet bean (*Mucuna pruriens*) to protect soil because this made the difference between making a small profit or a financial loss (Stocking & Tengberg 1999). In these situations, the aim of the soil conservationist is to work with the farmer to enhance the performance of the measures and optimize their benefits. Where such benefits are not perceived, farmers will even remove conservation work to which they already have access, such as terracing installed on food-for-work programmes (Holden & Shiferaw 1999).

7.3.5 External factors

The causes of soil erosion rarely relate only to the existing physical and socio-economic conditions of a farming system but arise also from its inability to adapt to changing circumstances brought about by drought, increases in population, fragmentation or enlargement of farms, economic pressures to grow cash crops rather than food crops and new technologies. Erosion in much of Europe reflects the breakdown of farming systems formerly based on a rotation of cereals, root crops and grasses, or a combination of tree crops, cereals and pastures. With economic incentives to grow continuous cereals and roots, all essentially soil-degrading crops, the loss of grass has led to a decline in organic matter and a deterioration in the aggregate stability of the soils. The penetration of a cash economy into areas of traditional subsistence farmers and pastoralists frequently marginalizes those sections of the community who do not have sufficient economic resources to take advantage of the new income opportunities (Franke & Chasin 1981). Pastoralists are often displaced from the better quality land and forced to utilize and therefore overgraze poorer areas. Small-scale farmers who plant new cash crops, which are often more soil-degrading, lose their food security and in years when the market price for the crop is low are unable to provide the necessary inputs to maintain soil fertility and conservation structures. To obtain money to buy food, they may have to relinquish control of their land and lose their security of tenure. With more of the land area taken up in cash crops, food crops are in short supply and more expensive. Similarly, surplus labour is taken up by the larger-scale farmers, reducing its seasonal availability. As a result, many farmers can no longer adjust their farming system to meet the needs of soil conservation.

If soil erosion as a response to changes in farming systems is to be avoided, soil conservation must be a dynamic process capable of adjusting to future change. It is insufficient to develop a strategy that merely solves an immediate problem. The need for adaptability in soil conservation systems is likely to increase because the global impacts on erosion and agriculture expected to arise from climatic change will be far greater than those hitherto associated with changes in land use.

7.3.6 Technological transfer

Though the principles of soil conservation are transferable, conservation strategies developed in one area will not necessarily work elsewhere. What is feasible on a 100–200 ha farm with a high level of mechanization in the Midwest of the USA is unlikely to be applicable to a smallholding of less than 0.5 ha in the tropics. The non-transferability of soil conservation measures is perhaps best illustrated by the failure of channel terraces within the small-scale farming systems of Africa. Borrowed from the USA, where they were successfully developed and implemented in the 1930s and 1940s, such terraces formed the nub of soil conservation programmes proposed by the European colonial administrations in much of central and eastern Africa. Since the Crown Land made available to European settlers was not under pressure, there was little incentive to practise soil conservation and the colonial governments were forced to support the implementation of terraces and other structural measures by legal means to ensure that the works were maintained and appropriate soil management, such as contour ploughing, followed. With the measures being generally effective on the large settler farms, the governments tried to impose them on the land reserved for the local population on which erosion was becoming particularly severe. Despite much exhortation and attempts at compulsion, the policy was a failure. The measures were seen as too labour-intensive for the expected economic return, took too much scarce land out of cultivation, were not supported by the same system of subsidies and loans that were made available to the settlers and were regarded as an illustration of what was increasingly seen as the unfairness of colonial rule (Temple 1972; Wood 1992). Today, there are many examples throughout Africa where channel terraces and bench terraces implemented by governments, pre- and post-independence, have not worked, whereas indigenous, often highly labour-intensive systems, have been successful (Roose 1992; Cheatle & Njoroge 1993; Tiffen et al. 1994).

As long as soil conservation and extension services were promoting technologies modelled on those developed in the USA, indigenous conservation practices were either ignored or dismissed as ineffective. The fact that many were simply good farming practices of soil and water management that also happened to be helpful in reducing erosion was disregarded. Although these included traditional bench terrace systems, like those built by the Incas in Peru, they also encompassed contour bunds to promote infiltration of water and increase soil moisture, mulching to conserve water and soil fertility, burying of weeds and trash to improve fertility and weed control, crop rotation and weed fallows (Critchley et al. 1994; Reij et al. 1996; Tengberg et al. 1998). A characteristic of these indigenous systems is that they are not standardized; they vary from area to area and, often, from farm to farm, and their designs do not conform to the technical standards proposed by soil conservation services (Humbert-Droz 1996). Indigenous systems are not without problems, however. Structural measures require large inputs of labour to construct and maintain. Trash lines and mulching enhance the survival and spread of insect pests and disease pathogens (Tengberg et al. 1999). Nevertheless, indigenous systems provide a base for the soil conservationist to work with the local farmers to improve the local technology.

7.3.7 Economic evaluation

A major problem with applying economic analysis to soil conservation is that many aspects, such as benefits related to minimizing adverse environmental impacts, are difficult to quantify in monetary terms. Although this may be of limited importance in a farm budget, it can be vital at a project scale for deciding on the size of financial incentives given to farmers to promote soil conservation for the benefit of the community. An analysis directed at the community should take account of the costs of not controlling the erosion, including downstream or downwind sedimentation, and the benefits in terms of crop production and employment. However, at this level the analysis generally suffers from inadequate information on: how to price labour when much of it is in the form of unpaid inputs from farmers and their families; how to take account of the effects of different currency exchange rates where a black currency market exists alongside the official exchange; and how to cost the differences between a project with soil conservation and one without when the impact of continuing erosion on productivity cannot be quantified (Bojö 1992). Detailed information on the economics of households and their decision-making processes can be difficult to obtain, with many farmers unwilling to provide it or supplying misleading data (Ellis-Jones & Mason 1999).

Different views exist on whether projects should be evaluated in terms of: their net present value (NPV), i.e. the sum of discounted benefits less the sum of discounted costs; their internal rate of return (IRR), i.e. the rate of interest that the project can afford to pay to cover the resources required; or their benefit–cost ratio, defined as the sum of discounted benefits divided by the discounted costs. Decisions on the appropriate discount rate can seriously affect the results of NPV analysis. Disagreement also exists on whether the evaluation should be made over 10, 20, 30 years or longer. Extending the time horizon to 50 years can have a detrimental effect on the outcome, especially if discount rates of 10 per cent or more are used (Bojö 1992).

If the analysis is applied to an individual farmer, costs and benefits can be more easily quantified and yearly cash flows calculated (Hedfors 1983). Benefit–cost ratios to farmers in the Acelhuate Basin, El Salvador, from adopting a system of contour cultivation, contour grass strips and contour ditches to grow maize, sorghum and beans on steep slopes ranged from 1.0 to 2.8, depending on slope steepness and the depth of soil (Wiggins 1981). Greatest returns were predicted for the gentler-sloping land with deeper soil, whereas benefits were very limited or neutral for farmers working slopes of 25–51° with no top soil and renting the land on annual leases. It is clear that these last farmers have no economic accessibility to soil conservation. At both 5 and 20 per cent discount rates contour grass barriers were found to be uneconomic for farmers in the Cochabamba region, Bolivia, unless they brought about a 20 per cent increase in productivity (Ellis-Jones & Mason 1999). Economic analysis of small farms in the Machakos District of Kenya showed that soil conservation measures gave a return of KSh5 per man-day where terracing and some tree- and grass-planting were used, and KSh21 where all the terrace edges were planted with fodder grasses and fruit trees (Tjernström 1992). Studies in the USA (Ervin & Washburn 1981; Mueller et al. 1985) showed that in the 1970s and early 1980s, with the effects of erosion on productivity being masked by higher-yielding crop varieties and improved soil and water management, the cost of adopting a conservation technique was often higher than any benefit gained. As farmers came under greater economic stress, they abandoned existing soil conservation practices to reduce costs and avoid bankruptcy (Napier 1988). More recent economic analysis based on partial budgets of the additional benefits and costs show that adopting a no-tillage practice with nutrient management can bring about net benefits to a farmer of US$17.4 per hectare for maize, $7.7 for wheat and $3.6 for soya bean compared to conventional tillage (Natural Resources Conservation Service 1999). Obtaining higher yields from the new practice is an important contributor to these benefits. Were the yields to remain the same, there would be a net benefit of $8.9 per hectare for maize and $2.4 for wheat but a net disbenefit of $7.3 for soya bean. Installing a mixed grass and forest riparian barrier to trap eroded sediment and help reduce off-site water pollution would produce a net disbenefit to the farmer of $19.8 per hectare, even after allowing for government payments of $34.4 per hectare for enrolling the land under the Conservation Reserve Program and an additional $1.2 per hectare for management of the barrier, and after amortizing the establishment costs over ten years at 8 per cent interest (Natural Resources Conservation Service 1999). The question arises as to whether further public support should be provided to obtain the additional benefits to river water quality.

As the last example shows, soil conservation can have many benefits beyond the farm level. Since these cannot always be easily expressed in monetary terms, alternative methods to cost–benefit analysis need to be used if they are to be taken into account. One approach, multi-criteria analysis, uses cost–benefit for those items that can be properly represented in this way and then modifies the results by a series of weightings, designed to assess the other costs or benefits. For the Konto River Watershed Development Project, Indonesia, weightings were developed to express: the expected income to the local population, particularly the landless; the changes in rates of runoff and erosion; and the preservation of natural forest (de Graaf 2001). In practice, multi-criteria analysis can be somewhat complex, with difficulties in deciding what the weightings should be and how their values should be determined, especially when, unlike measures of erosion and runoff, they cannot be easily expressed numerically. A better approach, which has yet to be widely used in soil conservation studies, would appear to be cost-effectiveness analysis. Here, the objectives are decided in advance and expressed as performance criteria. The lowest cost of achieving these objectives is then determined. This approach has the potential for including social, ecological and environmental benefits.

Approaches to soil conservation

7.4.1 Cultivated lands

A risk of erosion exists on cultivated land from the time trees, bushes and grasses are removed. Erosion is exacerbated by attempting to farm slopes that are too steep, cultivating up-and-down hill, continuous use of the land for the same crop without fallow or rotation, inadequate use of fertilizers and organic manures, compaction of the soil through the use of heavy machinery and pulverization of the soil when trying to create a seed-bed. Least protection of the soil is afforded by crops grown in rows, tall tree crops and low-growing crops with large leaves (section 3.4). As a result, crops like maize, rubber, oil palm, grape vines, cassava and sugar beet can all give moderate to serious erosion problems. Small grain cereals, such as wheat and barley, afford better protection provided they are planted at sufficient density. Erosion control is dependent upon good management, which implies establishing sufficient crop cover and selecting appropriate tillage practices. Thus soil conservation relies strongly on agronomic methods combined with sound soil management, with mechanical measures playing only a supporting role (Fig. 7.1).

Conservation strategies are aimed at establishing and maintaining good ground cover. The feasibility of this is determined by what crops are grown and how quickly, under the local climatic and soil conditions, they attain 40–50 per cent canopy cover. Rapid establishment of crops is important, particularly in those parts of the world where erosion risk is high at and immediately after planting – for example, where the first rains of a wet season are highly erosive. In the Indore area of India, timely planting of sorghum early in the rainy season followed by a crop of good yield can produce a 40 per cent canopy cover by about 20 July. With late planting and a

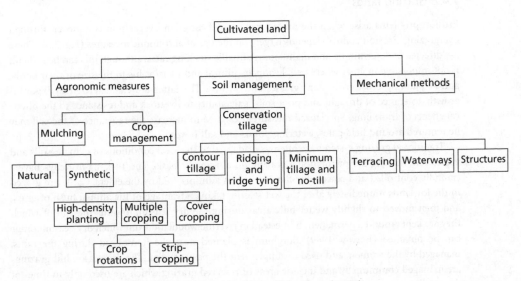

Fig. 7.1 Soil conservation strategies for cultivated land (after El-Swaify et al. 1982).

poor yield, 40 per cent cover is not attained until 30 September, thereby extending the period of erosion risk throughout the summer monsoon (Shaxson 1981). Aina et al. (1977) found that the time to establish 50 per cent canopy cover from the date of planting varied from 38 days for soya beans to 46 days for pigeon peas and 63 days for cassava. Soil loss under these crops was proportional to the time needed for the canopy to develop. Thus quick-growing crops may be viewed as soil-conserving crops. Where climatic conditions permit, early season cover can be provided by planting off-season crops of small grains or mustard and destroying them with herbicide prior to drilling the main season crop into the residue. To be successful, this approach requires adopting conservation tillage, using appropriate equipment to manage the residue, herbicides to control weeds and, ideally, precision agriculture to apply fertilizers at variable rates across fields according to local variations in soil.

As implied above, soil conservation measures must be both technically sound and socially and economically acceptable to the farmers. In southern Mali in the late 1970s, a system of graded terraces, diversion drains and grass waterways, supported by contour cultivation, and the use of mulches and manures, was tested in the village of Fonsébougou. Although the measures successfully reduced the erosion, the engineering works took up 10–14 per cent of the farm land and the terraces and the grass cover in the waterways were poorly maintained. Only a limited number of farmers were involved in decision-making or the execution of the work. Farmers gained the impression that once the system was installed, their problems were solved and no further work was required. As a result of this experience, an alternative approach was tried in the village of Kaniko based on existing farming systems, and integrating the ideas of the farmers with those of the researchers and extension staff. Greater emphasis was placed on contour grass strips and the use of trees and hedges rather than terraces (Fig. 7.2; Kleene et al. 1989; Hijkoop et al. 1991). Increasingly, it is recognized that strategies for soil conservation must rely on improving traditional systems (Roose 1992; Table 7.3), instead of imposing entirely new techniques from outside, and on enhancing land husbandry (Hudson 1993b; Hudson & Cheatle 1993; Critchley et al. 1994).

7.4.2 Grazing lands

Erosion problems arise when the protective cover of rangeland vegetation is removed through overgrazing. Erosion control depends largely on the use of agronomic measures (Fig. 7.3). These are directed at determining and maintaining suitable stocking rates, although this can be difficult if not impossible in areas where people attach cultural and social value to the size of their herds, and at planting erosion-resistant grasses and shrubs. The latter are characterized by vigorous growth, tolerance of drought and poor soils, palatability to livestock and resistance to the physical effects of trampling. Specialized measures designed to increase the resistance of the soil may be required around field gates, watering points and salt boxes.

Traditional grazing systems are often well adapted to the local conditions of climate, soils and vegetation, making use of rotational grazing on a nomadic basis. The Turkana in Kenya have a carefully controlled and reasonably flexible system administered by the elders. Cattle are grazed in the lowlands immediately after the wet season, taking advantage of the annual flush of grass, and then moved to slightly wetter hilly areas during the dry season, towards the end of which they are kept around a permanent homestead in riverine woodland from which dry-season forage can be obtained (Barrow 1989). Sorghum is planted near the homestead during the rains, managed by the women, and used to supplement the pastoral diet. The dry-season hill-grazings are managed communally and include areas of reserved grazing which are used only in times of drought.

(a) Toposequence

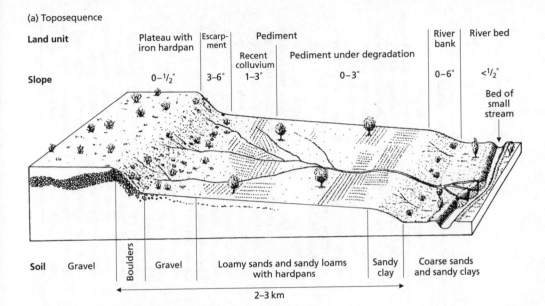

Land unit	Plateau with iron hardpan	Escarp- ment	Pediment		River bank	River bed
			Recent colluvium	Pediment under degradation		
Slope	0–¹/₂°	3–6°	1–3°	0–3°	0–6°	<¹/₂°

Bed of small stream

| Soil | Gravel | Boulders | Gravel | Loamy sands and sandy loams with hardpans | Sandy clay | Coarse sands and sandy clays |

2–3 km

(b) Proposed land use and management

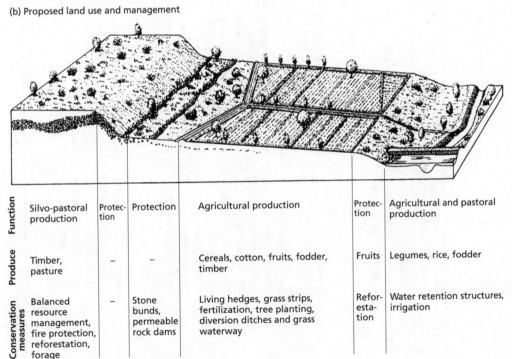

	Silvo-pastoral production	Protec- tion	Protection	Agricultural production	Protec- tion	Agricultural and pastoral production
Function						
Produce	Timber, pasture	–	–	Cereals, cotton, fruits, fodder, timber	Fruits	Legumes, rice, fodder
Conservation measures	Balanced resource management, fire protection, reforestation, forage	–	Stone bunds, permeable rock dams	Living hedges, grass strips, fertilization, tree planting, diversion ditches and grass waterway	Refor-esta-tion	Water retention structures, irrigation

Fig. 7.2 Proposed soil conservation scheme on cultivated land in southern Mali (after Hijkoop et al. 1991).

Table 7.3 Variation in traditional and proposed soil conservation systems with climate in West Africa

Area	South Soudanian	North Soudanian	South Sahelian	North Sahelian
Rainfall	>1000 mm	700–1000 mm	400–700 mm	150–400 mm
Conservation objectives	Dispose of excess rain while conserving soil	Hold rain *in situ* while conserving soil	Concentrate rain by water harvesting from a catchment area while conserving soil on cropped land	Concentrate rain by water harvesting from a catchment area while conserving soil on cropped land
Soils	Ferralitic	Leached ferruginous + vertisols + brown on basic rocks	Leached ferruginous + vertisols + brown on basic rocks	Sandy sheet on brown sub-arid soil
Vegetation	Tree savanna	Tree savanna	Bush savanna	Steppe, bush
Farming system	Yams on large mounds. Maize + intercropping in ridges. Millet + ground nut + various on ridges. Intercropping + agroforestry. Drainage between plots.	Superficial tillage. Cropping on the flat. Two weedings. Sorghum, cotton or millet + sorghum or ground nut + cowpeas. Hillslope runoff on waterways. Total infiltration or rain in the field. Stone bunds and walls. Grass, stone or brush barriers on contour.	Superficial tillage. Cropping on the flat. Two weedings. Sorghum or millet then ground nut or cowpeas. Mulching, stone bunds, grass or brush barriers on contour. Tied ridging on sandy soils (sometimes).	Sowing on the flat. Two weedings. Millet on sandy soils. Sorghum on clay soils on low ground. Grazing on hillslopes. Gardens in low-lying areas. Retreat flooding.
Proposed modern management	Afforestation of laterite screes. Grassed buffer strips. Gully restoration. Protection of rice fields.	Improvement of grassland. Living hedges + stone bunds or grassed buffer strips. Grass waterways.	Improvement of grassland. Pounds for cattle + supplementary irrigation of gardens. Stone bunds protected by grass. Gully management.	Shrub forage plantings in lunettes or ditches. Grassland management. Pounds for cattle. Trapping runoff on hillslopes. Stone bunds on pediment.

Source: after Roose (1992).

Virtually all traditional grazing systems are under pressure today. First, human population numbers, as a result of better health, and livestock numbers, as a result of better veterinary services, are both increasing. Second, there is a conflict between the individual and family ownership of livestock and the communal ownership of the pasture. The individual derives a positive utility of almost one for every additional animal owned and grazed on communal land but experiences a negative utility of only a small fraction of one as a result of ensuing overgrazing. The maximization of individual benefit at the expense of the community, termed by Hardin (1968) 'the tragedy of the commons', is one of the biggest challenges facing soil conservationists on rangeland. The conflict becomes most marked when incentives for commercial livestock production are so strong that individuals, usually those owning the larger numbers and better quality stock, break away and take over much of the land, enclosing it by fencing, displacing other members of the community and increasing the pressure on the remaining range. Third, the provision of additional watering points, often located without consideration of the traditional movement of stock, has meant that availability of forage rather than water has become the limiting factor on livestock numbers, creating additional pressure on the land. Fourth, the concentration of people in settlements to provide health, education and water more efficiently and to promote cash cropping with irrigation has caused people to abandon their nomadic tradition, with the result that stock are kept all year on pastures close to the homestead. Sometimes new settlement schemes will be sited on seasonal grazing areas, thereby removing them from the system. Fifth, education of the youth and changed political systems are gradually eroding the authority of the elders.

Grazing can be considered as the removal of biomass from rangeland. The rate of removal depends upon the number of animals and their daily intake of forage. Soil conservation is therefore aimed at controlling grazing numbers so that a sufficient vegetation cover is sustained over time to protect the soil. The loss of vegetation increases the rate of runoff and erosion and decreases the amount of water in the soil, which, in turn, reduces the amount of vegetation growth. In reality, the vegetation–erosion–grazing interaction is more complex because of the need to consider soil fertility, loss of soil nutrients, production of litter, the palatability and digestive value to stock of the different species in the plant community and the ability of the different species to survive under changing grazing, moisture and nutrient conditions. At present, the interaction is poorly understood. On the one hand, there is considerable evidence that many tropical rangelands are overstocked. It is suggested by Pratt and Gwynne (1977) that the threshold stocking rate on the semi-arid lands of northeast Kenya is about 1 Tropical Livestock Unit (TLU) per 25 ha whereas actual stocking rates range from 1 TLU per 4 ha to 1 TLU per 30 ha (Peden 1987). In the Middle Veld of Swaziland, stocking rates are about 1 TLU per 1.3 ha compared with sustainable rates of 1 TLU for every 2.5 to 3.5 ha (Nsibandze 1987; Dlamini & Maro 1988). On the other hand, application of a simple grazing model developed by Biot (1990) to the rangelands of Botswana suggests that at a grazing intensity of 1 TLU per 5 ha, the system would remain reasonably stable for the next 2000 years. Clearly with such discrepancy between conventional wisdom and model predictions, it is difficult to determine the safe level of grazing, particularly when short-term fluctuations in climate also need to be considered. Further, there is often a conflict between designing a strategy that sustains the productivity of the rangeland and maximizes economic benefit over the medium to long term, and a strategy that maintains erosion at a much lower tolerable level to minimize environmental impacts.

7.4.3 Forest lands

Forests provide excellent protection of the topsoil against erosion. They maintain high rates of evapotranspiration, interception and infiltration and therefore generate only small quantities of

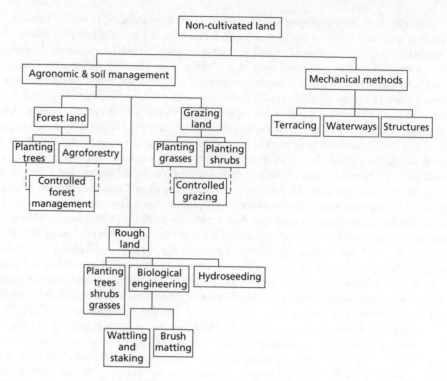

Fig. 7.3 Soil conservation strategies for non-cultivated land (after El-Swaify et al. 1982).

runoff. Low runoff rates and the protective role of the litter layer on the surface of the soil produce low erosion rates. Increases in erosion occur where the land is permanently or, in the case of shifting agriculture, temporarily cleared for agriculture. While the forest cover remains largely intact, the most important erosion problems are associated with: cropping of trees for firewood; destruction of the trees and surrounding shrub and ground cover by grazing; and logging operations. An estimated 40 per cent of the world's population use wood as the primary fuel. In Sudan, Colombia, Ethiopia, Nigeria and Indonesia, fuel wood accounts for 80 per cent or more of annual timber removals. Afforestation schemes that include rapid-growing tree species that can be cropped for firewood are therefore an important feature of erosion-control strategies (Fig. 7.3).

Livestock grazing is frequently detrimental to the survival of forests. The animals trample and compact the soil, injure roots close to the surface and browse on the tree seedlings. At the time of settlement between AD 875 and 930, about 65 per cent of the land surface of Iceland was covered with vegetation (Thorsteinsson et al. 1971), with between 25 and 40 per cent of the country under low-growing birch forest with a luxuriant undergrowth of forbs and grasses. Overgrazing and stripping of the bark from the trees has prevented the regeneration of the forest, much of which has also been cut for fuel. As a result, the area under forest is now reduced to only 1 per cent of the area of the country (Arnalds 1987).

Logging causes limited disturbance because erosion is confined to the area of land where the trees have been removed. With good management, the vegetation cover regenerates quickly so that high erosion rates are restricted to the first and sometimes the second and third years after felling. The level of disturbance is related to the method of clearance. Studies of three practices in Nigeria (Lal 1981) showed that more erosion followed mechanical clearance using crawler tractors with tree-pusher and root-rake attachments than manual clearance using chain saws. Least erosion followed manual clearance with machete and axe. Erosion immediately after clearance is generally associated with surface runoff. Another effect of forest removal, however, is the gradual loss of shear strength of the soil following the decay of the root systems (O'Loughlin 1974; O'Loughlin & Watson 1979). This induces a risk of landslides, which is greatest about five years after clearance (Bishop & Stevens 1964; Beschta 1978), although some researchers suggest a slower deterioration, with maximum slide hazard being reached 15 years after logging (Rice & Krammes 1970).

The main erosion problems in logged areas are associated with skid trails and roads, which are frequently areas of bare compacted soil. Studies in Pahang, Malaysia, showed that traffic along skid trails and logging roads increased the bulk density of the ultisol soil to 1.5 and 1.6 Mg m^{-3} respectively, compared to 1.1 Mg m^{-3} on undisturbed land, with resulting respective decreases in porosity of 62 and 69 per cent (Baharuddin et al. 1996). Erosion rates at log-landings and on haulage roads on slopes of less than 10° in Sabah, Malaysia, are some three times higher than those from primary rain forest on much steeper slopes (Clarke et al. 2002). As reported by Swanson and Dyrness (1975), who measured 2,300,000 t ha^{-1} on a roadside in western Oregon, USA, annual erosion rates from logging roads can be very high. Roads, tracks and paths can contribute up to half of the total sediment yield in forested catchments (Reid et al. 1981). Cut slopes are particularly vulnerable, being up to ten times more erodible than fill slopes (Riley 1988). Erosion can be minimized by locating roads on ridges and gentle slopes to avoid the need for cut slopes.

Another source of sediment, widely reported from afforestation schemes in Wales (Newson 1980; Robinson & Blyth 1982; Murgatroyd & Ternan 1983; Francis & Taylor 1989), occurs where the land has to be drained by surface ditches. The ditch banks erode as the channels adjust from their relatively deep and narrow cut-form to a shallower, wide one. The increased runoff from the ditches also causes bank erosion in the rivers immediately downstream.

7.4.4 Land clearance

The methods used for clearance of forest for agricultural development not only affect the amount of soil loss that takes place but also influence the subsequent crop yields. The mechanical clearance of forest on an alfisol in Nigeria (Lal & Cummings 1979) significantly increased the bulk density and decreased the infiltration capacity of the soil compared with removal by slash-and-burn. Bulk densities were increased from 0.9 Mg m^{-3} on uncleared land to 1.12 Mg m^{-3} with slash-and-burn and 1.25 Mg m^{-3} with mechanical clearance. The respective infiltration capacities were 112, 52 and 19 mm h^{-1}. Clearance of forest on 30–40° slopes in Malaysia on an ultisol soil reduced the infiltration capacity from 214 mm h^{-1} on uncleared land to 149 mm h^{-1} using slash-and-burn and 43 mm h^{-1} using crawler tractors with straight blades; the respective bulk densities were 1.06, 1.33 and 1.48 Mg m^{-3} (Jusoff & Majid 1988).

The response of crop yield to different types of land clearance depends on the degree of disturbance and the subsequent method of cultivation. Potentially, the situation is worse with mechanical clearance because much of the litter layer is scraped away, removing the immediate protection of the soil against erosion and the source of organic material. Yet, on an ultisol soil on a 2° slope in northern Thailand, mechanical clearance of the forest by tractor with a tree-pusher blade followed by construction of contour bunds, no-tillage and inclusion of legumes in a rotation system led to higher rice yields compared to other treatments (Boonchee et al. 1988). In contrast, on ultisols in Peru (Seubert et al. 1977) and alfisols in Nigeria (Lal 1981), significantly higher yields were obtained using conventional tillage practices or chisel ploughing following mechanical clearance compared to using no tillage or mulching. The latter techniques initially resulted in rather low yields, even when fertilizer was added, because they did not overcome the increased compaction of the soil.

If disturbance of the surface is kept to a minimum and a system of land management imposed for eight to ten years after clearance to overcome any increases in compaction of the soil, forest land can be successfully cleared for agriculture. If the land cannot be cropped immediately because the disturbance has been too great, the planting of a cover crop will help to restore the soil close to its original bulk density and infiltration capacity (Hulugalle et al. 1984), after which a range of management practices, including no tillage, is possible. Once under agricultural use, the appropriate soil conservation practices for cultivated land must be adopted.

7.4.5 Rough lands

Areas of rough ground remaining in their natural habitats because they are too marginal for other forms of land use include hilly and mountainous terrain with shallow stony soils and steep slopes, alpine grasslands, arctic tundras and sand dunes. Since they are often areas of spectacular scenery, they attract recreational use. Their marginality, however, means that they are sensitive to any disturbance, and their ability to withstand recreational impacts is low. The main problems of land degradation are associated with footpaths, tracks created by horses, motor cycles and off-road vehicles, and very intensive use at car parks and camp sites.

Approaches to controlling erosion range from exclusion of people, appropriate in areas of special scientific interest or in nature reserves, to use of erosion-resistant plant species, supported where necessary by drainage to reduce soil moisture content and increase soil strength, and artificial strengthening or reinforcement of footpaths. Although few plants can withstand prolonged and heavy pressure from walkers and horses, studies of footpath erosion in the English Lake District show that grasses, including *Nardus stricta*, *Agrostis* spp. and *Festuca* spp. resist trampling better than *Calluna* heathland (Coleman 1981). One way of comparing the vulnerability of different habitats to trampling is to determine the number of passes by individual walkers they can withstand before the plant cover is reduced from 100 to 50 per cent. Sand dune pasture can withstand between 1100 and 1800 passes, alpine plant communities less than 60 and arctic tundra only about eight passes (Liddle 1973). Trampling of a grass cover on a clay soil with a moisture content close to field capacity resulted in runoff rising rapidly between 0 and 50 passes, then levelling off before rising again between 700 and 900 passes. The rate of soil loss, however, increased linearly with increasing number of passes. Since patches of bare ground did not appear until 250 passes and even after 900 passes between 25 and 50 per cent of the cover remained, the critical period when increased runoff and erosion are initiated seems to occur before any decrease in the vegetation cover is observed (Quinn et al. 1980).

In addition to being able to survive in the climatic and soil environment, the plant species selected for revegetation of rough areas should ideally be local, so as not to introduce new species into the ecological system, and should meet the following criteria: short or prostrate growth form, flexible rather than brittle or rigid stems and leaves, basal or underground growth points, rapid growth, ability to withstand burial by soil and rocks and ability to withstand exposure of the root system (Coppin & Richards 1990). Often an ecological succession can be planned involving a mixture of species, both grasses and shrubs, to give an initial rapid growth and cover of the ground followed by longer-term development of erosion-resistant plants.

7.4.6 Urban areas

Urban development frequently results in intensive erosion. The exposure of bare soil during the construction phase results in higher volumes of peak runoff, shorter times to peak flow, higher and more frequent flood flows and rapid increases in erosion by overland flow, rills and gullies, producing high sediment concentrations. Residential development started in the Anak Ayer Batu catchment in Kuala Lumpur in the mid-1960s. By 1970 the river was choked with sediment. Flooding and deposition of sand occurred regularly on the flood plain, with suspended sediment concentrations in the river water of 4–81,230 mg l^{-1} (Douglas 1978). By 1977 the situation had changed little with sediment concentrations ranging from 3.7 to 15,343 mg l^{-1} compared to between 7 and 1080 mg l^{-1} in similar channels in rain forest areas. Similar sediment concentrations with peaks between 15,000 and 49,000 mg l^{-1} were recorded during construction work on the new campus of the University of Singapore at Kent Ridge and on the old campus at Bukit Timah (Gupta 1982).

Strategies for erosion control in urban areas depend on scheduling developments to retain as much plant cover as possible, but since this is generally feasible to only a limited extent, a much greater reliance is placed on mechanical methods than is the case with other types of land use. Erosion control in the final phase when urban development is complete requires rapid establishment of plant cover and permanent use of purpose-designed waterways and embankment-stabilizing structures (Fig. 7.4). On-site erosion is very low at this stage because most of the land is covered in concrete but runoff from impervious surfaces of streets, pavements, roofs, gutters and sewers can cause bank erosion in rivers downstream of the urban area. The instability of roadside slopes often presents an additional problem.

7.4.7 Mining land

The main purpose of erosion control on land previously used for mining is to create a stable environment for vegetation establishment and growth, often to reclaim the land for agriculture or recreation, and to minimize off-site damage. Erosion can start very quickly on banks of unconsolidated mine spoil and, if rills and gullies form, may reach rates of 100–500 t ha^{-1} annually (Porta et al. 1989). Revegetation requires designing a plant succession that will give adequate surface cover and increase the fertility of the soil. The succession should include: rapid-growing grasses to give ground cover as quickly as possible and stabilize the surface; legumes, such as clovers and

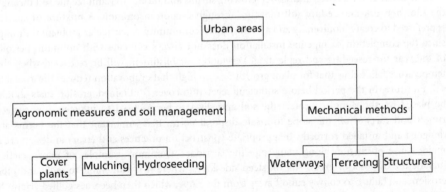

Fig. 7.4 Soil conservation strategies for urban areas (after El-Swaify et al. 1982).

Strategies for erosion control

vetches, to fix nitrogen; and other grasses and shrubs to provide the long-term cover. Species selection will depend on local soil and climatic condititions but should aim to produce a uniform rather than clumpy pattern of vegetation to avoid concentrations of runoff and localized erosion (section 3.4.2). The ideal condition is difficult to achieve, however, where the soils have low water-retention capacity or are toxic, or where cold, drought or exposure inhibit plant growth. Choosing vegetation that has beneficial effects is important. In order to reduce the costs of reclamation of spoil mounds from tin mining on the Jos Plateau, Nigeria, the Mine Lands Reclamation Unit chose *Eucalyptus* spp., which could be cropped as a source of fuel and pole timber, but these have resulted in significant declines in the base status and pH of the soil (Alexander 1990) to the detriment of other plant growth. Among the mechanical measures, terraces and waterways will be required to control runoff and remove excess water safely from the site, usually to silt-traps and soak-aways at the base of the slope, and revegetation may need to be supported by the use of geotextiles to provide an immediate protection to the soil and prevent erosion washing or blowing away the seeds. Contour wattling, contour brush-layering or terrace construction may be additionally employed to supplement the vegetation for longer-term protection of the soil.

7.4.8 Pipeline corridors

Restoration work along pipeline rights-of-way is usually directed at re-establishing the original vegetation cover or returning the land to agriculture. Erosion can inhibit restoration by washing out seeds and young plants. In addition, sediment washed off the right-of-way can cause environmental damage on surrounding land and rill and gully erosion can expose the pipe. The potential for erosion is high because in order to install the pipe, the top soil has been removed, temporarily stored and then replaced by spreading it on to land which has been heavily compacted by machinery. Infiltration rates through the compacted layer are low, so that runoff is easily generated and the top soil is erodible because it has lost much of its cohesion through disturbance. A compromise has to be sought between compacting the top soil sufficiently to increase its strength while keeping compaction low enough so that it does not inhibit plant emergence or root development. Bulk densities in excess of $1.3\,\mathrm{Mg\,m^{-3}}$ on clay soils, $1.4\,\mathrm{Mg\,m^{-3}}$ on silty and loamy and $1.6\,\mathrm{Mg\,m^{-3}}$ on sandy soils will limit plant growth (Daddow & Warrington 1983).

Similar to the situation described above for restoring mine spoil, a plant succession needs to be designed based on quick-growing native species, adapted to the local soils and climate, to provide dense ground cover, and slower-growing shrubs and bushes to reinforce the soil through the root network and reduce soil moisture through evapotranspiration. A mixture of species is preferred to create biodiversity and a healthy plant community. A particular problem is timing, since the completion of pipeline installation does not always coincide with optimum periods of the year for reseeding and replanting. Vegetation establishment will be restricted when the temperature falls below that for plant growth or is so high that evaporation reduces the available soil moisture. In the period before sufficient vegetation cover is obtained, erosion mats should be placed on the slope to protect the soil and diverter berms used to intercept any surface runoff and carry it off the slope to a safe disposal point. It is important that the berms are designed and installed correctly. Inappropriate construction practices and errors in design are a major cause of erosion along many pipeline corridors. Common examples of poor practice include the disposal of excess soil on steep sideslopes, where it is vulnerable to gully and pipe erosion, and failure to convey runoff away from the slope, which then becomes gullied (Hann & Morgan 2003).

7.4.9 Road banks

As indicated in section 7.4.3, road banks on either cut or fill slopes are frequently major sources of sediment. Poor disposal of runoff is an important cause of erosion associated with roads. Runoff upslope of the road should be collected in roadside drains or ditches, which are then led into culverts under the road, usually discharging into existing valleys. It is common, however, to discharge the runoff collected from several small streams into a single culvert, which means that the catchment area of the recipient channel downslope of the road is increased. Where adequate protection measures are not put in place, the increased runoff can enlarge the channel into a gully (Nyssen et al. 2002), which, over time, will extend back upslope and damage the road. The road shoulder, the land between the road surface and the side drain, is also vulnerable to erosion, particularly where the road surface is metalled and the shoulder is unprotected (Freer-Hewish 1991; Kamalu 1991).

The methods used for erosion control range from engineering structures such as revetments and retaining walls to stabilization of the slope by vegetation. Grasses will protect the slope against erosion by raindrop impact and runoff, and also trap moving sediment, while shrubs and trees will increase the strength of the soil through root reinforcement. Trees will also help to support the slope by buttressing (section 3.4.4). Vegetation increases the infiltration of water into the soil, but this can cause problems where rainfall amounts and intensities are very high. While the resulting reduction in runoff will help to control surface erosion, the increased moisture content of the soil may exacerbate the risk of mass soil failure. Whether this occurs will depend on the extent to which root reinforcement has increased the cohesive strength of the soil and root penetration across the potential slide plane has anchored the soil mass to the underlying material. Where soil saturation is a danger, the vegetation may need to be supported by cut-off drains above the slope to prevent the influx of surface water, and trench or herringbone rubble drains to aid drainage of the slope itself. In the initial stages, until the vegetation cover is established, wattling, brush layering or erosion mats may be needed to provide temporary erosion control and stability. For surface erosion and for soil failures extending to depths of up to 2 m, vegetation has the advantages over engineering structures of being cheaper to install, aesthetically more pleasing and, in the long-term, self-repairing, with an indefinite life. It will, however, require regular maintenance and occasional repair. Although they have not yet been adopted as standard practice worldwide, civil and highway engineers are showing increasing interest in both bioengineering, in which vegetation is used on its own as an engineering material, and biotechnical engineering, in which the engineering properties of vegetation are combined with inert structures. A detailed manual has been developed for Nepal describing the properties of some 13 local grass, 13 shrub, 15 tree and six clumping bamboo species suitable for bioengineering (Howell 1999).

When considering strategies for erosion control related to roads, it is important to take account of the location of the road within the landscape and, in particular, the nature of the land use on the slope above the road bank. In the Himalaya where the road banks are frequently oversteepened cut sections of the hillside, seepage from irrigation terraces on the slope above can increase instability. Inclusion of forestry, either as agroforestry or as community-managed forests, may help to stabilize both the slope above and the road bank itself and be more satisfactory than an engineering solution. If the forest can provide fodder, fuel and fruits, its long-term management may be vested in the individual farmers and the community as a whole.

Box 7

Planning a soil conservation strategy

Soil conservation design most logically follows a sequence of events (Fig. B7.1) beginning with a thorough assessment of erosion risk using the techniques described in Chapter 4. This is followed by the design of a sound land-use plan, based on what the land is best suited for under present or proposed economic and social conditions and what is compatible with the maintenance of environmental stability.

Land capability classification

By adopting land capability classification as a methodology for land-use planning a distinction can be made, as seen in section 4.2.1, between areas where erosion is likely to occur through mismanagement and areas where erosion occurs through misuse. In the first case, erosion can be reduced to a tolerable level by suitable conservation practices but, in the second case, erosion control is expensive and difficult, if not impossible. The land is likely to become so degraded that eventually it will have to be abandoned. Much money can be spent attempting to reclaim land where soil degradation has become virtually irreversible but the low productivity of the land means that the investment is highly questionable since it can never be repaid.

Defining conservation needs

The ultimate success of soil conservation schemes depends on how well the erosion problem has been identified, the suitability of the conservation measures selected to deal with the problem and the willingness of the farmers and others to implement them. These features of an erosion-control strategy are just as important as the ability of the engineer to design the conservation structures required. Correct identification of the major areas of erosion and therefore the main sources of sediment is also essential. Too often expensive construction work has been carried out on areas of land that were thought to be important sources of sediment but in fact were not

(Perrens & Trustrum 1984). By contrast, treating a small but carefully targeted area of land can be very effective. Soil loss in the Wuding Valley, China, was reduced by half by treating 20.5 per cent of the catchment (Jiang et al. 1981).

Existing farming systems

The farming systems already operating in the area should be analysed to see how well they protect the land, particularly at the critical periods of the year when erosion risk is highest. This is best achieved by relating the farming calendar to the monthly pattern of erosivity. For example, in eastern Thailand (Prinz 1992), erosivity reaches its peak in the first half of the wet season when the crops of kenaf, groundnut and upland rice do not yet provide adequate cover. On the loamy soils of north-eastern France (Monnier & Boiffin 1986), the danger of erosion under winter-sown cereals is high, even when the soil is covered and rainfall is relatively low, because of the marked reduction in infiltration rates between December and March as a result of crusting of the soil in the autumn (section 2.1). Studies of the farming systems can also provide information on how the farmers perceive erosion, the conservation measures already adopted and the social, economic, political and institutional constraints on the take-up of further measures.

Selection of conservation strategies

Based on assessments of the severity of erosion and its off-site impacts, a decision can be made on whether soil conservation measures are required in order to meet a desired soil loss tolerance. If so, a further decision is needed on the performance criteria they should achieve. These can be wider than meeting a given soil loss tolerance and include protection of watersheds and areas of scientific or heritage interest. They may also cover socio-economic objectives such as relieving poverty and producing sustainable rural liveli-

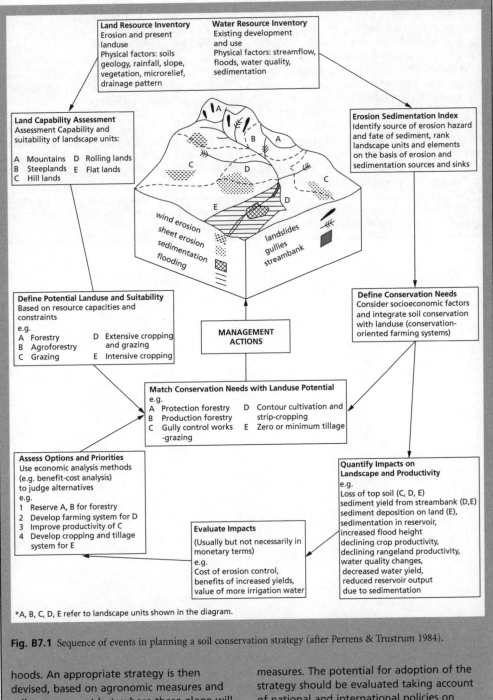

Fig. B7.1 Sequence of events in planning a soil conservation strategy (after Perrens & Trustrum 1984).

hoods. An appropriate strategy is then devised, based on agronomic measures and soil management but, where these alone will not be sufficient, supported by mechanical measures. The potential for adoption of the strategy should be evaluated taking account of national and international policies on agriculture and the environment.

Continued

Strategies for erosion control

Impact assessment

The next stage is to quantify the impacts of the proposed land use and associated conservation strategy. Simple environmental impact assessments can be made using the predictive models described in Chapter 6. Models have also been developed for forecasting the impact of erosion on the productivity of the soil. Most, such as the Productivity Index (PI) (Larson et al. 1985) and the Erosion Productivity Impact Calculator (EPIC) (Williams et al. 1984) are only valid in areas such as the USA and western Europe, where soil fertility is not a limiting factor because nutrients are added to the soil through fertilizers. At present, the models can only serve as guides to the likely impacts because they have not been widely validated owing to a lack of adequate data. To obtain the necessary information on changes in the nutrient status and physical properties of a soil as a result of erosion requires a minimum period of ten years under natural conditions. Trying to speed up the process by scalping the top soil to different depths results in an underestimation of the effect on nutrient removal by some five to ten times (Stocking & Peake 1987). Impact analysis can include economic assessments at the farmer level, using cost–benefit analysis, as well as wider assessments taking account of costs and benefits that cannot be described in monetary terms (section 7.3.7).

CHAPTER 8

Crop and vegetation management

Agronomic measures for soil conservation use the protective effect of plant covers to reduce erosion (section 3.4). Because of differences in their density and morphology, plants differ in their ability to protect the soil. Generally, row crops are the least effective and give rise to more serious erosion problems. This is because of the high percentage of bare ground, particularly in the early stages of crop growth, and the need to prepare a seed bed. In the design of a conservation strategy based on agronomic measures, row crops must be combined with protection-effective crops. Agronomic measures also form an important component of conservation tillage (Chapter 9) and of strategies to control erosion in plantations, forests and non-agricultural areas.

8.1
Rotation

The simplest way to combine different crops is to grow them consecutively in rotation. The frequency with which row crops are grown depends upon the severity of erosion. Where erosion rates are low, they may be grown every other year, but, in very erodible areas, they may be permissible only once in five or seven years. A high rate of soil loss under the row crop is counteracted by low rates under the other crops so that, averaged over a six- or seven-year period, the annual erosion rate remains low. The long-term effects of rotation versus continuous monoculture can be quite dramatic as shown by data from the experiments carried out at Sanborn Field, Missouri. On 4° slopes with a silt-loam soil, the annual erosion rate under continuous maize was $44\,t\,ha^{-1}$ compared with $6\,t\,ha^{-1}$ from a maize–wheat–clover rotation. After 100 years, land under continuous maize had only 44 per cent as much top soil as land kept permanently under grass, whereas the land under the rotation system had 70 per cent (Gantzer et al. 1990).

Suitable crops for use in rotations are legumes and grasses. These provide good ground cover, help to maintain or even improve the organic status of the soil, thereby contributing to soil fertility, and enable a more stable aggregate structure to develop in the soil. The effects are often sufficient to reduce erosion and increase yield during the first year of row-crop cultivation, but they rarely extend into the second year. For this reason, two continuous years of planting with a row crop should be avoided. Hudson (1981) showed that a rotation of tobacco–grass–grass–tobacco–grass–grass was more effective in Zimbabwe than one of two consecutive years of tobacco followed by four years of grass. The respective mean annual soil loss rates were 12 and $15\,t\,ha^{-1}$.

8.1.1 Shifting cultivation

Shifting cultivation is a traditional method of reducing soil erosion in the tropics by rotating the location of the fields. An area of forest is cleared by slash-and-burn, the soil loosened by hand hoeing and a crop planted. Where two crops a year can be obtained, a second crop is grown after the harvest of the first; otherwise the land is left in a weed fallow. The same area may be cropped for a second year before being allowed to revert to scrub and secondary forest. Typical crops grown are cassava, maize, upland rice and yam. The residual effect of the forest on the organic content and aggregate stability of the soil generally lasts for the first year of farming so that soil loss remains low. Erosion rates rise rapidly if the land continues to be cropped in subsequent years. Soil loss from land under upland rice in Mindanao, Philippines, averaged $0.38\,g\,m^{-2}$ per day on a new clearing but rose to $14.91\,g\,m^{-2}$ per day on a clearing in its twelfth year of cultivation (Kellman 1969).

The practice of shifting cultivation will maintain soil fertility and reduce erosion to tolerable levels provided the associated conditions of low crop yields and low ratio of population to land area remain socially and economically acceptable. The critical factor in the system is the length of the fallow period. This is traditionally between 7 and 20 years in West Africa (Okigbo 1977) but increasing population densities and the desire of people to raise their standard of living by changing from subsistence to cash cropping put pressure on the land, resulting in a reduction and sometimes the elimination of the fallow period. Serious erosion problems are created when the fallow period is reduced and alternative farming practices to the shifting cultivation system are required to solve them, using intercropping, mulching and agroforestry.

8.1.2 Row-crop cultivation

Particular problems are associated with maize, which, when grown as a row crop with conventional tillage and clean weeding, results in an annual soil loss on 2–5° slopes of between 10 and $120\,t\,ha^{-1}$, taking data from Zimbabwe (Hudson 1981), Malaysia (Sulaiman et al. 1981) and India (Singh et al. 1979). These rates are well above most soil loss tolerance levels. Where maize is grown in a more traditional manner with cultivation being restricted to ploughing and with no weeding, it presents fewer difficulties but the yields are rather low, often only $1\,t\,ha^{-1}$ compared with over $5\,t\,ha^{-1}$ regularly achieved in the USA. There is an urgent need to find a way of increasing the yield without increasing the erosion.

Soya beans are often used in rotation with maize because of their apparent ability to reduce soil loss by intercepting a high percentage of the rainfall. At 90 per cent canopy cover, soya beans intercept 58 per cent of the rainfall, compared with 40 per cent for maize interplanted with cassava and only 28 per cent for cassava on its own. Respective annual soil losses are 4.0, 6.9 and $11.0\,t\,ha^{-1}$ in the Ibadan area of Nigeria (Aina et al. 1979). However, studies in the Midwest of the USA indicate that soya beans can result in as much if not more erosion than maize (Laflen & Moldenhauer 1979). The average annual soil loss on a 4° slope with a silt-loam soil over seven years was $7.0\,t\,ha^{-1}$ for continuous maize, $6.5\,t\,ha^{-1}$ for maize following soya beans and $9.6\,t\,ha^{-1}$ for soya beans following maize. The main reasons for the increased erosion under soya beans are that the crop affords less protection than maize in the stage between canopy cover and harvest, and produces less residue after harvest.

In addition to controlling erosion, the inclusion of grasses and legumes in a rotation can increase yields of the main crop. Yields of maize greater than $5\,t\,ha^{-1}$ were obtained at Ibadan following a year with the land under *Centrosema pubescens*, *Setaria splendida* or *Stylosanthes gracilis*

(Okigbo & Lal 1977). Unless the fallow crops can be used for grazing or fodder, however, they have no immediate value to the farmer. Thus, crop rotations with grasses and legumes are rarely practised in the main cereal-growing areas of the world and are unlikely to be acceptable anywhere if their inclusion gives no income to the farmer. Under these conditions, an alternative approach to erosion control is required based on minimizing the period of bare ground – for example, by leaving crop residue on the land after harvest and delaying ploughing until the following spring, or, where winter cereals are grown, sowing as early as possible to ensure good cover before winter temperatures inhibit plant growth. Late sowing of winter cereals was a major cause of soil erosion in the 1980s over much of southern and eastern England (Boardman & Evans 1994) and northern France (Monnier & Boiffin 1986).

8.1.3 Grazing land management

Rotation is commonly practised on grazing land, moving the stock from one pasture to another in turn, to give time for the grass to recover. Generally, rangelands should not be exploited to more than 40–50 per cent of the annual production of their most palatable species (Fournier 1972; Thorsteinsson 1980a), a condition that is reached when about 10–20 per cent of the annual growth of the woody species has been removed. The vegetation should be allowed to regenerate sufficiently to provide a 70 per cent ground cover at times of erosion risk. Grazing has to be very carefully managed. While overgrazing can lead to deterioration of the rangeland and the onset of erosion, undergrazing can result in the loss of nutritious grasses, many of which regenerate more rapidly when grazed.

Overgrazing of the summer pastures in Iceland led to a decline in the proportion of grasses within the plant community from 30 to 12 per cent and rises in the proportions of sedges and rushes from 18 to 26 per cent, and mosses from 15 to 24 per cent (Thorsteinsson 1980a). The annual yield of woody species on overgrazed land is only $150\,kg\,DM\,ha^{-1}$ compared with $800\,kg$ $DM\,ha^{-1}$ on protected land. The plant density is often low with open scars in the vegetation cover. An increase in the numbers of sheep and a decrease in the length of the growing season from 240 to 210 days between 1944 and 1968 created pressure on the better pastures of the Peak District in the UK, leading to a breakdown in the turf mat, soil erosion and a change in sward type from *Agrostis–Festuca* to *Nardetum* (Evans 1977), with a consequent lowering of the sheep-carrying capacity. This change in vegetation type and also that from heather to grassland has been observed generally in the UK following overgrazing of upland pastures over periods of five to ten years (Fenton 1937; Jeffers 1986). Where undergrazing occurs, grassland and heather give way to birch scrub and woodland over a period of 10–50 years but are also in danger of invasion by bracken, in which case the succession to trees may be delayed and take centuries to occur (Jeffers 1986).

Controlled burning, preferably practised on a rotational basis, is essential for the removal of undesirable species, whereas uncontrolled burning can prevent plants from re-establishing and, by increasing the extent of bare ground, result in serious erosion. Generally, soil loss is highest in the first year after fire (Shakesby et al. 1993) and returns to background levels within three to four years. The absolute increase in erosion may be quite low, however, on soils with a stable granular structure (Kutiel 1994). On other soils, a high soil loss occurs because vaporization and readsorption of stable organic substances during the burn makes the soil hydrophobic (De Bano et al. 1970), which decreases the infiltration rate, increases the runoff and enhances detachability by raindrop impact (Terry & Shakesby 1993; Doerr et al. 2002). The severity of the effect depends upon the temperature of the burn. In an experimental study under Mediterranean *maquis* vegetation, annual erosion was five times higher than that of an unburned area following a fire with a mean soil surface temperature of 180°C, but was 50 times higher after a burn with a mean soil surface temperature of 475°C (Giovannini et al. 1988). The higher temperature burn caused the finer particles in the subsoil to form sand-size cement-like aggregates with low porosity (Giovannini 1994), which encouraged the generation of more runoff. Depending upon the soil, rainfall, temperature of burn and degree of insulation provided by duff (Dimitrakopoulos et al. 1994), it takes some three years for the surface soil to be restored to pre-burn condition and some seven to twelve years for the plant community to re-establish following a fire (Naveh 1975; Trabaud & Lepart 1980), which implies that burning every seven to ten years is probably required to control invasion of the rangeland by undesirable species.

Prescribed burning modifies the density, stature and composition of brush stands within a plant community but kills only a few species outright. Most brush species sprout vigorously after fire to the detriment of grasses. Grasses, however, fill the gaps between the brush and reach their maximum development in the second and third years following a fire. Thus, fire can be used to topkill and open up mature stands of brush so that a seed bed can be created for the establishment of grasses. Once the brush resprouts, however, it has a competitive advantage over the young grass and needs to be controlled by herbicides until the grass is strong enough to carry fire. After this stage has been reached, burning can be used periodically to control brush and to maintain a more diverse plant community than would occur if the ecological succession were allowed to develop uninterrupted (Kutiel 1994).

A critical factor in a rotational grazing system is the quality of the poorest rangeland. The traditional sheep-grazing system in Iceland relied on cultivated pastures close to the farm, which were grazed in the spring and autumn, and on communal upland pastures, which were grazed in the summer, while two crops of hay were cut from the cultivated lands for winter feed. During the 1960s the farmers reduced the time that sheep spent in the uplands with the result that only one crop of hay was obtained and winter feed had to be imported. However, with imported feed, more stock could be kept during the winter. With increasing sheep numbers, the pressure on the summer grazings intensified, leading to overgrazing and erosion. The availability of summer pastures thus became the limiting factor in sheep production (Thorsteinsson et al. 1971). Avoiding overgrazing and erosion in the uplands through the adoption of appropriate stocking rates was therefore vital. Government policies were therefore introduced to subsidize environmental projects undertaken by farmers in return for reducing sheep numbers. As a result, the number of sheep has halved since 1979, bringing about improvements to the quality of the rangeland, better carcass weights and economic benefits to the farmers (Arnalds 1999).

There is no universal formula for determining stocking rates and the carrying capacity of pastures is usually imprecisely defined. The determination is more difficult in regions with high variability in rainfall from year to year, so that overgrazing is almost inevitable when several years of drought follow in succession. The sheep-grazing capacity of the pastures in Iceland was assessed by mapping the vegetation of the country from aerial photography and field survey and classifying it into 94 different types (Steindórsson 1980). Based on the average annual production of dry matter, the palatability of the plant species to sheep and limiting the annual removal to 50 per cent of the annual growth of the palatable species, the carrying capacity of each type was calculated (Thorsteinsson 1980a, b). In reality, such surveys can only determine an average carrying capacity. The actual carrying capacity at a given moment will depend upon local soil and vegetation conditions, variations in climate and the extent to which trees and shrubs are cut for fuel and timber. Regular inspection and judgement of the quality of the pastures are thus needed to determine when stock should be moved from one area to another in a rotation system. The carrying capacity is also affected by the browsing of wildlife. The migration of wildlife on to a ranch

(a) Scheme for a four-block balanced rotational grazing system, as used for the rehabilitation of overgrazed land in Kenya, showing the movement of livestock as each block is grazed in turn (G) for a period of four months. The complete grazing cycle lasts four years, at the end of which the cycle is repeated. In the areas where this system has been used, the wet season lasts from April to September, as shown by the shaded area.

(b) A variation of the 4-block grazing system. The grazing is still balanced between blocks, but grazing periods last 3 months and rest periods are of different lengths. Provision is included for burning each block once in 4 years, with a full growing season's rest before and after burning, but one of the other rest periods during the cycle does not include part of a growing season. However, under a bimodal rainfall pattern, with rains in April–June and October–December, the latter objection would not arise and there would be more opportunities for burning.

(c) A second variant of the 4-block grazing system with longer grazing periods, each of 6-months' duration, which eliminate the short rests which appear in Fig. (b). Two opportunities for burning are shown in each block, though the one that is preceded by a short rest period should only be used in years of favourable rainfall. Alternatively, grass not burnt might be used as reserve grazing.

(d) A balanced 4-block grazing system, in which the grazing applied to each block comprises one period of 4 months and one of 8 months. This is an alternative to Fig. (c), though the second opportunity for burning during the long rest period would seldom be used since it is followed soon after by 8 months grazing.

Fig. 8.1 Rotational grazing system for semi-arid lands in Kenya (after Pratt & Gwynne 1977; Mututho 1989).

in the Narok area of Kenya in the rainy season causes the stocking intensity to rise to 1.23 TLU ha^{-1} compared with a wet-season carrying capacity of 0.9 TLU ha^{-1}. When the wildlife move away during the dry season, the stocking rate falls to 0.5 TLU ha^{-1}, which is very close to the dry-season carrying capacity (Mwichabe 1992). Clearly, if the land user is to obtain optimal stocking, wildlife must be kept out of the pastures.

Rotational grazing and burning are most appropriate for large-scale ranches of 4000 ha or more since this size allows for greater flexibility in the design of suitable systems with respect to variability in land quality, water supply, levels of degradation and climate. Figure 8.1 shows examples of rotational grazing systems used in the Arid and Semi-Arid Lands (ASAL) project in Kenya (Pratt & Gwynne 1977; Mututho 1989) to manage communal grazing.

Crop and vegetation management

8.1.4 Forest management

The principles of rotation can be applied to other forms of land use. The timber resources of forests are often exploited commercially by clear-felling patches of land on a rotational basis. Erosion rates are highest in the years immediately following logging but decline in subsequent years with either regrowth of the natural vegetation or replanting so that, averaged over 12 or more years, they may be little different from those of undisturbed land. In western Oregon, annual sediment yield from Deer Creek catchment averaged $0.97\,t\,ha^{-1}$ for the period 1959–65 before roads were constructed in late 1965 for the patch-cutting of 25 per cent of the catchment in late 1966 (Beschta 1978). The sediment yield in 1966 increased by 150 per cent but in 1967 it was only 40 per cent above the pre-cutting average. Sediment production in subsequent years then fell to the pre-cutting average until, in 1972, it suddenly increased by 170 per cent because of several mass failures along roadsides in the patch-cut area (section 7.4.3). The 1973 sediment yield was again similar to the pre-cutting average. Thus, provided erosion on roadside embankments and cuttings can be controlled, an area can recover quickly from the effects of patch-cutting. Clearance of much larger areas can create longer-term problems. In the neighbouring Needle Branch watershed, annual pre-cutting sediment yield averaged $0.53\,t\,ha^{-1}$. Following clearance of 82 per cent of the catchment in 1966, sediment yield was 50 per cent above the pre-cutting average in 1966 and 260 per cent above in 1967, and did not return to pre-cutting levels until 1973.

Forest fires represent an additional hazard once an area has been cut because, as shown by observations following wildfire in Pine Creek, Boise National Forest, Idaho, the erosion directly attributable to burning is greater in cut than in uncut areas (Megahan & Molitor 1975). Responses to fire in terms of hydrophobicity of the soil, generation of runoff and erosion are similar to those described for burning on rangeland (section 8.1.3). The severity of the burn is important. On the island of Lesvos, Greece, post-fire erosion in 30 rainfall events between October 1992 and March 1993 was: $12.2\,t\,ha^{-1}$ following a severe burn that destroyed the litter layer and all but 0–10 per cent of the crown canopy; $1.3\,t\,ha^{-1}$ following a moderate burn where the litter layer was consumed but 45–60 per cent of the canopy remained; and $0.2\,t\,ha^{-1}$ from an unburned area (Dimitrakopoulos & Seilopoulos 2002). Replanting needs to take place quickly after clear felling before the loss of top soil and plant nutrients through erosion reduces the quality of the land. If planting is delayed, it is desirable to establish ground cover using some of the cover crops referred to in section 8.2.

Although with careful management and limiting clearance to small areas at a time patch-cutting may be an acceptable technique, in very erodible areas soil loss may still be too severe to permit its use. In these areas, selective felling, in which only mature trees are removed and the other trees remain to provide a plant cover, is a better conservation practice. The chosen trees should be removed by directional felling to avoid damage to seedlings, young trees and rivers. A filter strip, with a minimum width of 20 m, should be maintained on the banks of perennial streams. Harvesting by tractors fitted with shear blades and using skid trails should be permitted only on slopes up to 30° (Wan Yusoff 1988).

8.2
Cover crops

Cover crops are grown as a conservation measure either during the off-season or as ground protection under trees. In the USA they are grown as winter annuals and ploughed in to form a green manure prior to sowing the main crop. Typical crops used are rye, oats, mustard, hairy vetch, sweet clover and lucerne in the north, and Austrian winter peas, crimson clover, crotalaria and lespedeza in the south. To be effective, the cover crop must be quick to establish, provide an early canopy cover, be aggressive enough to suppress weeds and possess a deep root system to improve the macroporosity of the soil. The broadcast sowing of winter rye at a rate of 120–130 kg ha^{-1} as early as possible in the autumn is practised to control wind erosion on the sandy soils of the northern Netherlands where sugar beet, potatoes and maize are grown (Eppink & Spaan 1989). Early sowing is essential because a good crop cover must be obtained by the end of December, since the climate is too cold for growth in February and March, the period when the soil surface is dry and strong winds with low humidity are most common. The crop is spray-killed with herbicide at a suitable time in the spring, usually before the drilling of sugar beet or before the emergence of potatoes and maize. Maintaining a cover crop over winter instead of bare soil is also beneficial in reducing the leaching of nitrogen to groundwater.

Ground covers are grown under tree crops to protect the soil from the impact of water drops falling from the canopy. They are particularly important with tall crops such as rubber where the height of fall is sufficient to cause the drops to approach their terminal velocity (section 3.4.1). With bare soil underneath the trees, erosion rates greater than 20 t ha^{-1} have been recorded under oil palm in Malaysia, even with bench terracing (Lim 1988). The erosion can be reduced to less than 0.5 t ha^{-1} with cover crops, although the harvesting paths still remain vulnerable (Maene et al. 1979). The most common cover crops used are *Pueraria phaseoloides*, *Calapogonium mucunoides* and *Centrosema pubescens*. Although they grow rapidly and retain nutrients in the soil that would otherwise be removed by leaching, their use can sometimes give problems. First, as the main crop becomes established and ground conditions change from strong, open sunshine to shade, there is a risk that a satisfactory cover will not be attained. Second, the cost of growing the cover crops may outweigh the benefits an individual farmer receives. Most covers give no income and this restricts their use on smallholdings where farmers do not have sufficient cash reserves to wait for the tree crop to mature. Research is required to find suitable cover crops, particularly varieties of beans and peas, that the smallholder can grow. Third, ground covers compete for the available moisture and, in dry areas, may adversely affect the growth of the main crop. Studies in rubber plantations in eastern Java show that cover crops may reduce the soil moisture by up to 50 per cent during the dry season compared with clean-weeding (Williams & Joseph 1970). An alternative conservation measure is required in these circumstances.

When used in vineyards in the Beaujolais region of France, a permanent grass cover, managed by mowing, reduced erosion to less than 7 per cent of that in vineyards with bare soil. There was no competition with the vines for moisture during the hot dry summer since the increased infiltration promoted by the grass limited the water loss by runoff during storms and thereby offset any potential water deficit (Gril et al. 1989). In the Mosel Valley, Germany, grass reduced the available water in the top 40 cm of soil in the summer when rainfalls were about average but improved the moisture status during dry summers, with no significant difference in yield compared with maintaining a bare tilled soil under vines (Husse 1991).

8.3

Strip-cropping

With strip-cropping, row crops and protection-effective crops are grown in alternating strips aligned on the contour or perpendicular to the wind (Fig. 8.2a). Erosion is largely limited to the row-crop strips and soil removed from these is trapped within and behind the next strip

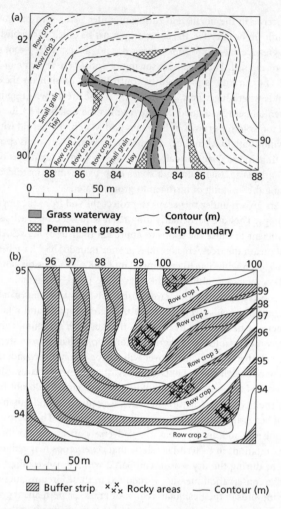

Fig. 8.2 Contour strip-cropping designs for (a) a five-year crop rotation and (b) the use of buffer strips. The contour lines are in metres above an arbitrary datum. Note the use of a grass waterway to evacuate excess runoff from the strip-cropped area and the inclusion of rocky areas that cannot be farmed within the buffer strips (after Troeh et al. 1980).

downslope or downwind, which is generally planted with a leguminous or grass crop. Row-crop widths vary according to the degree of erosion hazard (Table 8.1). The protective strips are usually about 4.5 m wide when comprised of grasses or a grass and legume mix, and 9 m wide when made up solely of leguminous species (Natural Resources Conservation Service 1999). When the protective strips comprise one of the crops used in a rotation, they are known as annual strips and their position on the hillslope can vary from one year to the next. Strip-cropping is best suited to well drained soils because the reduction in runoff velocity, combined with the low rate of infiltration on a poorly drained soil, can result in waterlogging and standing water.

On steep slopes or on very erodible soils, it may be necessary to retain the strips as permanent vegetation. These buffer strips are usually 2–4 m wide and are placed at 10–20 m intervals (Fig.

Table 8.1 Recommended strip widths for strip cropping

Water erosion (width of cropped strips on soils with fairly high water intake)	
2–5 per cent slope	30 m
6–9 per cent slope	25 m
10–24 per cent slope	20 m
15–20 per cent slope	15 m
Wind erosion (strips perpendicular to wind direction)	
Sandy soil	6 m
Loamy sand	7 m
Sandy loam	30 m
Loam	75 m
Silt loam	85 m
Clay loam	105 m

Source: after FAO (1965), Cooke and Doornkamp (1974).

8.2b). On slopes around 5°, the strips act by retarding the flow, causing ponding of water upslope of the barrier, encouraging infiltration of runoff and sedimentation (Melville & Morgan 2001). The gradual build-up of soil behind the barrier leads, in time, to the formation of bench-type terraces. On a contour grass strip system with a tufted grass, *Setaria anceps*, on a clay nitisol soil near Nairobi, Kenya, up to 120 mm depth of soil accumulated above the strips within two years, reducing the local slope from 6 to 5° (Wolde & Thomas 1989). In the Cochabamba area of Bolivia, increases of 20–450 mm in riser height were recorded within two years behind barriers of *Phalaris tuberoarundinacea* grass (Sims et al. 1999). However, sediment deposition also leads to a rise in height of the ground surface above the strip relative to that below, thereby locally increasing the slope steepness. Unless the soil within the strip is properly protected by the vegetation, water leaving the strip, with its sediment load filtered out and flowing over the steeper slope, may initiate erosion (Ghadiri et al. 2001). Erosion rates can then be higher with contour grass strips than without (Emama Ligdi & Morgan 1995). On steeper slopes, 12–16°, the strips filter out the sediment from the flow but have little effect on runoff (Boubakari & Morgan 1999). The maximum slope at which grass strips remain effective has not been determined. Although they may be inadequate as the sole conservation measure on slopes above 8.5° (FAO 1965), they have been adopted by farmers on slopes up to 30° (Wolde & Thomas 1989) or even 50° (Sims et al. 1999).

The main disadvantage of strip-cropping is the need to farm small areas, which limits the kind of machinery that can be operated. The technique is therefore not compatible with highly mechanized agriculture. Although this is a less relevant consideration on smallholdings, the difficulty here is that much land is taken up with protection-effective crops of limited value. Contour grass strips were introduced into Swaziland by a decree of King Sobhuza II in 1948 and by the 1950s some 112,000 km of strips were laid out, protecting virtually all the arable land surrounding individual homesteads. The strips were originally 2 m wide but by the 1970s there was evidence that farmers were gradually allowing the cropped areas to encroach on the strips. However, the strip system appears to be relatively flexible regarding strip width and there may be little additional benefit in having strips wider than 1.5 m. Wolde and Thomas (1989) found that strips of 0.5 and 1.0 m wide reduced soil loss to 36 per cent of that from an unprotected bare plot but a 1.5 m wide strip gave only a further 18 per cent reduction. In south Limbourg, the Netherlands, 1 m wide

Crop and vegetation management

strips reduced sediment discharge by 50–60 per cent, 5 m wide strips by 60–90 per cent and 10 m wide strips by 90–99 per cent (van Dijk et al. 1996).

The plants chosen to form permanent buffer strips are usually grasses. They should be perennial, quick to establish and able to withstand periods of both flood and drought. They should have deep-rooted systems to reinforce the soil and reduce scouring, a uniform density of top growth to provide a filter for sediment and reduce flow velocity; their growth points should be close to the ground or below the soil so that they are not grazed out and can recover from damage after fire; and they should either be sterile or propagate very slowly so that they do not become weeds in the adjacent cropped strips. Rhizomous species should be avoided since they spread very rapidly on to surrounding land. Tussocky grasses, such as Kikuyu grass (*Pennisetum clandestinum*), should be avoided because they concentrate the flow (section 3.4.2). Grasses with an erect growth habit of interwoven stems which act like a porous filter are more effective in reducing erosion than those with slender creeping rhizomes and a more horizontal growth form (Lakew & Morgan 1996; Melville & Morgan 2001). Suitable species include Napier grass (*Pennisetum purpureum*), Guatemala grass (*Tripsacum laxum*), Makarikari grass (*Panicum coloratum*), Canary grass (*Phalaris canariensis*), oat grass (*Hyparrhenia* spp.), wheat grass (*Agropyron* spp.) and lyme grass (*Elymus* spp.).

Considerable publicity has been given to vetiver grass (*Vetiver zizanioides*). Vetiver grass establishes quickly from splits and does not compete with adjacent crops. It is adapted to a wide range of climatic and soil conditions with tolerance to both heavy metal and saline toxicity (Pease & Truong 2002). Over an 11-month period on an oxisol under cassava at the Santander de Quilichao Research Station, near Cali, Colombia, soil loss was 1.3 t ha^{-1} with vetiver grass contour barriers, 4.0 t ha^{-1} with Napier grass strips and 8.3 t ha^{-1} with no conservation measures (Laing 1992). Over a three-year period on the Punjabrao Krishi Vidyapeeth University Farm, Manoli, India, annual soil loss averaged 3.3 t ha^{-1} with vetiver grass strips on the contour compared with 11.4 t ha^{-1} using across-slope cultivation for a rotation of green gram–pigeon pea–safflower, pearl millet–safflower and pearl millet (Bharad & Bathkal 1991). While vetiver grass has the advantage that it is not palatable to livestock and therefore any strips formed from it are likely to remain undamaged, it provides no income to the farmer, whereas Napier grass can be cut and fed to cattle. Clearly the ideal situation is where the economic value of the grass strips equals or exceeds that lost by taking the land out of agricultural production.

Strip systems can also be used to provide in-field shelter for wind-sensitive crops and protect the soil against erosion by wind. In the fens of eastern England, live barley strips are used to protect onions, sugar beet and carrots from wind damage and to control erosion on the lowland peat soils. The barley is sown in February or March, allowing time for it to emerge before the main crop is drilled in mid to late April, the critical period for erosion. Once the main crop has sufficient biomass to perform a protective role, the barley is killed with a selective herbicide. Live barley meets the design requirements of a plant for infield shelter in having bladed rather than ovate leaf forms to minimize the risk of streamlining in strong winds, rigid leaves to reduce flutter and thereby minimize the 'wall effect' (section 3.4.3) and an upright growth habit with at least 0.65 kg DM m^{-3} of uniform biomass in the lowest 5 cm to provide an adequate filter (Morgan 1989). The barley strips should be spaced at distances of eight to ten times their height, which is every 2 m if the barley is 20 cm tall at the time when the risk of erosion is highest.

Barrier strips or buffers can be used to protect water bodies from pollution by sediment and the chemicals adsorbed to it, particularly phosphorus. For this purpose, grass barriers need to be placed in critical positions in the landscape, such as: alongside rivers, lakes and reservoirs, where they form riparian barriers; on the convexities of hillslopes to reduce runoff velocities and prevent the initiation of rills; and across pathways of water and sediment movement such as unsurfaced tracks. Since they need to trap all the sediment, including the fine clay particles, they need to be wider than those described above. Riparian barriers are often 5–10 m wide. A 9-m wide strip is usually sufficient to trap 80 per cent of the sediment (Natural Resources Conservation Service 1999). Studies in Denmark indicate that a 12-m wide buffer is sufficient to trap all sediment and particulate phosphorus transported from surrounding hillslopes and prevent it entering the water course (Kronvang et al. 2000). It should be noted that their ability to trap sediment means that they are 'designed to fail' since they become less effective over time as sediment builds up within and upslope of the strip. When the sediment accumulates to a depth of about 150 mm, it is recommended that it is removed by tillage and the barrier re-established. Otherwise there is a risk of the runoff being diverted around the barrier. Mowing of the barrier leads to more vigorous growth of the grass, making it more effective in nutrient uptake and carbon sequestration.

Forests can be used for riparian barriers with the additional advantages of creating shade for the aquatic environment and a diverse habitat for wildlife. Typically, they comprise three zones: (i) a zone of mixed grass species forming a filter adjacent to agricultural land; (ii) an intermediate zone of managed forest; and (iii) a zone of undisturbed forest alongside the river. The first zone, usually 10 m wide, reduces runoff by 56–72 per cent and the transport of sediment by 78–83 per cent (Sheridan et al. 1999). Where the surrounding land is pasture and not crop, this zone can be dispensed with. Where the riparian barrier is installed alongside a river on a flood plain that is wider than 100 m, the width of zones (ii) and (iii) combined should be at least 30 m. For narrower flood plains, the combined width should be 13.7 m. For channels without a flood plain, a combined width of 10.7 m is recommended (Natural Resources Conservation Service 1999).

8.4

Multiple cropping

The aim of multiple cropping is to increase the production from the land while providing protection of the soil from erosion. The method involves either sequential cropping, growing two or more crops a year in sequence, or intercropping, growing two or more crops on the same piece of land at the same time. Many schemes involve a mixture of the two. Multiple cropping has been traditionally practised in the Caribbean in the kitchen gardens, which, averaging 0.2 ha in size, provide the subsistence component of fruits and vegetables for many families in rural areas. Those in Grenada combine fruit trees, with banana the most ubiquitous but also including coconut, cocoa, mango and breadfruit, with vegetables, especially root crops such as sweet potatoes and yams. Pigeon peas, groundnuts, dasheen, okra, pepper and tomatoes are also grown. The home garden thus mimics the natural, multilayered ecosystem (Hoogerbrugge & Fresco 1993). Although the garden has a confused appearance (Fig. 8.3), the organization of the cropped land is systematic and the better farmers carry out regular weeding, selective application of fertilizers and irrigation (Brierley 1976).

When using sequential cropping, attention must be paid to the order in which the crops are grown. Lal (1976) found that on a 6° slope near Ibadan, Nigeria, a crop of maize followed by one of cowpeas with no tillage produced a loss of only 0.2 t ha^{-1}. Growing cowpeas followed by maize but with tillage gave a soil loss of 6.2 t ha^{-1}. Although it is difficult to isolate the effects of zero tillage, it seems that a maize–cowpeas sequence produces less erosion than a cowpeas–maize one because maize is a soil-depleting crop and, when grown second, is planted into an already partially exhausted soil. Soil loss is also greater under maize than cowpea (Lal 1977a) and the

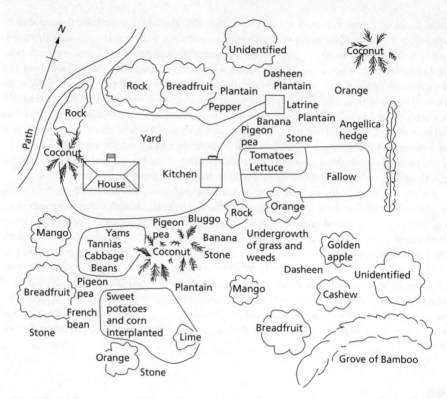

Fig. 8.3 Land use in kitchen garden, near Concord, St John's, Grenada, showing an example of multiple cropping (after Brierley 1976).

difference between the two crops is likely to be enhanced when maize is the second crop. Crop yields were 4.9 t ha^{-1} for maize and 4.3 t ha^{-1} for cowpea in the maize–cowpeas sequence but 5.0 t ha^{-1} for cowpea and only 2.1 t ha^{-1} for maize in the cowpeas–maize sequence.

The intercropping of maize with cassava offers the advantages of a two-storey canopy, giving a higher interception capacity and reducing the detachment of soil particles by raindrop impact to 35 and 60 per cent of the respective values from cassava and maize alone (Lal 1987). On a 6° slope, mixed maize–cassava reduced annual soil loss to 86 t ha^{-1} compared with 125 t ha^{-1} for cassava as a monoculture (Lal 1977a). Both values, however, are well above most soil loss tolerance levels, as are those found in similar experiments involving cassava with groundnuts, maize, cowpeas and peppers (16.2 t ha^{-1}) near Benin City, Nigeria (Odemerho & Avwunudiogba 1993). Generally, multicropping needs to be combined with other practices.

8.5 High density planting

High density planting is used to try to obtain the same effect for a monoculture that multiple cropping achieves with two or more crops. Again the technique may need to be supplemented by other practices. Hudson (1981) showed in Zimbabwe that increasing the planting density of maize from 25,000 to 37,000 plants per hectare and using a trash mulch at the higher density reduced annual soil loss from 12.3 to 0.7 t ha^{-1}. Respective crop yields were 5 and 10 t ha^{-1}. The arrangement of the plants also has an effect. With a plant population of 55,000 plants per hectare, the lowest soil loss from maize in the Doon Valley, northern India, was obtained by increasing the distance between the rows and increasing the plant density within the rows. Thus changing from a row spacing of 450 mm and a plant spacing of 400 mm to a row spacing of 900 mm and a plant spacing of 200 mm reduced annual soil loss under maize from 22.6 to 13.9 t ha^{-1} (Bhardwaj et al. 1985).

High density planting can be effective even when targeted only on those parts of the landscape that are most vulnerable to erosion. In a study on the Belgian loess under winter-sown triticale, a second sowing was carried out along the valley floor where runoff concentrates and ephemeral gullies form. This increased the plant density by 50 per cent and resulted in a 42 per cent reduction in erosion (Gyssels et al. 2002). Clearly, the need to make a second sowing would increase costs but this would not be necessary with modern precision agriculture, where the sowing rate could be automatically adjusted to position in the field.

8.6

Mulching

Mulching is the covering of the soil with crop residues such as straw, maize stalks, palm fronds or standing stubble. The cover protects the soil from raindrop impact and reduces the velocity of runoff and wind. It is most useful as an alternative to cover crops in dry areas where insufficient rain prevents the establishment of a ground cover before the onset of heavy rain or strong winds, or where a cover crop competes for moisture with the main crop. In the semi-humid tropics, the side-effects of a mulch in the forms of lower soil temperatures and increased soil moisture are beneficial and may increase yields. Elsewhere the effects can be detrimental. In cool climates, the reduction in soil temperature shortens the growing season, while in wet areas, higher soil moisture may induce gleying and anaerobic conditions. Crop yields may also be reduced because the mulch competes with the main crop for nitrogen as it decomposes.

The effectiveness of mulching in reducing erosion was demonstrated by the field experiments of Borst and Woodburn (1942), who found that, on a silt-loam soil on a 7° slope, annual soil loss was 24.6 t ha^{-1} from uncultivated bare land but only 1.1 t ha^{-1} from land covered with a straw mulch applied at 5 t ha^{-1}. Lal (1976) found that covering an alfisol on a 6° slope with 6 t ha^{-1} of straw mulch resulted in an annual soil loss of 0.2 t ha^{-1} compared with 23.3 t ha^{-1} on bare soil. Although such low soil losses are not always achieved, the extent to which a mulch will reduce erosion is generally impressive. A grass mulch at 4 t ha^{-1} applied to a silt-loam soil on a 4.5° slope near Dehra Dun, India, reduced annual soil loss under maize to 5.0 t ha^{-1} compared with 22.4 t ha^{-1} for maize tilled conventionally up-and-down slope (Khybri 1989). A grass mulch at 15 t ha^{-1} on a sandy loam soil with a 3° slope near Pakhribas in eastern Nepal gave an annual soil loss under maize of 16.9 t ha^{-1} compared with 32.8 t ha^{-1} for the traditional practice of maize planting followed by two hoeings (Sherchan et al. 1990).

Mulches can also be used under tree crops. Using pruned fronds to cover harvesting paths in an oil palm plantation in Johor, Malaysia, reduced annual soil loss to 4.2 t ha^{-1} from 14.9 t ha^{-1} recorded on unprotected paths (Maene et al. 1979). A grass mulch applied between rows of young tea on a 6° slope on an ultisol near Kericho, Kenya, reduced average annual soil loss over three years to 0.7 t ha^{-1} compared with 210.8 t ha^{-1} for bare hand-tilled soil between the rows (Othieno 1978).

Most applications of mulching involve spreading crop residues on the surface of the soil but this method creates problems in drilling and planting through the mulch, unless specialized equipment is used. An alternative approach is to incorporate the mulch in the soil. Although this reduces the overall surface cover, the mulch elements help to bind the soil, simulating the effect of plant roots, and increase infiltration rates. Duley and Russel (1943) found that incorporated crop residue reduced annual soil loss from 35.7 t ha^{-1} on a bare soil to 9.9 t ha^{-1}. The effect of incorporated mulch depends upon the material used. Wheat straw is particularly effective because it produces a large number of individual mulch elements, resulting in a more uniform pattern of incorporation, good contact between the mulch and the soil, and sufficient material on the surface to form miniature dams, behind which water ponds, and protect the soil against crusting. In contrast, maize stalks perform badly because their large size means that the surface is sparsely covered, the contact with the soil is poor and the individual elements are easily washed out and transported downslope (Abrahim & Rickson 1989). Overall, incorporated mulches are less effective than surface mulches.

When using mulches to control wind erosion, standing stubble is required because of the danger of spread mulches blowing away. A mulch of 1 t ha^{-1} of standing wheat stubble or 2 t ha^{-1} of flattened wheat straw will reduce annual wind erosion rates to a tolerable level of 0.2 t ha^{-1}. To achieve the same effect with sorghum stubble requires a mulch of 6.7 t ha^{-1} (Chepil & Woodruff 1963). In a review of research carried out in the Sahel, Bielders et al. (1998) showed that 2 t ha^{-1} of millet stover could reduce soil loss by between 42 and 92 per cent, depending on wind velocities.

There is considerable experimental evidence (Wischmeier 1973; Lal 1977b; Laflen & Colvin 1981; Norton et al. 1985; Bekele & Thomas 1992) to show that the rate of soil loss decreases exponentially with an increase in the percentage area covered by a mulch. The mulch factor (MF), defined as the ratio of soil loss with a mulch to that without, is related to the percentage residue or mulch cover (RC) by the expression (Laflen & Colvin 1981):

$$MF = e^{-a.RC} \tag{8.1}$$

where a ranges in value from 0.01 to 0.07, depending on the degree of soil disturbance by tillage. For a mouldboard plough, the value is 0.03, and for no tillage, 0.06 (Norton et al. 1985). Hussein and Laflen (1982) found that the exponential relationship applied only to rill erosion and that the rate of interrill erosion decreased linearly with increasing residue cover. Since for most soils the contribution of interrill erosion to total soil loss is quite small for slope lengths in excess of 25 m, eqn 8.1 can be used to obtain mulch factor values that, when multiplied by the C-factor values, allow the effects of mulching to be included in the Universal Soil Loss Equation (section 6.1.1).

A mulch should cover 70–75 per cent of the soil surface. With straw, an application rate of 5 t ha^{-1} is sufficient to achieve this. A lesser covering will not adequately protect the soil, while a greater covering may delay plant emergence and suppress plant growth. Denser mulches, giving at least 90 per cent cover, are sometimes used under tree crops to control weeds. An estimate of the required application rate to control erosion can be made for a preselected set of conditions using the Manning equation for flow velocity (eqn 2.18) and the relationship between Manning's n and mulch rate for straw (Foster et al. 1982):

$$n_m = 0.071M^{1.12} \text{ interrill erosion} \tag{8.2}$$

Table 8.2 Estimating the required density of a maize stalk mulch to control water erosion

Estimation is made for the following conditions:
Sandy soil, desired maximum flow velocity = 0.75 ms^{-1}
Flow depth in small channels $\qquad$ = 100 mm
Slope $\qquad$ = 5°

Estimating the required value for total Manning's _n_:
From the Manning equation:

$$n = \frac{r^{0.67} s^{0.5}}{v}$$

For simplicity, assume the hydraulic radius (_r_) is approximated by the depth of flow and that slope (_s_) can be represented by the tangent of the slope angle. Thus

$$n = \frac{0.1^{0.67} \times 0.087^{0.5}}{0.75}$$

$$n = 0.0843$$

Estimating required value for Manning's _n_ due to mulch (n_m):
According to Foster et al. (1982):

$$n_m = \left(n^{3/2} - n_s^{3/2}\right)^{2/3}$$

Taking a value of Manning's _n_ due to the soil (n_s) = 0.02, gives

$$n_m = \left(0.843^{3/2} - 0.02^{3/2}\right)^{2/3}$$

$$n_m = 0.0767$$

Estimating required mulch application rate (_M_):
Rearranging eqn 8.3 gives

$$M = \left(\frac{n_m}{0.0105}\right)^{1.06}$$

$$M = \left(\frac{0.0767}{0.0105}\right)^{1.06}$$

$$M = 0.72 \text{kg m}^{-2}$$

Note: Theoretically this procedure can be used to determine application rates for other mulches but in practice the relationships between _M_ and n_m have not been established. Eqn 8.3 cannot therefore be used for wheat straw, soya bean residue, palm fronds or other mulch materials.

$$n_m = 0.105 M^{0.84} \text{ rill erosion} \tag{8.3}$$

where n_m is the value of _n_ due to the mulch and _M_ is the mulch rate (kg m^{-2}) (Table 8.2). No similar procedure exists for determining mulch rates to control wind erosion, but the effectiveness of standing crop residues depends upon the number of standing stalks per unit area and their size.

As indicated above, controlling erosion with a mulch poses special problems for the arable farmer because tillage tools become clogged with the residue, weed control and pest control are

Crop and vegetation management

more difficult, planting under the residue is not always successful and crop yields, especially in humid and semi-humid areas, are sometimes lower. Mulching on its own is not always an appropriate technique but where it is combined with conservation tillage, many of these problems can be overcome and it has tremendous potential as a method of erosion control.

Rock fragments on the surface can act in the same way as a mulch. Relationships between rate of erosion and percentage stone cover frequently follow that described by eqn 8.1 (Poesen 1992). Therefore stone mulches might be appropriate for erosion control on non-agricultural areas such as road banks and construction sites. However, where the stones are embedded in a sealed or crusted soil, interrill erosion can be enhanced and will increase with percentage stone cover because the runoff generated on the impermeable stone elements is unable to infiltrate the surrounding soil (Poesen & Ingelmo-Sanchez 1992). On an erodible soil, rock fragments can also induce local turbulence, leading to scouring on the upstream side of the stones and deposition on the downstream side. With small stones or pebbles around 15 mm in size, scouring can cause sediment yield to increase as cover increases from zero to 20 per cent but, with further increases in percentage cover, the protective effect prevails. On covers of 60 per cent and above, scour is almost impossible. For cobbles around 86 mm in size, however, there is no reduction in sediment production and the soil loss increases exponentially with increasing percentage cover (Bunte & Poesen 1994). Since rock fragments can both decrease and increase erosion according to local circumstances, their use as a mulch to control water erosion cannot be recommended.

A different picture emerges for wind erosion control. Pebble and gravel mulches have been used in the semi-arid regions of northwest China for over 300 years (Gale et al. 1993) to help to preserve moisture from the limited rainfall and trap wind-blown material. A pebble mulch reduces the rate of wind erosion by 84–96 per cent (Li et al. 2001) and can trap between 1.6 and 1.8 times more sediment than the surrounding land. Pebbles of 25 mm diameter are more effective than those of 100 mm diameter, trapping 2.8 times as much, probably because the larger pebbles create more turbulence (Li & Liu 2003). Since the nitrogen and organic content of the dust is two to three times higher than that of the local soil, the accumulated sediment adds to the fertility of the gravel-mulched fields. Over time, the spaces between the pebbles fill with soil and the surface becomes smoother, thereby reducing the effectiveness of the technique. Tillage, however, will help to return the gravel to the surface.

8.7

Revegetation

Vegetation plays a major role in the process of erosion control on gullied areas, landslides, sand dunes, road embankments, construction sites, mine spoils and pipeline corridors. Rapid revegetation is also necessary for replanting forest in areas cleared by patch-cutting and for covering land cleared of forest in favour of agriculture. The first-listed cases represent marginal environments for plant growth where the risk of vegetation failing to re-establish is high. The second-listed cases are less marginal and the objective is to minimize damage during clearance and to establish cover quickly before the environment has time to deteriorate.

When developing a plan for revegetating an area, a soil test should be carried out to establish pH, nutrient levels, moisture status, salinity levels and the presence of toxic ions, all of which will influence the range of species that will grow. Climatic conditions should also be studied, including the frequency of drought and waterlogging. Topographic influences on the local climate are important – for example, differences in temperature and moisture between sunny and shady slopes, and frost hollows. Topography also determines the location of dry and wet sites through its effect on movement of water through the soil. Plant species should be selected for their properties of rapid growth, toughness in respect of diseases and pests, ability to compete with less desirable species and adaptability to the local soil and climatic conditions. Wherever possible, native species should be chosen. A study of neighbouring sites often gives a good indication of what species are most likely to survive and thrive. The use of introduced or exotic species should not be ruled out, however, especially where the local environment has deteriorated beyond that of adjacent sites or where numbers of local species are limited. The revegetation plan should allow for plant succession to take place naturally. In many cases, the objective is to establish pioneer species to give immediate cover and improve the soil, permitting native species to come in and take over as the colonizing plants decline. Generally, a mix of plant species is required because it is impossible to predict the success of any one species in marginal environments. A monoculture is also more susceptible to disease. The species mix should include grasses, forbs and woody species, both bushes and trees, except where specific requirements make such a mix undesirable, as with certain types of gully reclamation or along pipeline corridors. Detailed coverage of the various methods of vegetation establishment can be found in Gray and Leiser (1982) and Coppin and Richards (1990).

8.7.1 Restoration of gullied lands

Revegetation is used in gully erosion control as a method of increasing infiltration and reducing surface runoff. Provided about 30–35 per cent vegetation cover remains on the land, closing an area on the gullied loess in the Xingzihe Basin by fencing and prohibiting its use for grazing can lead to a natural increase in cover to 70–90 per cent within three to four years. However, with supplementary planting of grasses and shrubs, the same effect can be achieved within two years (Tang et al. 1987). Generally, the area around the gullies needs to be treated with grasses, legumes, shrubs and trees or combinations thereof, aided in the early stages by mulching or the use of geotextiles. Research carried out at the Suide Soil and Water Conservation Experiment Station of the Huang He Conservancy Commission shows that afforestation can reduce runoff in the gullied loess areas by 65–80 per cent and soil loss by 75–90 per cent. Growing grass reduces runoff by 50–60 per cent and soil loss by 60–80 per cent (Gong & Jiang 1977). The planting of trees and herbs on the steep slopes of the gully sides and raising crops on bench terraces on the gentler slopes of the divides (Fig. 8.4) can stabilize the land and prevent sediment from entering the gullies. By supporting these measures with check dams and reservoirs along the gully beds, the volume of sediment entering the Huang He from gullies in the Wuding Valley was reduced by 44 per cent over the period 1971–8 (Jiang et al. 1981).

In the studies mentioned above, the main objective was to increase infiltration. Achieving this will help to reduce gullying when surface processes are involved, but where the gullies are fed by subsurface pipes or tunnels (section 2.6.1), it is necessary also to promote infiltration in a uniform pattern. Since trees and grasses have different root densities and their root networks extend to different depths, their mixture may result in more infiltration under the trees, which may, in turn, feed water into a pipe system. Even if the pipe network has been previously broken up by subsurface ripping, concentrations of water in the soil may encourage pipes to reform. This is more likely to occur if tree species with long tap roots are planted. Thus, the best vegetative treatment for tunnelled areas is to establish a dense, uniform grass cover. Where conditions are too marginal for grass to grow, shrubs and trees will have to be used but species with a good system of lateral rather than vertical roots should be chosen.

Crop and vegetation management

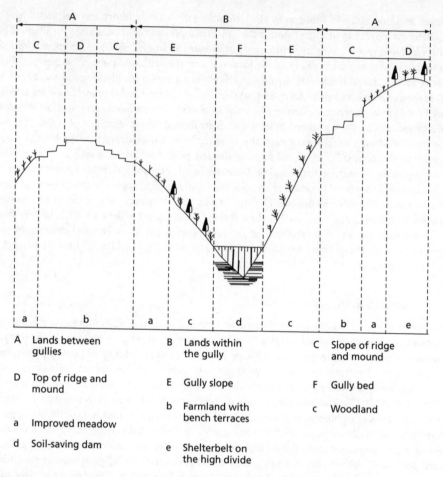

Fig. 8.4 Cross-section showing soil and water conservation measures adopted in the gullied loess area of China (after Jiang et al. 1981).

A	Lands between gullies	B	Lands within the gully	C	Slope of ridge and mound
D	Top of ridge and mound	E	Gully slope	F	Gully bed
		b	Farmland with bench terraces	c	Woodland
a	Improved meadow				
d	Soil-saving dam	e	Shelterbelt on the high divide		

8.7.2 Restoration of landslide scars

Tree planting is recommended as a method of stabilizing slopes prone to mass movement. Although the addition of trees to a hillside may sometimes induce sliding because of the increase in weight (De Ploey 1981), this effect is generally offset by an increase in cohesion associated with the binding of the soil within the root network and by the tensile strength of the roots themselves (section 3.4.4). Further, the surcharge effect can be minimized by placing trees at the bottom rather than the top of the slope. Living tree roots can contribute up to 20 kPa to the soil shear strength (O'Loughlin & Watson 1979). It is believed that the lateral roots contribute most to binding because of their greater density, whereas the vertical roots add most of the tensile strength and, where they cross a potential slide plane, help to anchor the soil to the slope. Deep-rooted species are thus preferred for stabilizing the slope and increasing its resistance to sliding. Bishop and Stevens (1964) show that large trees can increase the shearing stress required for sliding by 2.5 kPa, which is why, as seen in section 7.4.3, their removal can promote landslides. Grass roots can increase the shear strength of soils by between 0.1 and 11.6 kPa (Lawrance et al. 1996; Preston & Crozier 1999).

Closing an area, particularly from livestock but also from wild animals, will allow vegetation to colonize landslide scars naturally. The rate of recovery, however, is slow. Herbs come in first, followed by grasses, but only after about four years do perennial grasses dominate the cover and shrubs start to appear. After seven years the cover on scars in the Mgeta Valley, Tanzania, was only 25 per cent (Lundgren 1978). Recovery of soil slip scars in the Wairarapa hill country, New Zealand, takes 20 years but even then the productivity of the pasture is only 80 per cent of that on uneroded land. Despite recolonization, the quality of the land deteriorates (Trustrum et al. 1984). Tree planting was attempted on the slide scars in the Mgeta Valley but the species used, *Acacia mearnssii*, *Cupressus lusitanica* and *Eucalyptus maidenni*, proved unsuitable. All the *Cupressus lusitanica* seedlings died and the survival of the others was threatened by gully erosion (Lundgren 1978). This example emphasizes the need for careful selection of species in relation to the environmental conditions.

8.7.3 Afforestation

Many countries have afforestation programmes aimed at arresting erosion and regulating floods. Most schemes involve closing the land to other uses. It should be recognized, however, that the success of such schemes depends on the methods used to prepare the land for replanting and that it takes some years before the all-important litter layer develops on the soil surface. As seen in section 7.4.3, erosion often increases in the early years of reforestation work.

In the Vasad and Kota areas of Gujarat, India, the closing of gullied lands to grazing allowed the establishment first of a good grass cover of desirable species and then an increase in tree numbers through natural colonization. Afforestation trials were also successfully implemented, with bamboo, teak, sissoo and eucalyptus as the most promising species (Tejwani 1981). The principal species in forest plantations in Kenya, *Cupressus lusitanica*, *Pinus patula*, *Pinus radiata* and *Eucalyptus saligna*, are selected because of their rapid growth rather than their value in soil and water conservation (Konuche 1983). Although the pines are generally satisfactory, the cypress results in a bare forest floor liable to erosion by overland flow unless the stands are pruned and thinned. The role of eucalypts is unclear because of evidence that when planted in wet areas they result in a reduction of water supply in springs and rivers (Gosh et al. 1978), yet when grown in drier areas they do not consume large quantities of water (Konuche 1983). Since eucalypts develop their root systems rapidly and promote infiltration and subsurface drainage, they may induce mass movements when planted on steep slopes with shallow soils (section 3.4.4). This is because of the impedance to subsurface water movement at the soil–rock interface and the reduction in the shear strength of the soil as the moisture content increases to saturation (De Ploey & Cruz 1979).

8.7.4 Restoration for pasture

Revegetation of lands for pasture is the major activity of Landgræðsla Ríkisins, the State Soil Conservation Service of Iceland. Work is concentrated on restoration of bare moving sands and gravels using aerial seeding. The land is first fenced to keep out livestock. *Elymus arenarius* is then planted on the moving dunes in strips at right angles to the erosive winds and fertilized annually until serious sand movement has been halted. This plant thrives well in drifting sand

and collects and fixes its own dunes. Once the sand has been stabilized, the *Elymus* dies out and *Festuca rubra*, *Poa pratensis* and *Phleum pratense* are aerially seeded. Imported seeds are used because the local climate is too severe to provide a reservoir of locally available seed. Fertilizer is applied aerially each year for the next two to four years, by which time a reasonable vegetation cover has been obtained. No further work is carried out and, in the absence of continued fertilizer application, the seeded grasses die out and the vegetation cover becomes poor. Sufficient stability and organic matter have been achieved, however, to allow native vegetation to recolonize. The land remains protected for 30 years (Runólfsson 1978; Arnalds et al. 1987) before being returned to the farmer.

In many countries of the world, revegetation is limited not only by the infertility of the soil but also by lack of water. In addition to supplying fertilizers in the early stages of reclamation, it is necessary to provide water-conserving structures. Mututho (1989) modified a traditional pitting practice of the WaMatengo people in southern Tanzania to form semi-circular pits, 5–12 m² in area, surrounded by a 15–30 cm high bank to trap water on rangeland reclamation schemes in the Kitui District, Kenya. The sites were excluded from grazing and the pits planted with indigenous perennial grasses. The pits, however, require considerable inputs of labour for their construction. A modified version, known as the Katumani pit, has been developed (Gichangi et al. 1992) that is only 1.5–3.0 m² in area. These pits are also planted with indigenous grasses but, in the first year after construction, a crop of beans and cowpeas can be obtained to give a short-term income until the grazing land has been rehabilitated.

8.7.5 Embankment and cut slopes

Rapid establishment of a grass or legume cover is essential on embankment and cut slopes to minimize surface erosion and enhance slope stability. This is commonly achieved by hydroseeding a seed and fertilizer mix. Where immediate erosion control is necessary, mulches or geotextiles should be used to protect the soil and prevent the seeds from being washed away. A straw mulch, applied at 4 t ha⁻¹ at the time of seeding and fertilizing the soil in the autumn, was found to reduce erosion to acceptable levels on 45° roadside slopes in western Oregon (Dyrness 1975) and allow the establishment of more than 70 per cent vegetation cover by the end of the following summer. Maximum cover, mainly rye grass, bent and fescue, was achieved after some three years but then lack of nitrogen caused the vegetation to decline to only 10 per cent cover after eight years. By this time the slope was protected by a dense litter of dead grass. Addition of more fertilizer quickly revived the vegetation, which developed to 90 per cent cover within one year. Most erosion occurred in the first year of the treatment, after which soil loss was virtually zero. An alternative way of establishing cover is to plant natural turf over the slope, as is standard practice in the urban areas of Singapore (Ramaswamy et al. 1981).

Grasses, mainly *Cynodon dactylon* and *Pennisetum clandestinum*, are used to control surface erosion on 30–65° road banks in eastern Nepal (Howell et al. 1991) but the increased infiltration that results enhances the risk of shallow slides (section 7.4.9). In order to reduce the risk, deeper-rooted grasses, such as *Eulaliopsis binata*, *Neyraudia reynaudiana*, *Saccharum spontaneum* and *Cymbopogon microtheca*, are recommended to anchor the soil and root mat to the underlying weathered material. However, these grasses are clumpy in habit and cause runoff to concentrate. Moreover, their roots do not always adhere well to the coarse debris and pull-out when the material starts to move. To counteract this and provide deeper stability, plant successions are planned for the revegetation work, allowing the grasses to be supported by shrubs and trees. Suitable tree species include *Pinus wallichiana*, *Pinus roxburghii*, *Acacia catechu*, *Dalbergia sissoo* and *Alnus nepalensis*. In order to sustain the grass cover, the forest must remain 50 per cent open, so long-term management will be required to prevent the closure of the forest canopy (Howell 1999). Despite the versatility of vegetation for erosion control on road banks, the steepness of many slopes and the harshness of the environment often mean that vegetative solutions need to be integrated with structural work.

8.7.6 Pipeline corridors

Pipeline corridors comprise strips of land, usually 20–40 m wide, extending for many hundreds of kilometres across country. As indicated in section 7.4.8, there is a risk of erosion between the time the pipe is installed and the regrowth of the vegetation. Severe erosion can hinder the restoration work by washing out seeds and young plants. In many situations, the most cost-effective method of combining revegetation and erosion control is to reseed, cover the slope with a surface-laid geotextile and then plant saplings through the geotextile material. Where a suitable local source is available, a mulch may be used instead of the geotextile, but it is important that at least a 70 per cent ground cover is maintained throughout the vulnerable period. The success of the restoration work, particularly where the aim is to re-establish the local plant community, can be enhanced by careful management of the top soil during its removal, storage and return (Coppin & Richards 1990) in order to preserve as much of the original soil structure and cohesion as possible and not to damage the seed bank contained in the soil. Since the seed bank contains representative material of the local plant community, it should be viewed as the foundation for re-establishing the local ecology; reseeding and replanting are used to supplement the seed bank. In dry climates or where revegetation work is carried out prior to the dry season, the seeds and plants must be watered regularly. Water conservation measures, such as pits and lunettes (Hudson 1987), can be used to trap and store surface runoff from occasional rain storms. As is the case with road banks, it is important to plan a plant succession. When deciding which species to use for reseeding and replanting, emphasis should be given to the colonizing ones in order to establish cover as rapidly as possible. Other species will emerge from the seed bank and by invasion from the surrounding land. Once a vegetation cover has been re-established, the plant succession must be managed in order to ensure that a uniform cover of grasses and shrubs is maintained and that trees do not take hold, since their roots can damage the pipe.

8.7.7 Sand dune restoration

The first stage in sand dune restoration is to stabilize the moving sand in order to create a suitable environment for plants to grow. Fences are erected to reduce wind speed, trap sand and enclose the area to keep out livestock and people. The sand surface is fixed using soil stabilizers or geotextiles. The main plants used in stabilizing sand dunes in the coastal areas of Europe and the USA are marram grass (*Ammophila arenaria*) and American beach grass (*Ammophila breviligulata*). These have strong extensive root networks in both lateral and vertical directions, which enable them to bind the soil while the grass acts as a sediment trap (Hesp 1979). They thrive well in moving sand with low nutrient availability. Grasses should be planted as 100–150-mm wide culms rather than seeded because of the risk of the seeds being blown away and the young plants damaged by sand blasting. Once stabilized, the dunes can be planted with shrubs and trees.

Similar approaches are used to stabilize desert dunes. Within eight years of the establishment of the fast-growing *Tamarix aphylla* on sands near the Al-Hasa oasis in Saudi Arabia, the depth of the organic horizon had increased from zero to 10 mm and the calcium carbonate content decreased from 30 to 15 per cent (Stevens 1974). The long-term effect of such schemes is not clear, however, because, as the trees mature, the soil moisture within the dune is depleted. Survival of the vegetation then depends on the ability of the tree roots to seek out groundwater. If this turns out to be salty, the trees may die and the dunes may be remobilized (Gupta 1979). Generally, where precipitation annually exceeds 250 mm, dunes can be stabilized by vegetation within five to ten years. Where annual precipitation is between 100 and 250 mm, vegetational stabilization is possible but may take 20–30 years; moreover, woody species cannot grow and only shrubs and grasses can be used. Where precipitation is less than 100 mm per year, a vegetation-based approach is not feasible except with irrigation over very small areas.

8.7.8 Restoration of recreational areas

Closing the land and replanting with trees, shrubs and grasses is frequently adopted to renovate areas eroded through recreational use. Examples are the Tarn Hows project in the English Lake District (Barrow et al. 1973), the reclamation of gullies in the Box Hill area of the North Downs in southeast England (Streeter 1977), the footpath restoration work of the Three Peaks area of the Yorkshire Dales National Park (Rose 1989) and the revegetation of the Heavenly Ski Resort in the Tahoe Basin, California (Morse 1992). In addition to ensuring that the species used for replanting are compatible with the physical environment and resistant to trampling, plant selection may be influenced by aesthetic considerations so as to enhance landscape quality and by the need to create varied and interesting wildlife habitats.

8.8

Agroforestry

Trees can be incorporated within a farming system by planting them on land that is not suitable for crop production. Where trees are deliberately integrated with crops or animals or both to exploit expected positive interactions between the trees and other land uses, the practice is defined as agroforestry (Lundgren & Nair 1985). Trees help to preserve the fertility of the soil through the return of organic matter and the fixation of nitrogen. They improve the soil's structure and help to maintain high infiltration rates and greater water-holding capacity. As a result less runoff is generated and erosion is better controlled. Trees are also attractive to the farmer where they provide additional needs, especially fuel, fodder and fruits. Multipurpose trees and shrubs are thus fundamental to agroforestry.

Agroforestry is encouraged in many countries as a way of modifying existing farming systems to promote soil fertility, erosion control and a diversified source of income. Three types of practice may be defined. Agrisilviculture involves the combination of trees and crops. Silvopastoralism is the combination of trees and animals. Agrosilvopastoralism combines trees with crops and animals. Within these systems, trees can be used to supplement existing erosion-control measures – for example, by being added to contour grass strips and terraces – or they can be used to control erosion direct – for example, in multiple-cropping systems such as kitchen gardens or with contour-aligned hedgerows in intercropping. Although the latter act as live barriers and trap sediment, leading to the formation of terraces over time, as described in section 8.3, an important feature of the hedgerows is that they increase infiltration by three to eight times because of their root systems (Kiepe 1995) and thereby reduce runoff. Much attention has been given to alley cropping systems, in which multipurpose trees are grown as contour hedges separated by strips of cropland. On a 16° slope at the Butare Research Station, Rwanda, the annual soil loss over four years from cassava in an alley cropping system with 5-m wide strips of the leguminous shrub *Calliandra calothyrsus* grown on microterraces was 12.5 t ha^{-1} compared with 111 t ha^{-1} for traditional cultivation of cassava. Mixed cropping of cassava in the alleys with beans, maize and sweet potatoes, alternating with leguminous cover crops, reduced soil loss to 1 t ha^{-1} (König 1992). In a range of alley systems on an alfisol on a 4° slope near Ibadan, Nigeria, annual soil loss over two years for a maize–cowpea rotation was 1.6 t ha^{-1} with *Leucaena* hedges at 4 m spacing, 0.15 t ha^{-1} with *Leucaena* hedges at 2 m spacing, 0.88 t ha^{-1} with *Gliricidia* hedges at 4 m spacing and 1.7 t ha^{-1} with *Gliricidia* hedges at 2 m spacing compared with 8.7 t ha^{-1} with conventional cultivation and 0.025 t ha^{-1} with no tillage (Lal 1988). Crop yields of the maize were similar in all systems. Studies in many countries show that agroforestry systems reduce erosion by at least ten times compared to having no soil conservation measures (Young 1998).

Agroforestry systems require careful selection of both crops and tree species if a beneficial interaction is to be obtained. Indeed, several studies bring into question whether such interaction can be achieved. With alley cropping systems at Bellary, India, yields of sorghum, safflower and Bengal gram were reduced by 16, 53 and 53 per cent respectively when grown between *Leucaena leucocephala* alleys at 4.5 m spacing (Srivastva & Rama Mohan Rao 1988), mainly as a result of soil moisture depletion on land close to the hedgerows. Sorghum yields were also reduced when grown in combination with *Acacia nilotica*, *Azadirachta indica* and *Eucalyptus* hybrid. Yield reductions in sorghum and pearl-millet were observed in agrisilvicultural systems near Karnal, northern India (Kumar et al. 1990). Clearly, with these adverse effects, the acceptability of the system will depend on whether the grain yields are still sufficient for survival and whether additional income can be obtained from the trees. According to the Natural Resources Conservation Service (1999), alleys at 12 m spacing will allow the cropping of maize and soya bean for between five and ten years before yields reduce; alleys at 24 m spacing will allow cropping for 20 years or more.

The most important tree species are: *Leucaena leucocephala*, which is a quick-growing fodder tree but also provides timber for fuel and pulpwood; *Prosopis juliflora* and *Prosopis chilensis*, which are drought-resistant and provide wood for fuel and poles; *Acacia albida*, which is well adapted to sandy soils and produces good fodder; *Acacia nilotica*; and *Sesbandia grandiflora*. Alternative systems, however, which might generate more income, could include fruit trees and more shade-loving crops.

Box 8

Selecting vegetation for erosion control

A dense and uniform ground cover of vegetation is the most effective way of controlling erosion. Revegetation using native grasses, trees and shrubs is thus widely recommended as the most appropriate long-term method of restoring gullied land, landslide scars, road banks, pipeline rights-of-way, mining spoil and eroded land in recreation areas. When designing a revegetation scheme it is important to draw up in advance clear specifications of what is required so that these can be agreed with the contractors who will carry out the work (Table B8.1).

In choosing vegetation species, attention should be given to:

■ native species, adapted to local climatic and soil conditions, to ensure integrity of the local ecology;
■ the availability of seeds and plants;
■ species with low fire risk;
■ species with appropriate engineering properties for erosion control (Table B8.2);

Continued

Table B8.1 Guidelines for specification of revegetation work

Section	Item	Coverage
Materials	Seeds	Types (species, quality, storage, sources, mixtures)
	Plants	Types (cuttings, container-grown, bare-rooted), size, handling and storage, source/provenance, pre-treatment and preparation
	Fertilizers	Content of N, P and K and other nutrients, lime, special requirements such as slow-release
	Mulches	Material, source, size or grade
	Composts/manures	Organic bulk materials: source, nutrient content and/or organic matter content, moisture content
	Soils	*In situ* or imported. Use or not of top soil. Texture, pH, fertility, organic content. Source.
	Geotextiles	Trade name, product, required properties, storage, testing, source
	Soil stabilizers	Trade name, product, quantities, method of application, source
	Herbicides	Formulation, active ingredient, type (liquid or granular)
Earthworks	Soil stripping	Depth and location of horizons to be stripped, timing, moisture content
	Soil stockpiling	Size and shape of stockpile, temporary protection (e.g. grassing)
	Compaction	Density, method of compaction
	Soil handling	Methods, timing (especially in relation to wet periods), moisture content
	Soil preparation	Deep, shallow or no cultivation, timing, direction in relation to slope, moisture content
Seeding	Method	Drilling, broadcasting, hydroseeding, hand application, uniformity
	Timing	Acceptable seeding periods in relation to predicted and actual weather conditions
	Site preparation	Weed clearance, stone picking, cultivation
	Aftercare	Follow-up fertilizing, overseeding, weeding, cutting
Planting	Method	Planting technique: excavation, preparation, backfilling, supports. Size/type of plants, mulching, watering, irrigation
	Timing	Acceptable planting periods according to weather and ground conditions
	Site preparation	Weed clearance, cultivation
	Aftercare	Fertilizing, pruning, weeding, watering, checking, adjustment and removal of tree supports
Management	Trees	Thinning, coppicing, replacement planting, trimming (hedges), checking for disease and pest damage
	Grasses	Mowing (frequency, height, method, disposal of cuttings); fertilizing, overseeding, pest and weed control
	Fencing	Check and repair

Source: after Coppin and Richards (1990).

Table B8.2 Desirable properties of plants for control of soil erosion by water

Plant component	Required properties
Roots	Vigorous development close to the surface to reinforce resistance to erosion, i.e. dense, shallow horizontal root structure
Shoots	Dense uniform growth close to soil surface to impart roughness to flow and reduce its velocity; resistant to damage in high flows
Growth cycle	Provision of sufficient shoot and root growth to protect the soil when erosion risk is greatest; may need to consider ability to protect the soil when heavy rain occurs after die-back at end of dry or cool season; ability to self-repair by rapid regrowth
Growth habit	Uniform growth habit close to the ground to protect the soil from raindrop impact and stresses exerted by running water; avoid clumpy or tussocky species and species with canopies taller than 0.5 m unless mixed with other low-growing species
Growth rate	Rapid establishment, especially for 'pioneer' species; low maintenance

■ mixtures of species to create biodiversity and a more healthy plant community;
■ mixtures of species with a range of establishment rates, including rapidly establishing 'pioneer' species to colonize the area and stabilize the surface and slower establishing species that will form the mature vegetation cover;
■ the required pattern of plant succession, taking account of when particular species will establish and when they will die out.

CHAPTER 9

Soil management

The aim of sound soil management is to maintain the fertility and structure of the soil. Highly fertile soils result in high crop yields, good plant cover and, therefore, in conditions that minimize the erosive effects of raindrops, runoff and wind. Soil fertility and the land husbandry to support it can therefore be seen as the key to soil conservation.

9.1 Organic content

One way of achieving and maintaining a fertile soil is to apply organic matter. This improves the cohesiveness of the soil, increases its water retention capacity and promotes a stable aggregate structure (section 3.2). Organic material may be added as green manures, straw or a manure that has already undergone a high degree of fermentation. The effectiveness of the material varies with the isohumic factor, which is the quantity of humus produced per unit of organic matter (Table 9.1; Kolenbrander 1974). Green manures, which are normally leguminous crops ploughed in, have a high rate of fermentation and yield a rapid increase in soil stability. The increase is short-lived, however, because of a low isohumic factor. Straw decomposes less rapidly and so takes longer to affect soil stability but has a higher isohumic factor. Previously fermented manures require still longer to influence soil stability but their effect is longer lasting because these have a still higher isohumic factor (Fournier 1972).

Ekwue et al. (1993) found that ploughing in groundnut haulm on a range of soil types in northern Nigeria was more effective than cow dung in immediately reducing soil detachment by raindrop impact. Bonsu (1985) also found that cow dung was not fully effective in the savanna regions of West Africa because the extremely high temperatures bring about volatilization of nitrogen and desiccation of the manure. It was better to combine cow dung with a straw mulch to increase the protection of the surface soil against erosion and enhance the moisture content of the soil. A combined application of cow dung at $5\,t\,ha^{-1}$ and wheat straw mulch at $4\,t\,ha^{-1}$ gave the lowest soil loss and the highest yields of grain sorghum compared with a range of other practices. The dung and mulch have to be applied every year because termites attack the straw during the dry season. Combined cow dung and maize straw mulch were similarly effective in reducing erosion in the humid tropical region of Ghana.

To increase the resistance of an erodible soil by building up organic matter is a lengthy process. Before any effect on stability is observed, the organic content must be raised above 2 per cent

Table 9.1 Farmyard manure equivalents of some organic materials

Material	Isohumic factor	FYM equivalent
Plant foliage	0.20	0.25
Green manures	0.25	0.35
Cereal straw	0.30	0.45
Roots of crops	0.35	0.55
Farmyard manure	0.50	1.00
Deciduous tree litter	0.60	1.40
Coniferous tree litter	0.65	1.60
Peat moss	0.85	2.50

Source: after Kolenbrander (1974).

(section 3.2). For soils with less than 1 per cent organic content, a large supply of organic material is required. Ploughing in maize residue at $5–10\,t\,ha^{-1}$ was found in Nigeria to increase the organic carbon content of the soil in absolute terms by only 0.004–0.017 per cent, while application of farmyard manure at $10\,t\,ha^{-1}$ was sufficient only to maintain, not to increase, an existing level of organic content (Jones 1971). A three-year grass ley, however, was equivalent to an annual application of farmyard manure at $12\,t\,ha^{-1}$. In the UK, Ekwue (1992) found that grass leys were more effective in reducing soil detachment by raindrop impact on a sandy loam soil than either farmyard manure or straw. Numerous field experiments worldwide show that grass leys are the only effective way of building up the organic content. Unfortunately, with the trend in many parts of the world towards larger mechanized arable farms where there is no demand for grass, clover or alfalfa, the addition of organic matter becomes uneconomic.

The value of organic matter is enhanced by the presence of base minerals in the soil as these bond chemically with the organic materials to form the compounds of clay and humus which make up the soil aggregates (section 3.2). The base minerals are thus retained in the soil rather than removed by leaching or subsurface flow. Where these minerals, which provide the essential nutrients for plant growth, are absent, they should be added to the soil as fertilizer in the amounts normally recommended for the crops being grown. However, mineral fertilizers cannot improve the aggregate structure of a soil on their own; they need organic support. The continual use of mineral fertilizers without organic manures may lead to structural deterioration of the soil and increased erodibility.

There is much evidence to show that the long-term use of the land for cropping reduces the organic carbon content of the soil unless conservation measures are taken. Soil erosion probably accounts for only a small component of the reduction, with biological mineralization of the carbon from the soil *in situ* being the most important process, particularly in semi-arid areas where bare soil summer fallows are used to conserve moisture (Rasmussen et al. 1998). However, recent research suggests that between 30 and 46 per cent of the carbon in the eroded material may also be mineralized (Jacinthe et al. 2002). Inversion ploughing sustains the process by regularly mixing the subsoil and top soil and bringing carbon reserves from depth to the surface. Following the conversion of prairie grassland to arable agriculture in the Midwest of the United States in the 1870s and 1880s, organic carbon levels in the top 200 mm of the soil fell rapidly over a 20- to 40-year period from 3–5 per cent to 0.5–1 per cent. They then remained at this level until the 1960s, when conservation practices based on returning crop residues to the land were adopted, raising levels to 2–3 per cent (Huggins et al. 1998). Decreases of 1–2 per cent in absolute values

Soil management

of organic carbon occurred over 30–70 years after parts of the Great Plains of the USA and Canada were brought into cultivation (Haas et al. 1957; Martel & Paul 1974).

As indicated in Chapter 1, in addition to the concern over the effects of carbon loss on the fertility of the soil, there is the recognition that the loss is a contributor to carbon dioxide in the atmosphere and may therefore play a part in any global warming (Flach et al. 1997). As a result, much research on soil management today is focused on the potential of using soils to sequester carbon. Practices such as the elimination of bare fallows, the use of no tillage and return of crop residues can bring about small increases in organic carbon in soils quite rapidly (Huggins et al. 1998; Janzen et al. 1998; McConkey et al. 2003) but the effect is just as quickly lost should the practices be discontinued. According to Lal (2002), the adoption of conservation tillage has the potential to sequester carbon at rates of 500–800 kg ha^{-1} yr^{-1} in humid temperate areas, 300–500 in arid temperate areas, 200–400 in the humid tropics and 100–200 in the arid tropics. Equivalent values obtained by the use of cover crops are 400–600, 200–400, 150–300 and 50–150 kg ha^{-1} yr^{-1} respectively. By far the most effective method of carbon sequestration is to establish a permanent vegetation cover. Five years after establishing grass on a sandy loam soil formerly under continuous wheat-fallow cropping at Keeline, Wyoming, soil organic carbon levels in the top soil had returned to those found on native rangelands (Reeder et al. 1998). A strategy of growing bio-energy crops on the surplus arable land in the United Kingdom would sequester carbon at an annual rate of 3.5 Tg, which is equivalent to 2.2 per cent of the country's 1990 carbon emission (Smith et al. 2000). Establishing a vegetation cover on the estimated 250 million hectares of severely eroded land across the world has the potential to sequester carbon at annual rates of 0.42–0.83 Pg (Lal 2002). The rate of sequestration peaks some 10–15 years after soil restoration and net accumulations are usually small after 25 to 30 years. Over this period, however, a combination of erosion control and restoration of degraded soils globally could sequester some 50–70 Pg C, which would provide some temporary mitigation to the greenhouse effect on world climates, while other policies on carbon emissions are put in place. However, without the international political will to promote erosion control and also prevent the reversion of restored land to agricultural production, soil conservation as a basis for carbon sequestration can remain only a potential.

9.2

Tillage practices

When managed so as to maintain their fertility, most soils retain their stability and are not adversely affected by standard tillage operations. Indeed, tillage is an essential management technique: it provides a suitable seed bed for plant growth and helps to control weeds. The tillage tools pulled by a tractor are designed to apply an upward force to cut and loosen the compacted soil, sometimes to invert it and mix it, and to smooth and shape its surface. When the moisture content of the soil is below the plastic limit (section 3.2), the soil fails by cracking, with the soil aggregates sliding over each other but remaining unbroken. The soil ahead of the loosening tool moves forwards and upwards over the entire working depth. Soil loosening is effective where the confining stresses resisting upward movement are less than those resisting sideways movement of the soil. Vertical confining stress is obviously zero at the surface and increases with depth until, at a critical depth, it equals the lateral confining stresses. Below this depth, the soil moves only forwards and sideways and failure occurs, with a risk of compaction (Godwin & Spoor 1977).

Table 9.2 Vulnerability of soils to compaction

(a) Texture and packing density

Texture class	Packing density (Mg m^{-3})		
	Low (<1.40)	Medium (1.40–1.75)	High (>1.75)
Coarse	VH	H	M*
Medium (<18% clay)	VH	H	M
Medium (>18% clay)	H	M	L
Medium fine (<18% clay)	VH	H	M
Medium fine (>18% clay)	H	M	L**
Fine	M[†]	L[‡]	L[‡]
Very fine	M[†]	L[‡]	L[‡]
Organic	VH	H	

Susceptibility classes: L = low; M = moderate; H = high; VH = very high.
* Except for naturally compacted or cemented sandy materials which already have a low (L) susceptibility. ** These soils are already compact. [†] These packing densities are usually found only in recent alluvial soils or in soils with >5 per cent organic carbon. [‡] Fluvisols in these categories have a moderate (M) susceptibility.

(b) Wetness

Susceptibility class	Wetness			
	Wet	Moist	Dry	Very dry
VH	E (E)	E (E)	V (E)	V (V)
H	V (E)	V (E)	M (V)	M (M)
M	V (E)	M (V)	N (M)	N (N)
L	M (V)	N (M)	N (N)	N (N)

N, not particularly vulnerable; M, moderately vulnerable; V, very vulnerable; E, extremely vulnerable.
Classes outside parentheses refer to situations with a strong firm top soil layer over a stronger subsurface layer or pan. Classes within parentheses refer to situations with a loose weak topsoil without a strong subsurface layer.
Source: after Spoor et al. (2003)

The effect of wheeled traffic and tillage implements on a soil depends upon its shear strength, the nature of the confining stresses and the direction in which the force is applied. The main effect of driving a tractor across a field is to apply force from above and compact the soil. This may result in an increase in shear strength through an increase in bulk density, but this is often offset by decreased infiltration and increased runoff, so that wheelings are frequently zones of concentrated erosion. The pattern of compaction depends upon tyre pressure, the width of the wheels and the speed of the tractor, the latter controlling the contact time between the wheel and the soil. Compaction generally extends to the depth of the previous tillage, up to 300 mm for deep ploughing, 180 mm for normal ploughing and 60 mm with zero tillage (Pidgeon & Soane 1978). Table 9.2 summarizes the risk of soils to compaction in relation to texture and wetness (Spoor et al. 2003). The best ways of preventing or reducing compaction are to avoid vehicle movements on the land when the soil is too wet or to confine wheels or tracks to permanent, sacrificial strips

Soil management

across a field (Chamen et al. 2003). Since plough pans help to protect a soil from compaction, they should only be broken up where they significantly impede root development, aeration or drainage (Spoor et al. 2003).

9.2.1 Conventional tillage

Over the years a reasonably standard or conventional system of tillage, involving ploughing, secondary cultivation, with one or more disc harrowings, and planting has been found suitable for a wide range of soils. Ploughing is carried out with the mouldboard plough, although, on stony land or with soils that do not fall cleanly off mouldboards, the disc plough is often used. Mouldboard ploughs invert the plough furrow and lift and move all the soil in the plough layer, usually to a depth of 100–200 mm. Secondary cultivation to form the seed bed and remove weeds is carried out by either disc or tine cultivators. With disc cultivators the soil is broken up by the passage of saucer-shaped metal discs mounted on axles. The most common tine cultivator is the chisel type, which consists of a series of metal blades mounted on a frame. The blades vary in width from narrow, 50 mm, to wide, 75 mm, but can be up to 300 mm wide. Ploughing produces a rough cloddy surface with local variations in height of 120–160 mm. Secondary cultivation reduces the roughness to 30–40 mm, while drilling and rolling decrease it still further (section 2.1; Table 2.2). Roughness is also reduced over time by raindrop impact and water and wind erosion. Soil loss (SL) by water erosion decreases with increasing roughness (R) according to the relationship (Cogo et al. 1984):

$$SL \propto e^{-0.5R} \tag{9.1}$$

which means that small increases in roughness from a virtually smooth surface can have substantial effects on reducing erosion but much larger increases will be needed to have the same effect with an already rough surface. Tillage can often be used successfully to roughen the surface to control wind erosion in an emergency, using a chisel to produce ridges and furrows across the path of the prevailing wind (Woodruff et al. 1957). Ridging was found to decrease wind erosion by between 15 and 26 per cent in experiments carried out in Niger (Bielders et al. 1998).

9.2.2 Contour tillage

Carrying out ploughing, planting and cultivation on the contour can reduce soil loss from sloping land compared with cultivation up-and-down the slope. The effectiveness of contour tillage varies with the length and steepness of the slope. It is inadequate as the sole conservation measures for lengths greater than 180 m at 1° steepness. The allowable length declines with increasing steepness to 30 m at 5.5° and 20 m at 8.5°. Moreover, the technique is only effective during storms of low rainfall intensity. Protection against more extreme storms is improved by supplementing contour farming with strip cropping (section 8.3).

On silty and fine sandy soils, erosion may be further reduced by storing water on the surface rather than allowing it to run off. Limited increases in storage capacity can be obtained by forming ridges, usually at a slight grade of about 1:400 to the contour, at regular intervals determined by slope steepness. Contour ridging is generally ineffective on its own as a soil conservation measure on slopes steeper than 4.5°. Greater storage of water and more effective erosion control can be achieved by connecting the ridges with cross-ties over the intervening furrows, thereby forming a series of rectangular depressions that fill with water during rain. Because crop damage can occur if the water cannot soak into the soil within 48 h, this practice, known as tied ridging, should only

Table 9.3 Tillage practices used for soil conservation

Practice	Description
Conventional	Standard practice of ploughing with disc or mouldboard plough, one or more disc harrowings, a spike-tooth harrowing and surface planting.
No tillage	Soil undisturbed prior to planting, which takes place in a narrow, 25–75 mm wide seed-bed. Crop residue covers of 50–100% retained on surface. Weed control by herbicides.
Strip tillage	Soil undisturbed prior to planting, which is done in narrow strips using rotary tiller or in-row chisel, plough-plant, wheel-track planting or listing. Intervening areas of soil untilled. Weed control by herbicides and cultivation.
Mulch tillage	Soil surface disturbed by tillage prior to planting using chisels, field cultivators, discs or sweeps. At least 30% residue cover left on surface as a protective mulch. Weed control by herbicides and cultivation.
Reduced or minimum tillage	Any other tillage practice that retains at least 30% residue cover.

be used on well drained soils. If it is applied to clay soils, waterlogging is likely to occur. Tied ridging increased yields of millet, maize and cotton in years of average and below average rainfall in Burkina Faso compared with open ridging or flat planting due to better water retention. In wet years, however, crops like cowpeas, which are sensitive to waterlogging, produced lower yields (Hulugalle 1988). Tied ridging with no till gave soil losses over a three-year period of less than $0.5 \, t \, ha^{-1}$ compared with up to $9.5 \, t \, ha^{-1}$ for conventional ploughing with a mouldboard under maize cultivation on erodible sandy soils in Zimbabwe (Vogel 1994). Maize yields increased with the system in the semi-humid region of the country but declined in the semi-arid region.

9.2.3 Conservation tillage

Numerous studies have taken place in recent years to examine the effects of different types of conservation tillage (Table 9.3) on soil erosion rates, soil conditions and crop yields. Conservation tillage can be defined as any practice that leaves at least 30 per cent cover on the soil surface after planting. In 1998, mulch tillage was used on 19.7 per cent of the planted area in the USA and 16.3 per cent was under no till. The success of the various systems is highly soil specific and also dependent on how well weeds, pests and diseases are controlled. The main barriers to adoption are the expense of the specialist equipment for managing cultivation in crop residues, problems of weed control and increases in pests, particularly rodents (Natural Resources Conservation Service 1999). Generally the better-drained, coarse- and medium-textured soils with low organic content respond best and the systems are not successful on poorly drained soils with high organic contents or on heavy soils where the use of the mouldboard plough is essential. Since the effectiveness of all the techniques depends on the amount of crop residue left on the surface at the time of greatest erosion risk, it is difficult to isolate the role of tillage from that of the residue, which acts as a mulch (section 8.6). Some plants produce toxic material during the breakdown of their residue, which can hinder the germination and establishment of the next crop. Since this is a particular problem in monocultures of maize and wheat, conservation tillage is also dependent on the use of crop rotations.

No tillage

No tillage describes the system whereby tillage is restricted to that necessary for planting the seed. Drilling takes place directly into the stubble of the previous crop and weeds are controlled by herbicides. Generally between 50 and 100 per cent of the surface remains covered with residue. The technique has been found to increase the percentage of water-stable aggregates in the soil compared to tine or disc cultivation and ploughing (Aina 1979; Douglas & Goss 1982; Schjønning & Rasmussen 1989; Suwardji & Eberbach 1998; Mrabet et al. 2001). It is not suitable, however, on soils that compact and easily seal because it can lead to lower crop yields and greater runoff.

No tillage reduced erosion rates under maize (Bonsu & Obeng 1979) and millet (Bonsu 1981) in Ghana to levels comparable with those achieved by multiple cropping but generally not to the levels obtained with surface mulching. Moreover, no tillage was not always effective in the first year of its operation because of the low percentage crop residues on the surface. At Ibadan, Nigeria, the technique reduced annual soil loss under maize with two crops per year to $0.07\,t\,ha^{-1}$, compared with $56\,t\,ha^{-1}$ for hoe and cutlass, $8.3\,t\,ha^{-1}$ for a mouldboard plough and $9.1\,t\,ha^{-1}$ for a mouldboard plough followed by harrowing (Osuji et al. 1980).

No tillage is viewed as the leading technology to control erosion on crop-livestock farms in the Midwest of the USA based on pig production and the growing of maize, soya bean and wheat in rotation. In a review of various studies of tillage systems, Moldenhauer (1985) showed that annual soil loss under no till on a range of erodible soils in the Corn Belt was 5–15 per cent of that from conventional tillage. No till is also recommended on the ultisol soils of the Southern Piedmont of the USA, which have become severely eroded after 150 years of continuous cropping. As alternatives to the conventional system of growing soya bean with a bare fallow over winter, the following no tillage systems were studied: soya bean with wheat as a winter cover and using in-row chisel tillage; soya bean with barley as a winter cover and using fluted coulters; and soya bean with rye as a green manure and using fluted coulters. The respective mean annual soil losses for the four systems are 26.2, 0.1, 0.1 and $3.4\,t\,ha^{-1}$ (Langdale et al. 1992). The use of no tillage on silty, crust-prone soils in northern France reduced erosion over two years (1993–4) to $0.04\,t\,ha^{-1}$ from $0.18\,t\,ha^{-1}$ with conventional tillage using a mouldboard plough, but increased runoff from 3.2 to 6.1 mm (Martin 1999), resulting in greater risk of erosion and flooding further downslope. There is some evidence, however, to suggest that such an adverse effect might be short-lived. After 16 years of no tillage, the saturated hydraulic conductivity of the surface layer of Oxic Paleustalf soils under winter wheat in the Wagga Wagga region of New South Wales was higher than that under conventional cultivation, an effect attributed to greater biological activity as witnessed by the greater earthworm population (Suwardji & Eberbach 1998). Increased earthworm numbers were also a feature of long-term no tillage on soils ranging from sands to silt loams in Germany (Tebrügge & Düring 1999).

Strip tillage

With strip tillage, the soil is prepared for planting along narrow strips, with the intervening areas left undisturbed. Typically, up to one-third of the soil is tilled in a single plough–plant operation. When used for maize cultivation on research plots of the University of Science and Technology, Kumasi, Ghana, the technique reduced soil loss from 23 storms totalling 452 mm of rain to

0.2 t ha^{-1} compared with 0.9 t ha^{-1} with a plough–harrow–plant sequence and 1.4 t ha^{-1} with traditional tillage using a hoe and cutlass (Baffoe-Bonnie & Quansah 1975). The plough–plant system caused least soil compaction, conserved most soil moisture and reduced losses of organic matter, nitrogen, phosphorus and potassium (Quansah & Baffoe-Bonnie 1981). Plough–plant systems have not become popular, however, because of problems of weed control and the slow speed of planting.

Mulch tillage

Difficulties with weed control, operating with large amounts of residue and, in many cases, lower yields have prevented the widespread take-up of stubble-mulch tillage. When tested to reduce water erosion on vertisols in Australia under continuous wheat production, the technique failed to increase the aggregate stability of the soil compared with conventional tillage (Marston & Hird 1978) and did not reduce erosion to an acceptable level (Marston & Perrens 1981). Nevertheless, the system has be used successfully to control wind erosion and conserve moisture in drier wheat-growing areas (Fenster & McCalla 1970).

Minimum tillage

Minimum tillage or reduced tillage refers to practices using chiselling or discing to prepare the soil whilst retaining a 15–25 per cent residue cover. With one disc cultivation prior to planting on a silty clay soil under continuous wheat production near Pisa, Italy, runoff was increased over that from conventional tillage. The minimum tilled plots retained moisture and this, in turn, reduced the cracking which plays a major role in promoting high infiltration of water in these soils. Despite the higher runoff, annual soil loss was lower under minimum tillage, at 1.6 t ha^{-1} compared with 4.1 t ha^{-1} from conventional tillage (Chisci & Zanchi 1981).

Chiselling in the autumn to produce a rough surface but retain residue cover, followed by disc cultivation in the spring to smooth the seed bed and cover the residue, is now widely practised in maize–soya bean agriculture in the Corn Belt of the USA. It can reduce soil loss by an order of magnitude over that recorded from conventional tillage (Siemens & Oschwald 1978; Johnson & Moldenhauer 1979). The technique works well when soya bean is planted in the chiselled maize residue but is not satisfactory when maize is planted in the chiselled soya bean residue because the latter deteriorates very rapidly and by the following spring there is insufficient cover to protect the soil. Alternative practices for this year of the rotation include growing a winter cover crop, killing it in the spring with a contact herbicide and planting maize in the residue of the cover, or planting maize with no tillage in order to minimize disturbance of the soil. After some 20 years of reduced tillage on loess soils in Germany, soil organic carbon was 5 Mg ha^{-1} higher than on similar conventionally tilled soils but the technique was found to create problems for farmers, with the accumulation of crop residues on the surface and infestations of weeds, fungal diseases and pests. Short-term inversion ploughing can be used to ameliorate these problems but its effects are so dramatic that only one year of conventional tillage is sufficient to remove all the benefits of reduced tillage with respect to carbon and organic matter (Stockfisch et al. 1999).

As with no tillage, there is concern that by not breaking up the soil, the technique will lead to a less porous surface with a resulting increase in runoff and erosion (Soane & Pidgeon 1975). However, Voorhees and Lindstrom (1984) found that on a silty clay loam, there was no difference in the porosity of the soil to 150 mm depth between no tillage and conventional tillage after four years and no difference to a depth of 300 mm after seven years. Kemper and Derpsch (1981) suggest that no tillage can be effective in restoring the porosity of oxisols and alfisols where it has been reduced by the development of a plough pan but that it will take ten to twelve years.

9.2.4 Alternatives to conservation tillage

An alternative approach to conservation tillage is to attempt, through careful timing of operations in relation to soil conditions, to use tillage to produce an erosion-resistant surface. Several farmers on sandy loam soils in the Midlands of England have adopted the Glassford system of ploughing and pressing the soil to produce a cloddy surface to control wind erosion on land devoted to sugar beet. When the soil is moist but not wet, a chisel is used to break up the crust and produce ridges and furrows at right angles to the direction of erosive winds. The land in the furrows is then rolled either in the same operation, by modifying the chisel plough to incorporate a roll press, or as soon as possible in a second operation before the soil dries out. This tillage is preferably carried out in January and the resulting surface of micro hills and valleys remains stable throughout the spring blowing period even after it has been broken up by drilling, which is carried out transverse to the press ridges. The Glassford system can thus be practised with standard farm equipment.

Conservation tillage has not proved appropriate for the management of the vertisols. These soils have a high percentage of smectitic clays, which undergo pronounced shrinking when dry, resulting in deep cracks that close only after prolonged wetting. In India, there is a problem of how to prepare a good seed bed on these soils, since sowing has to be carried out in the hard dry soil in advance of the rains in order to obtain a good crop cover to protect the soil from erosion in the rainy season. Experiments by ICRISAT showed that by preparing a surface of broad-based beds (950 mm wide) and furrows (550 mm wide and graded at 1:150) as soon in the rainy season as the soil becomes workable, annual soil loss was reduced over a six-year period to 1.2 t ha^{-1} compared with 6.6 t ha^{-1} with conventional tillage practice (Pathak et al. 1985). Steeper furrow grades resulted in too much erosion, while gentler grades did not provide sufficient drainage under wet conditions (Kampen et al. 1981). With better moisture control due to surface drainage along the furrows, crop yields were also increased (El-Swaify et al. 1985). However, the take-up of the technique by farmers is limited because of the need for specialized and costly equipment, namely a wheeled tool bar to make the raised bed, and increased labour requirements.

Subsoiling is used to break up impermeable soil layers, such as plough pans, at depth, although, as seen above (Spoor et al. 2003), this should only be done when the pan has an adverse effect on drainage and crop yield. Deep tillage using a crawler tractor to pull two chisels through ground to open up furrows about 100 mm wide and 500–700 mm deep is a recommended practice to break up subsurface pipes and tunnels (Colclough 1965; Crouch 1978). This treatment aids the establishment of grasses, forbs and legumes and is therefore carried out prior to reseeding the land for pasture. Control of tunnel erosion (section 2.6.1) is dependent upon the success of the revegetation in achieving a uniform pattern of infiltration because the effects of the ripping decline after three to five years (Aldon 1976).

9.3

Drainage

Drainage is used as a soil conservation measure to reduce runoff and therefore erosion on heavy clay soils. Heavy sticky soils with a moisture content above the plastic limit are difficult to manage because they fail compressively during tillage and become smeared (Spoor & Godwin 1979). Erodible soils with more than 20 per cent clay content will benefit from the installation of mole drains and from the break-up of compacted layers at depth by subsoiling. Pipe drainage was found to be very effective in reducing erosion on clay soils derived from Pliocene marine sediments in central Italy (Chisci et al. 1978; Zanchi 1989). However, subsurface drains can be important sources of sediment. On marine clay soils in southern Norway, the suspended sediment concentrations prior to autumn ploughing range from 0.13 to $0.15\,\mathrm{g}\,\mathrm{l}^{-1}$ in drainage waters and 2.7 to $3.2\,\mathrm{g}\,\mathrm{l}^{-1}$ in surface runoff; after ploughing, the respective values are 3.8–4.3 and $2.2–2.9\,\mathrm{g}\,\mathrm{l}^{-1}$. In years with no tillage, concentrations in drainage water never exceed $0.8\,\mathrm{g}\,\mathrm{l}^{-1}$ (Øygarden 1995).

9.4

Soil stabilizers

Improvements in soil structure can be achieved by applying soil conditioners. These may take the form of organic by-products, polyvalent salts and various synthetic polymers. Polyvalent salts such as gypsum bring about flocculation of the clay particles, while organic by-products and synthetic polymers bind the soil particles into aggregates. Although soil stabilizers are too expensive for general agricultural use, where the cost is warranted they are helpful on special sites like sand dunes, road cuttings, embankments and stream banks, to provide temporary stability prior to the establishment of a plant cover.

Gypsum has been used successfully to improve the structure of sodic soils in southeast Australia (Davidson & Quirk 1961; Rosewell 1970). These soils are particularly susceptible to tunnel erosion. Their high sodium content results in the dispersal of clay minerals when in contact with water, with consequent structural deterioration. The most effective treatment is to apply gypsum as a cation to replace the sodium. A good drainage system is also necessary to wash out the sodium from the soil. The treatment is extremely expensive over large areas and, unless accompanied by ripping to break up the tunnels and the sowing of grass, gives only temporary relief. Gypsum has also been used to reduce surface crusting and runoff on red-brown earths in the wheat-growing region of South Australia where the soils are unstable because of high contents of exchangeable magnesium (Grierson 1978).

Soil conditioners fall into two groups, those that render the soil hydrophobic and therefore decrease infiltration and increase runoff, and those that make the soil hydrophilic, increase infiltration and decrease runoff. Hydrophobic conditioners based on bitumen are generally effective in controlling erosion for only a few storms and are not always suitable for soil conservation purposes. They can, however, be employed to increase water yield; for example, to supply farm ponds (Laing 1978). Asphalt and latex emulsions seal the surface, thereby increasing runoff, but they are effective in stabilizing the soil and preventing erosion until the seal is broken. When applied to agricultural soils, for example, subsequent discing to a depth of 200 mm can partially destroy the seal and promote aggregate destruction (Gabriels & De Boodt 1978). This problem can be alleviated to some extent by incorporating the emulsion in the top 100–200 mm of the soil (Gabriels et al. 1977). The critical factor in this case is the size of the aggregates that are produced: if they are too small, infiltration rates remain low. For effective infiltration with hydrophobic conditioners, the aggregates should be at least 2 mm in size and ideally larger than 5 mm (Pla 1977).

Experience with polyacrylamide conditioners that are hydrophilic shows that high infiltration rates can be obtained regardless of aggregate size. Small plot studies with rainfall simulation indicate that, for best results, the conditioners should be sprayed directly on to the surface rather than mixed into the top soil (Wallace & Wallace 1986). However, on severely degraded fluvisols near Lake Baringo, Kenya, better results were achieved by applying the polyacrylamide conditioner to a tilled surface and then raking it into the top 20 mm of the soil (Fox & Bryan 1992). Studies on a 52° road bank with a silt loam soil near Thurston, West Virginia, showed that polyacrylamide conditioners reduced the sediment concentrations in runoff by between 22 and 82 per cent compared with untreated soils (Tobiason et al. 2001). Polyacrylamide conditioners are also effective in wind erosion control because they prevent abrasion of the soil surface by saltating particles (Armbrust 1999).

Considerable interest has been shown in soil conditioners that combine hydrophobic and hydrophilic components. Fullen et al. (1995) found that such a conditioner significantly increased the aggregate stability of loamy sand in the Midlands of England. In contrast, a similar conditioner used on a vertisol from Oahu, Hawaii, failed to increase aggregate stability although it did significantly reduce detachment of the soil by raindrop impact (Sutherland & Ziegler 1998). Polyurea polymers, which contain a mixture of hydrophilic ethylene oxide and hydrophobic propylene oxide in proportions according to the degree of hydrophilicity or hydrophobicity required, have been used successfully to stabilize sand dunes at Oulled Dhifallah, Tunisia. *Acacia cyanophylla* plants, used in the revegetation programme, had a higher survival rate and made faster growth on the stabilized areas (De Kesel & De Vleeschauwer 1981).

Box 9

Tillage erosion

Tillage erosion is the net downslope movement of soil brought about by tillage operations. As seen in section 9.2, the passage of a tillage tool loosens the soil to an extent dependent upon the soil strength, the type of implement and the depth of tillage. Where tillage occurs in a downslope direction, the soil is displaced downslope, rolling and sliding along the plough furrow under the influence of gravity. Where tillage takes place in an upslope direction, the soil is moved upslope but by a smaller distance because the movement is offset by the downslope movement due to gravity. With tillage on the contour, soil may be moved either upslope or downslope, depending on the alignment of the tool in relation to the slope direction; allowing for gravity, however, there is a net displacement of soil downslope. Overall, tillage results in a net transport of soil downslope. This is known as tillage displacement and it depends on the type of implement, the speed at which it moves through the soil, the angle of the slope and the resistance of the soil. The detachment and displacement of soil downslope can be described by a tillage transport coefficient (k; $kg\,m^{-1}$).

For upslope or downslope tillage, k is calculated from (Govers et al. 1994):

$$k = -D\rho_b b \tag{B9.1}$$

where D is the depth of tillage (m), ρ_b is the bulk density of the soil ($kg\,m^{-3}$) and b is the slope of the linear regression equation between the displacement of soil by tillage and the slope of the land ($m\,m^{-1}$). More comprehensive descriptions take account of the depth and speed of tillage (V). For a mouldboard plough, values of k are related exponentially to D for upslope and downslope tillage and linearly for contour tillage (van Muysen et al. 2002):

$$k = 2.026\rho_b\,D^{1.989}\,V^{0.406}, \text{ for upslope and downslope tillage} \tag{B9.2}$$

$k = 0.406\rho_b \, DV^{0.385}$, for contour tillage   (B9.3)

The data shown in Table B9.1 indicate that mouldboard ploughs displace more soil than chisel ploughs or rotary harrows. The tillage equipment used in the 1940s gives lower values than that of today because of the slower speed and shallower depth of ploughing (Mech & Free 1942). Low values are also found where the mouldboard is pulled by oxen (Thapa et al. 1999). Not surprisingly, manual cultivation by hoe gives very low values (Turkelboom et al. 1999).

When the rates of displacement of the soil by tillage are converted into annual rates of soil movement, they are seen to be of a similar magnitude to rates of soil erosion, indicating that tillage, particularly in intensively mechanized agriculture, can lead to major changes in the landscape. Tillage is more likely to explain the loss of soil on convexities and spurs where the landscape encourages divergent rather than convergent flow, and rill and gully erosion do not occur. Tillage can also move soil into hollows where, in its unconsolidated state, it is more likely to be eroded by concentrated flow.

Table B9.1 Displacement of soil by tillage

Implement	Conditions	Tillage speed (m s^{-1})	Tillage depth (m)	Tillage displacement (kg m^{-1} per pass)	Source
Mouldboard	Up- and downslope	2.1	0.24	330	Lindstrom et al. 1992
	Up- and downslope	1.25	0.28	234	Govers et al. 1994
	Up- and downslope	1.7	0.23	346	Lobb et al. 1999
	Up- and downslope	0.5	0.33	254	van Muysen et al. 1999
	Up- and downslope	0.75	0.15	70	van Muysen et al. 1999
	Up- and downslope	1.45	0.25	224	van Muysen et al. 2002
	Up- and downslope	1.54	0.21	169	van Muysen et al. 2002
	Up- and downslope	1.81	0.21	177	van Muysen et al. 2002
	Up- and downslope	1.25	0.2	153	Gerontidis et al. 2001
	Up- and downslope	1.25	0.3	383	Gerontidis et al. 2001
	Up- and downslope	1.25	0.4	670	Gerontidis et al. 2001
	Up- and downslope	1.0*	0.08	24	Mech and Free 1942
	Contour	2.1	0.24	363	Lindstrom et al. 1992
	Contour	1.25	0.2	134	Gerontidis et al. 2001
	Contour	1.25	0.3	252	Gerontidis et al. 2001
	Contour	1.25	0.4	360	Gerontidis et al. 2001
	Contour	1.37	0.26	184	van Muysen et al. 2002
	Ox-driven, up- and downslope	n/a	0.2	365–481	Thapa et al. 1999
	Ox-driven, contour	n/a	0.2	306–478	Thapa et al. 1999
Ridger	Ox-driven, contour	n/a	0.2	85–381	Thapa et al. 1999
Chisel	Up- and downslope			111	Govers et al. 1994
Rotary harrow + seeder	Up- and downslope	2.2	0.07	123	van Muysen & Govers 2002
Ard plough	Contour	n/a	0.08	68	Nyssen et al. 2000
Hoe	Manual cultivation	n/a	0.08	44	Turkelboom et al. 1999

* Value estimated by van Muysen et al. (2002).

CHAPTER 10

Mechanical methods of erosion control

Mechanical field practices are used to control the movement of water and wind over the soil surface. A range of techniques is available and the decision on which to adopt depends on whether the objective is to reduce the velocity of runoff and wind, increase the surface water storage capacity or safely dispose of excess water. Mechanical methods are normally employed to support agronomic measures and soil management.

10.1 Contour bunds

Contour bunds are earth banks, 1.5–2 m wide, thrown across the slope to act as a barrier to runoff, to form a water storage area on their upslope side and to break up a slope into segments shorter in length than is required to generate overland flow. They are suitable for slopes of 1–7° and are frequently used on smallholdings in the tropics where they form permanent buffers in a strip-cropping system, being planted with grasses or trees. The banks, spaced at 10–20 m intervals, are generally hand-constructed. There are no precise specifications for their design and deviations in their alignment of up to 10 per cent from the contour are permissible. In Wallo Province, Ethiopia, Hurni (1984) found that earth bunds were only effective on slopes up to 6°. An alternative technique on stony soils is to construct stone bunds, 250–300 mm high, set in a shallow trench on the contour. In order to enhance their ability to filter runoff and trap sediment, smaller stones should be placed on the upslope side and, if possible, gravels upslope of them (Hudson 1987). On steeply sloping land with erodible andosol soils devoted to potatoes and forage oats near Cochabamba, Bolivia, annual erosion from fields with stone bunds ranged from 38 to 124 t ha^{-1} compared with 118 to 164 t ha^{-1} on land without (Clark et al. 1999).

10.2 Terraces

Terraces are earth embankments constructed across the slope to intercept surface runoff, convey it to a stable outlet at a non-erosive velocity and shorten slope length. They thus perform similar functions to contour bunds. They differ from them by being designed to more stringent specifi-

Table 10.1 Design lengths and grades for terrace channels

Maximum length	Normal	250 m (sandy soils) to 400 m (clay soils)
	Absolute	400 m (sandy soils) to 450 m (clay soils)
Maximum grade	First 100 m	1 : 1000
	Second 100 m	1 : 500
	Third 100 m	1 : 330
	Fourth 100 m	1 : 250
	Constant grade	1 : 250
Ground slopes	Diversion terraces	Usable on slopes up to 7°; on steeper slopes the cost of construction is too great and the spacing too close to allow mechanized farming
	Retention terraces	Recommended only on slopes up to 4.5°
	Bench terraces	Recommended on slopes of 7–30°

Source: after Hudson (1981).

cations. Decisions are required on the spacing and length of the terraces, the location of outlets, the gradient and dimensions of the terrace channel and the layout of the terrace system (Tables 10.1 and 10.2).

Terraces can be classified into three main types: diversion, retention and bench (Table 10.3). The primary aim of diversion terraces is to intercept runoff and channel it across the slope to a suitable outlet. They therefore run at a slight grade, usually 1 : 250, to the contour. There are several varieties of diversion terrace. The Mangum terrace, formed by taking soil from both sides of the embankment, and the Nichols terrace, constructed by moving soil from the upslope side only, are broad-based, with the embankment and channel occupying a width of about 15 m. Narrow-based terraces are only 3–4 m wide and consequently cannot be cultivated. For cultivation to be possible, the banks should not exceed 14° slope if small machinery is used or 8.5° if large reaping machines are operated. Diversion terraces are not suitable for agricultural use on ground slopes greater than 7° because of the expense of construction and the close spacing that would be required. Closer spacings are feasible, however, on steeper slopes on road banks, mining spoil and along pipeline corridors.

Surprisingly, standard pipeline engineering practice is to grade the channel behind the terrace or berm at a slope of 9° so as to remove surface water as rapidly as possible without the channel overtopping (Marshall & Ruban 1983). Such a grade is far too steep, particularly since the risk of overtopping is extremely small. In the Tbilisi area of Georgia, sheet and gully erosion occurred on pipeline corridors on slopes of 18–26° as a result of installing channel terraces that were too steep and at spacings that were too wide, and a failure to extend them on to vegetated land beyond the right-of-way (Morgan & Hann 2003). This example shows that a poorly designed terrace system can actually exacerbate an erosion problem.

Retention terraces are used where it is necessary to conserve water by storing it on the hillside. They are therefore ungraded or level and generally designed with the capacity to store the runoff volume expected with a ten-year return period without overtopping. These terraces are normally recommended only for permeable soils on slopes of less than 4.5°.

Bench terraces consist of a series of alternating shelves and risers (Fig. 10.1) and are employed where steep slopes, up to 30°, need to be cultivated. The riser is vulnerable to erosion and should be protected by a vegetation cover or faced with stones or concrete. Unprotected risers can be the source of most of the erosion in terraced systems (Critchley & Bruijnzeel 1995). The basic bench

Table 10.2 Formulae for determining spacing of terraces

Approach

Many formulae have been developed for determining the difference in height between two successive terraces; this difference in height is known as the vertical interval (*VI*).

Theoretical or analytical formulae

For steady state conditions, the runoff (*Q*) at slope length (*x*) on a hillside can be expressed as:

$$Q = (R - i) \, x \cos\theta$$

where *R* is the rainfall intensity, *i* is the infiltration capacity of the soil and θ is the slope angle.

From the Manning equation of flow velocity:

$$Q = (R - i)x \cos\theta = \frac{r^{5/3}\sin^{1/2}\theta}{n}$$

The hydraulic radius (*r*) is expressed by:

$$r = \left(\frac{vn}{\sin^{1/2}\theta}\right)^{3/2}$$

Therefore:

$$(R - i)x \cos\theta = \left[\left(\frac{vn}{\sin^{1/2}\theta}\right)^{3/2}\right]^{5/3} \frac{\sin^{1/2}\theta}{n}$$

Rearranging for given values of *R* and *i*, say those for the 1-hour rainfall with a 10-year return period, and for a preselected value of *v*, say the maximum permissible velocity for the soil (*v$_c$*; Table 10.6), gives a slope distance x_{crit}, which can be used as the distance between the terraces downslope:

$$x_{crit} = \frac{v_c^{5/2}n^{3/2}}{(R - i)\sin^{3/4}\theta\cos\theta}$$

A value of *n* = 0.01 is recommended for bare soil.

For example, if the peak rainfall excess (*R* – *i*) on a sandy loam soil is 0.2 mm s^{-1}, and the selected value for *v$_c$* is 0.75 m s^{-1}, then for a slope of 3°:

$$x_{crit} = \frac{0.75^{5/2} \times 0.20^{3/2}}{0.0002 \times 0.1904 \times 0.9886}$$

$$x_{crit} = 22.55\,\text{m}$$

$$VI = x_{crit}\sin\theta = 1.18\,\text{m}$$

A similar approach was used by Mirtskhoulava (2001) to produce the formula:

$$x_{crit} = \frac{0.000034 v_c^{3.32}}{(R - i)m^{2.32}s^{1.16}n}$$

Table 10.2 *Continued*

where m is a coefficient describing the roughness related to the soil particles ($m = 1.0$ for silts, 1.1 for sands and sandy loams, 1.3 for loamy sands and clay loams and 1.5 for clays), and s = slope (mm⁻¹). This method tends to give wider terrace spacings on low-angled slopes and closer spacings on steep slopes. When the two methods were applied to the design of diverter berm spacings on pipeline rights-of-way near Tbilisi, Georgia, they gave very similar results over slopes ranging from 19 to 26° (Morgan et al. 2003).

Empirical formulae

Ramser method	$VI(m) = 0.305\left(\dfrac{S}{a} + 2\right)$	where $a = 3$ or 4, depending on the severity of erosion
United States Soil Conservation Service	$VI(m) = aS + b$	where a varies from 0.12 in the south to 0.24 in the north of the USA and b varies between 0.3 and 1.2 according to the erodibility of the soil
Zimbabwe	$VI(ft) = \dfrac{S+f}{2}$	where f varies from 3 to 6 according to the erodibility of the soil
South Africa	$VI(m) = \dfrac{S}{a} + b$	where a varies from 1.5 for low rainfall areas to 4 for high rainfall areas and b varies from 1 to 3 according to the erodibility of the soil
Algeria	$VI(m) = \dfrac{S}{10} + 2$	
Israel	$VI(m) = XS + Y$	where X varies from 0.25 to 0.3 according to the rainfall and Y is 1.5 or 2.0 according to the erodibility of the soil
Kenya	$VI(m) = \dfrac{0.3(S+2)}{4}$	
New South Wales, Australia	$HI(m) = KS^{-0.5}$	where K varies from 1.0 to 1.4 according to the erodibility of the soil
Marshall et al., Rocky Mountains, USA (used for spacing of diverter berms on pipeline rights of way	$HI(m) = 45$ $HI(m) = 30$ $HI(m) = 305/S$	slopes < 5% slopes 5–10% slopes > 10%

Bench terraces

Algeria/Morocco	$VI(m) = (260S)^{-0.3}$ $VI(m) = (64S)^{-0.5}$	for slopes of 10–25% for slopes > 25%
India	$VI(m) = 2(D - 0.15)$	where D is the depth of productive soil
Taiwan/Jamaica	$VI(m) = \dfrac{S.Wb}{100 - (S.U)}$	where Wb is the width of the shelf (m) and U is the slope of the riser (expressed as a ratio of horizontal distance to vertical rise and usually taken as 1.0 or 0.75)

Table 10.2 *Continued*

China	$VI(m) = \dfrac{Wb}{(\cos S - \cos \beta)}$	where β is the angle of slope of the riser (normally 70–75°)
Taiwan	$VI(m) = \dfrac{(Wb.S) + (0.1S - U)}{100 - (S.U)}$	for inward-sloping bench terraces
Fanya juu Kenya	$VI(m) = aS + b$	where $a = 0.075$ and $b = 0.6$

VI, vertical interval between the terraces; *HI*, horizontal interval between the terraces; *S*, slope (per cent).

Note: it is recommended that three or more of the formulae be used and that the design spacing be based on a consensus of the results.

Source: after Ramser (1945), Lakshmipathy and Narayanswamy (1956), Gichungwa (1970), Sheng (1972b), Bensalem (1977), Charman (1978), Chan (1981b), Fang et al. (1981), Hudson (1981), Marshall and Ruban (1983), Thomas and Biamah (1989), Mirtskhoulava (2001), Morgan et al. (2003).

Table 10.3 Types of terraces

Diversion terraces	Used to intercept overland flow on a hillside and channel it across slope to a suitable outlet, e.g. grass waterway or soak away to tile drain; built at slight downslope grade from contour.
Mangum type	Formed by taking soil from both sides of the embankment.
Nichols type	Formed by taking soil from upslope side of the embankment only.
Broad-based type	Bank and channel occupy width of 15 m.
Narrow-based type	Bank and channel occupy width of 3–4 m.
Retention terraces	Level terraces; used where water must be conserved by storage on the hillside.
Bench terraces	Alternating series of shelves and risers used to cultivate steep slopes. Riser often faced with stones or concrete. Various modifications to permit inward-sloping shelves for greater water storage or protection on very steep slopes or to allow cultivation of tree crops or market-garden crops.
Fanya juu terraces	Terraces formed by digging a ditch on the contour and throwing the soil on the upslope side to form a bank. If the soil is thrown downslope, it is called *fanya chini*.

terrace system can be modified according to the nature and value of the crops grown. Two kinds of system are used in Malaysia. Where tree crops are grown, the terraces are widely spaced, the shelves being wide enough for one row of plants, usually rubber or oil palm, and the long, relatively gentle riser banks being planted with grass or a ground creeper; this system is sometimes known as orchard terracing. With more valuable crops such as temperate vegetables grown in the highlands, the shelves are closely spaced and the steeply sloping risers frequently protected with

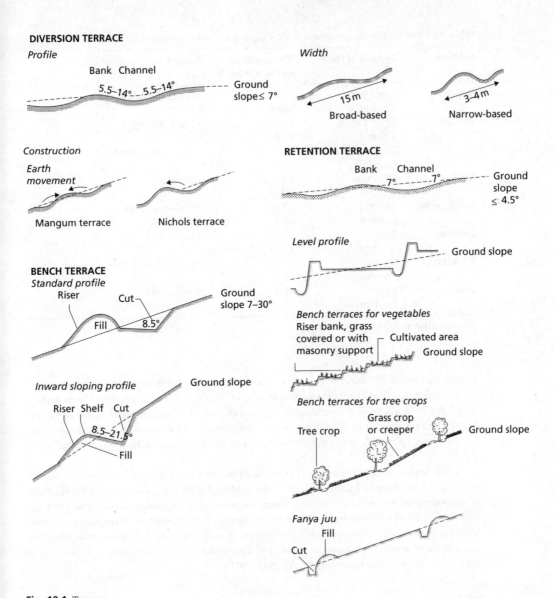

DIVERSION TERRACE

Profile

Bank   Channel

5.5–14°   5.5–14°   ---- Ground slope ≤ 7°

Width

15 m

Broad-based

3–4 m

Narrow-based

Construction

Earth movement

Mangum terrace   Nichols terrace

BENCH TERRACE

Standard profile

Riser   Cut   Ground slope 7–30°

Fill   8.5°

Inward sloping profile

Riser Shelf Cut   Ground slope

8.5–21.5°

Fill

RETENTION TERRACE

Bank   Channel   Ground slope ≤ 4.5°

7° ---- 7°

Level profile

Ground slope

Bench terraces for vegetables
Riser bank, grass covered or with masonry support   ┌ Cultivated area

Ground slope

Bench terraces for tree crops

Grass crop or creeper

Tree crop   Ground slope

Fanya juu

Fill

Cut

Fig. 10.1 Terraces.

masonry. Level bench terraces are used where water conservation is also a requirement, as in the loess areas of China (Fang et al. 1981). Bench terraces are unsuitable for shallow soils because their construction can expose infertile subsoil.

Fanya juu terraces are used in many parts of East Africa as an alternative to bench terraces. They consist of narrow shelves constructed by digging a ditch on the contour and throwing the soil upslope to form an embankment, which is later stabilized by planting grass (Thomas & Biamah 1989). During cultivation, vegetation and crop residues are spread over the shelves. Over time, redistribution of the soil within the inter-terrace area causes the inter-terrace slope to decline

Mechanical methods of erosion control

Table 10.4 Soil erosion rates on terraced land compared to similar unterraced land as a control

Location	Type of terrace	Mean annual soil loss (t ha⁻¹)	Source
Nepal	Level bench	20.0	Partap & Watson 1994
	Control	100.0	
Taiwan	Level bench	1.4	Liao & Wu 1987
	Control	11.6	
	Orchard terraces	0.01	Wu & Wang 1998
	Control	36.5	
Nigeria	Diversion	0.3	Lal 1982
	Control	3.3	
Sierra Leone	Bench	7.5	Millington 1984
	Control	48.0	
Burundi	Bench	5.0	Roose 1988
	Control	150.0	
Rwanda	Bench	12.3	Nyamulinda & Ngiruwonsanga 1992
	Control	72.0	
Jamaica	Bench	17.0	Sheng 1981
	Control	133.0	

Source: after Critchley et al. (2001).

in angle and bench-like features to develop. Since this decreases the storage area for runoff behind the embankment, maintenance is required to raise the height of the bank to prevent overtopping. Some degree of safety from overtopping is provided, however, because water flowing over the embankment is trapped by the ditch. Thomas and Biamah (1989) recommend the use of *fanya juu* on slopes up to 17°, although Hurni (1986) suggests that they can be used on slopes up to 26°.

Bench terraces have been used as a conservation measure for over 2000 years in China, Southeast Asia, Mediterranean Europe and the Andes. Although they can reduce erosion substantially compared to unterraced land, the steep slopes on which they are used means erosion rates can still remain above tolerable levels (Table 10.4). Their success depends on them being well constructed and, equally important, well maintained. In areas of rural depopulation or increased availability of alternative sources of income to agriculture, the labour is often not available to undertake the necessary repairs (Critchley et al. 2001).

10.3

Waterways

The purpose of waterways in a conservation system is to convey runoff at a non-erosive velocity to a suitable disposal point. A waterway must therefore be carefully designed. Normally its dimensions must provide sufficient capacity to convey the peak runoff from a storm with a ten-year return period. Three types of waterway can be incorporated in a complete surface water disposal system: diversion channels, terrace channels and grass waterways (Fig. 10.2). Diversions are placed upslope of areas of farmland to intercept water running off the slope above and divert it across the slope to a grass waterway. Terrace channels collect runoff from the inter-terrace areas and also

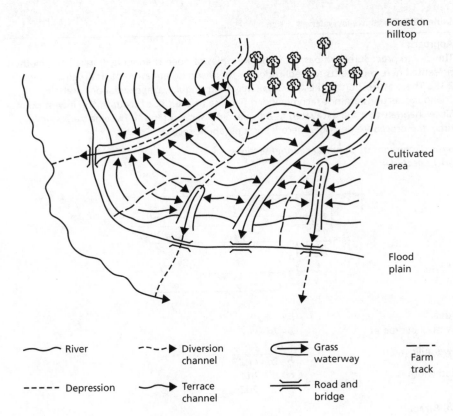

Forest on hilltop

Cultivated area

Flood plain

River	Diversion channel	Grass waterway	Farm track
Depression	Terrace channel	Road and bridge	

Fig. 10.2 Typical layout of waterways in a soil conservation scheme.

Table 10.5 Types of waterways used in soil conservation systems

Diversion ditches	Placed upslope of areas where protection is required to intercept water from upslope; built across slope at slight grade so as to convey the intercepted runoff to a suitable outlet.
Terrace channels	Placed upslope of terrace bank to collect runoff from inter-terraced area; built across slope at slight grade so as to convey the runoff to a suitable outlet.
Grass waterways	Used as the outlet for diversions and terrace channels; run downslope, at grade of the sloping surface; empty into river system or other outlet; located in natural depressions on hillside.

convey it across the slope to a grass waterway. Grass waterways are therefore designed to transport downslope the runoff from these sources to empty into the natural river system (Table 10.5); wherever possible, they are located in natural depressions or hollows. Design procedures for waterways are described in Table 10.6. A method of predicting the design runoff is given in Table 10.7.

Grass waterways are recommended for slopes up to 11°; on steeper slopes the channels should be lined with stones, acceptable for slopes up to 15°, or concrete. On hillsides with alternating

Table 10.6 Procedures for waterway design

Approach

The design procedures are based on the principles of open-channel hydraulics. The method presented here represents an application of the Manning equation of flow velocity (eqn 2.18). The cross-section of the waterway may be triangular, trapezoidal or parabolic. Triangular sections are not recommended because of the risk of scour at the lowest point. Since channels that are excavated as a trapezoidal section tend to become parabolic in time, the procedure described here is for a parabolic section.

Basic dimensions of common channel sections
(a) Trapezoidal

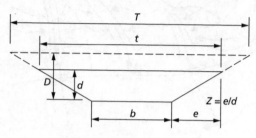

Area	$bd + Zd^2$
Wetted perimeter	$b + 2d\sqrt{1 + Z^2}$
Hydraulic radius	$\dfrac{bd + Zd^2}{b + 2d\sqrt{1 + Z^2}}$
Top width	$t = b + 2dZ$
	$T = b + 2dZ$

(b) Parabolic

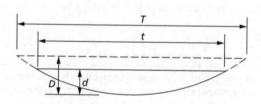

Area	$\dfrac{2}{3}td$
Wetted perimeter	$t + \dfrac{8d^2}{3t}$
Hydraulic radius	$\dfrac{t^2 d}{1.5t^2 + 4d^2} \approx \dfrac{2d}{3}$
Top width	$t = \dfrac{3a}{2d} \quad T = t\left(\dfrac{D}{d}\right)^{1/2}$

Table 10.6 *Continued*

The waterways used in soil conservation works on agricultural land are normally designed to convey the peak runoff expected with a 10-year return period without causing scour or fill. For other situations, such as construction sites, mining land or pipeline corridors, higher magnitude events may be chosen, particularly if there is a risk of destruction of assets or loss of life should the waterway overtop. The peak runoff can be estimated using the method shown in Table 10.7.

The design is based on the conditions expected to prevail two years after installation, i.e. after the grass vegetation lining has become established. Although the waterway is vulnerable to erosion in the interim period, providing a stable design for this time would result in overdesign for the rest of its life. It would also lead to an unnecessary reduction in the area that can be devoted to arable farming.

Designs are based on the concept of a maximum allowable or safe velocity of flow in the channel. Temple (1991) has shown that the influence of flow duration on the stability of the channel is small in the first 100 hours of flow unless the vegetation is destroyed or has a clumpy growth habit, in which case erosion of the bare areas can undermine the vegetation cover. Since, in these channels flow durations of 100 hours are unlikely, no allowance is made in the channel design for the duration of the flow. The design, therefore, is based on peak flows not exceeding the maximum permitted value.

Design problem
Design a parabolic grass waterway to convey a peak flow of $6\,m^3\,s^{-1}$ on a 1 per cent slope over an erodible sandy soil with Bermuda grass vegetation in a good stand cut to a height of 60 mm.

Procedure
Discharge (Q) $= 6\,m^3\,s^{-1}$ (given)
Slope (s) $= 0.01$ (given)
Velocity (v)
  select a maximum permissible velocity according to the proposed vegetation cover and local soil:
    $= 1.5\,m\,s^{-1}$

Maximum safe velocities ($m\,s^{-1}$) in channels based on covers expected after two seasons

Material	Bare	Medium grass cover	Very good grass cover
Very light silty sand	0.3	0.75	1.5
Light loose sand	0.5	0.9	1.5
Coarse sand	0.75	1.25	1.7
Sandy soil	0.75	1.5	2.0
Firm clay loam	1.0	1.7	2.3
Stiff clay or stiff gravelly soil	1.5	1.8	2.5
Geotextile mat	1.5	2.5	3.5
Coarse gravels	1.5	1.8	n/a
Shale, hardpan, soft rock	1.8	2.1	n/a
Hand-pitched stone	2.0	n/a	n/a
Hard cemented conglomerates	2.5	n/a	n/a
Rip-rap (D_{50}: 250–400 mm)	3.0	n/a	n/a
Box gabions, grouted stone	5.0	n/a	n/a
Concrete block systems	6.0	n/a	n/a

n/a, not applicable, since a medium or very good grass cover is unlikely to be obtained. Intermediate values may be selected.

Table 10.6 *Continued*

Roughness (*n*)

Select a suitable value according to the vegetation retardance class (*CI*); the value of *CI* can be estimated knowing the length of the plant stems (*m*) and the density of the stems per unit area (*M*); the latter can be estimated in turn from the grass type, a qualitative description of the stand and the percentage cover:

m = 0.06 (given)
M = 5380 stems m^{-2} (from table)
CI = $2.5(m\sqrt{M})^{1/3}$ (Temple 1982)
  = $2.5(0.06 \times \sqrt{5380})^{1/3}$
  = 6.75

Interpolating between values in table relating *n* to *CI*, select a value for:

n = 0.034

Properties of grass channel linings for good uniform stands*

Cover group	Estimated cover factor (*CF*)	Covers tested	Reference stem density (stems m^{-2})
Creeping grasses	0.90	Bermuda grass	5380
		Centipede grass	5380
Sod-forming grasses	0.87	Buffalo grass	4300
		Kentucky blue grass	3770
		Blue grama	3770
Bunch grasses	0.50	Weeping love grass	3770
		Yellow blue stem	2690
Legumes†	0.50	Alfalfa	5380
		Lespedeza sericea	3230
Annuals	0.50	Common Lespedeza	1610
		Sudan grass	538

* Multiply the stem densities by 1/3, 2/3, 1, 4/3 and 5/3 for poor, fair, good, very good and excellent covers respectively. The equivalent adjustment for *CF* remains a matter of engineering judgement until more data are obtained or a more analytical model is developed.
† For the legumes tested, the effective stem count for resistance (given) is approximately five times the actual count very close to the bed. Similar adjustment may be needed for other unusually large-stemmed and/or woody vegetation.

Values of Manning's n for vegetated channels

CI	Description	n
10.0	very long (over 600 mm) dense grass	0.06–0.20
7.6	long (250–500 mm) grass	0.04–0.15
5.6	medium (150–250 mm) grass	0.03–0.08
4.4	short (50–150 mm) grass	0.03–0.06
2.9	very short (less than 50 mm) grass	0.02–0.04

Table 10.6 *Continued*

Calculate the hydraulic radius (r) from the Manning equation:

$$r = \left(\frac{vn}{s^{0.5}}\right)^{1.5}$$

$$r = \left(\frac{1.5 \times 0.034}{0.01^{0.5}}\right)^{1.5}$$

$$r = 0.364 \, \text{m}$$

Calculate the required cross-sectional area (A) of the channel:

$$A = \frac{Q}{v}$$

$$A = \frac{6}{1.5}$$

$$A = 4 \, \text{m}^2$$

Calculate the design depth, which for a parabolic section can be approximated by:

$$d = 1.5r$$
$$d = 1.5 \times 0.364$$
$$d = 0.55 \, \text{m}$$

Calculate the top width, which for a parabolic section is expressed by:

$$t = \frac{A}{0.67d}$$

$$t = \frac{4}{0.67 \times 0.55}$$

$$t = 10.86 \, \text{m}$$

Check that the capacity given by the design criteria is adequate. For a parabolic section:

$$Q = Av = 0.67tdv$$
$$Q = 0.67 \times 10.86 \times 0.55 \times 1.5$$
$$Q = 6 \, \text{m}^3 \text{s}^{-1}, \text{ which is adequate}$$

Add 20 per cent free board to the design depth:

$$d = 0.55 + 0.11 = 0.66 \, \text{m}$$

Final design criteria:

depth   = 0.66 m
top width   = 10.86 m

Notes
1 This procedure can be used for all waterways in a terrace and waterway system. With terrace channels and diversion channels, however, the slope is not predetermined by the ground slope but should be selected from guidelines given in Table 10.1. Terrace channels are usually unvegetated except with broad-based terraces, where they may be cropped as part of the inter-terrace area. A value of $n = 0.02$ should be used for bare soil.

Table 10.6 *Continued*

2 When the above procedure is applied to small discharges, the design depths are sometimes greater than the design widths; since the channel then resembles a gully, it is undesirable. Moreover, the dimensions are too small for the channel to be constructed easily. To avoid these problems, a minimum size of 2.0 m wide and 0.5 m deep is recommended for terrace channels.

3 With very large discharges, the procedure gives channel widths which are very large and depths that are very shallow. Usually, it is not possible simply to increase the depth and decrease the width while retaining the required cross-sectional area, since this will increase the flow velocity. The only solution is to increase the resistance of the channel lining, which may mean using stones, rip-rap or concrete instead of grass.

4 With long grass waterways it is often necessary to allow them to cross farm roads. Where the roads are rarely used, it may be sufficient to allow the road to cross the waterway as a 'drift', i.e. the road surface goes down one bank of the waterway, across the channel floor and up the other bank; at this point the channel is effectively unlined, being formed on compacted bare soil for the width of the road. The alternative is for the waterway to pass underneath the road in a culvert. A rough estimate of the size of the pipe required is obtained by dividing the cross-sectional area of the grass waterway by 4. The pipe should not be larger than 0.5 m in diameter. If this is insufficient to give the required cross-sectional area, two or more smaller pipes should be used; the diameter of the pipes must add up to the total diameter needed. The pipes should be placed at least 0.5 m below the road surface, spaced one pipe diameter apart, on a bed of pea gravel and covered with compacted backfill. The length of the pipe should be equal to twice the pipe diameter plus twice the length of bank slope between the edge of the road and floor of the grass waterway plus the road width. Upstream and downstream of the culvert, the channel should be protected with rip-rap or stones for a distance of five times the total diameter of the pipes.

Source: after Schwab et al. (1966), Hudson (1981), Temple (1982), Hewlett et al. (1987), Escarameia (1998), Crowley (2003).

Table 10.7 Estimating the volume of peak runoff

Approach
Several methods have been developed for estimating the volume of peak runoff from small areas where no measured data exist. These include the rational formula and the United States Soil Conservation Service Curve Number. Both these require meteorological information that is not always readily obtainable. The procedure described here can be operated with the minimum of data. It was developed by Hudson (1981) for use in tropical Africa.

Problem
Estimate the volume of peak runoff for a roughly circular catchment of 50 ha of which (A) 10 ha comprises steeply sloping land with shallow rocky soils, (B) 15 ha is cultivated land with loamy soils and slopes of 6–9° and (C) 25 ha is flat land devoted to pasture on clay soils.

Procedure
The runoff generating characteristics for any catchment can be represented by an area-weighted score based on the vegetation, soil and slope conditions. Typical scores for each of these factors are shown in the table below.

Table 10.7 *Continued*

Catchment characteristics

Cover		Soil type and drainage		Slope	
Heavy grass or forest	10	Deep, well drained soils, sands	10	Very flat to gentle (0–3°)	5
Scrub or medium grass	15	Deep, moderately pervious soil, silts	20	Moderate (3–6°)	10
Cultivated lands	20	Soils of fair permeability and depth, loams	25	Rolling (6–9°)	15
Bare or eroded	25	Shallow soils with impeded drainage	30	Hilly or steep	20
		Medium heavy clays or rocky surface	40	Mountainous	25
		Impervious surfaces and waterlogged soils	50		

The value of the catchment characteristic (CC) for the problem catchment is calculated as follows:

Region	Factor values			Percentage area weighting	Total
	Cover	Soil	Slope		
A	25	+40	+20	×0.20	17.0
B	20	+25	+10	×0.30	18.0
C	10	+40	+5	×0.50	27.5
Catchment characteristic (CC)				=	62.5

From the table, read peak runoff with a 10-year return period for area (A) = 50 ha and CC = 62.5.

Interpolating gives peak runoff = $9.25\,\mathrm{m^3\,s^{-1}}$.

Peak runoff as a function of catchment characteristics and area

A\CC	25	30	35	40	45	50	55	60	65	70	75	80
5	0.2	0.3	0.4	0.5	0.7	0.9	1.1	1.3	1.5	1.7	1.9	2.1
10	0.3	0.5	0.7	0.9	1.1	1.4	1.7	2.0	2.4	2.8	3.2	3.7
15	0.5	0.8	1.1	1.4	1.7	2.0	2.4	2.9	3.4	4.0	4.6	5.2
20	0.6	1.0	1.4	1.8	2.2	2.7	3.2	3.8	4.4	5.1	5.8	6.5
30	0.8	1.3	1.8	2.3	2.9	3.6	4.4	5.3	6.3	7.3	8.4	9.5
40	1.1	1.5	2.1	2.8	3.5	4.5	5.5	6.6	7.8	9.1	10.5	12.3
50	1.2	1.8	2.5	3.5	4.6	5.8	7.1	8.5	10.0	11.6	13.3	15.1
75	1.6	2.4	3.6	4.9	6.3	8.0	9.9	11.9	14.0	16.4	18.9	21.7
100	1.8	3.2	4.7	6.4	8.3	10.4	12.7	15.4	18.2	21.2	24.5	28.0
150	2.1	4.1	6.3	8.8	11.6	14.7	18.2	21.8	25.6	29.9	35.0	40.6
200	2.8	5.5	8.4	11.7	15.3	19.1	23.3	28.0	33.1	38.5	45.0	52.5
250	3.5	6.5	9.7	13.2	17.2	21.7	27.0	32.9	39.6	46.9	55.0	63.7
300	4.2	7.0	10.5	14.7	19.6	25.2	31.5	38.5	46.2	54.6	63.7	73.5
350	4.9	8.4	12.6	17.2	23.2	30.2	37.8	46.3	53.8	62.5	71.5	81.0
400	5.6	10.0	14.4	19.4	25.6	33.6	42.2	51.0	60.0	69.3	79.5	90.0
450	6.3	10.5	15.5	21.5	28.5	36.5	45.5	55.5	65.5	76.0	86.5	97.5
500	7.0	11.0	17.0	23.5	31.0	40.5	51.0	62.0	73.0	84.0	95.0	106.5

Table 10.7 *Continued*

A is the area of the catchment in hectares, CC is the catchment characteristics from the previous table, and the runoff ($m^3 s^{-1}$) is for a 10-year return period.

Notes:

Rainfall intensity:	tropical (high)	multiply by 1.0
	temperate (low)	multiply by 0.75
Catchment shape:	long, narrow	multiply by 0.8
	square, circular	multiply by 1.0
	broad, short	multiply by 1.25
Return period:	2 years	multiply by 0.9
	5 years	multiply by 0.95
	10 years	multiply by 1.0
	25 years	multiply by 1.25
	50 years	multiply by 1.5

gentle and steep sections, a grass waterway with drop structures on the steeper slopes should be used. The selection of grasses for planting should take account of the local soil and climatic environment and the need to establish dense cover very rapidly. Commonly used grasses are *Cynodon dactylon* (Bermuda grass), *Poa pratensis* (Kentucky bluegrass), *Bromis inermis* (smooth bromegrass) and *Pennisetum purpureum* (Napier grass). It is recommended, however, to seek local agronomic or ecological advice before making the final selection.

Grass waterways can be replaced in the water disposal system by tile drains. Diversion and terrace channels are graded to a soak-away, normally located in a natural depression, which provides the intake to the drain. The tile system is designed to remove surface water over a period not exceeding 48 hours so that crop damage does not occur. Soil loss from tile-outlet terraces is much reduced because less than 5 per cent of the sediment delivered to the soak-aways passes into the drainage system (Laflen et al. 1972). The tile outlet consists of four parts: the inlet tube, the orifice plate, the conducting pipe and the outlet. The inlet tube is usually made of plastic and rises from a pipe below ground to a height 70–100 mm above the adjacent terrace bank; the tube has holes or slots at regular intervals above ground level and a removable cap to prevent the entry of debris and allow access. The orifice plate is positioned at the base of the tube, where it connects with the conducting pipe; it regulates the downward flow of water. The conducting pipe, also of plastic, carries water from one or more inlet tubes to the outlet, which is normally in a natural waterway. The terrace bank adjacent to the inlet must be level to reduce the risk of overtopping by ponded water. Although tile outlets are commonly used in the USA because they take up less cropland, they are more expensive. On a worldwide basis, grass waterways remain the cheapest and most effective form of terrace outlet.

The main reason why terrace and waterway systems reduce erosion is the way they manage the runoff. The terraces divide the hillside into inter-terrace areas, which should be small enough in area to generate only small quantities of runoff. The grass waterway reduces the speed of flow because of the retardance effects of the vegetation. The arrangement of the waterway network gives a high tributary (diversion and terrace channels) to main channel (grass waterway) ratio and a catchment that is elongate in shape rather than square or circular; both attributes contribute to a reduction in peak flow.

A terrace and waterway system must be designed to give the most efficient layout possible in terms of farming operations. This can be achieved by following a systematic design procedure and then making adjustments within certain tolerance limits to take account of local topography. Once a system has been constructed, regular maintenance is required to prevent it from deteriorating. This includes: cutting the grass in the waterways to maintain it at the height on which the channel design is based; regular applications of fertilizers to promote grass growth; closure of the waterways to animals and vehicles, especially when the soil is wet and damage could occur; and regular inspection and repair of breaks in the terrace banks. Although a terrace and waterway system will fail with the occurrence of a storm of much higher magnitude than that for which it is designed, by far the most common cause of failure is inadequate maintenance. Once a failure in vegetation cover occurs, the reduction in flow resistance is immediate and it may take two or more years for natural recovery to take effect (Temple & Alspach 1992).

10.4

Temporary measures

Temporary measures for erosion control are required on construction sites and also, until a vegetation cover has been established, on road banks and pipeline corridors. Generally, when the catchment area is less than 2 ha, sediment traps such as silt fences are sufficient, but for larger areas it is usually necessary to install a sedimentation pond or basin.

10.4.1 Silt fences

Silt fences are synthetic geotextile meshes attached to vertical posts driven into the ground to form a fence. They are placed across the slope approximately perpendicular to the direction of runoff. Their purpose is not to prevent erosion *per se* but to trap sediment and prevent it leaving the slope. It is important that the fence is high enough and anchored sufficiently well to support the hydraulic and silting loads. The maximum height of the fence (H_{max}; m) is defined by (Rankilor 1989):

$$H_{max} \geq 1.4 \sqrt{\frac{PtL}{s}} \qquad (10.1)$$

where P is the precipitation rate of the design storm ($m\,h^{-1}$), t is the duration of the storm, L is the slope length between successive fences (m) and s is slope ($m\,m^{-1}$). In order to support the fence, posts should be driven into the ground to a depth (G) that is at least three times H_{max} and the fence should be anchored in a trench that is at least $0.5\,H_{max}$. Posts should be spaced at a distance no greater than $2G$. Failure to anchor the fence and install the posts at sufficient depth and with suitable spacing is the main cause of failure of silt fences. Usually the limiting factor is the depth to which the posts can be driven. If the design depth is not feasible, the fences must be placed closer together so as to reduce L and, thereby, reduce H_{max}.

10.4.2 Burlap rolls, straw bales

Burlap rolls, also known as bolsters, are tubes of hessian or jute filled with soil or stones that can be placed across the slope on the contour. Generally, they are not successful. They need to be placed in a shallow trench to prevent runoff cutting channels underneath them because their cylindrical shape results in very limited contact with the soil. When used on road banks in Nepal, the hessian rotted rapidly, causing the structure to fail. On pipeline corridors in Georgia, they were quickly destroyed by cattle. Straw bales arranged as a barrier across the slope are similarly fragile and since 1992 the United States Environmental Protection Agency has not recognized bale barriers as an appropriate technique for reducing sediment in runoff.

10.4.3 Sedimentation ponds

Sedimentation ponds or basins are used to trap suspended sediment particles contained in runoff and prevent them leaving a site. Since it is almost impossible to contain all the runoff that may be discharged from a construction site, the aim is to capture the suspended sediment and then allow the clearer runoff to overflow through a drain to a safe outlet. The required volume of a sedimentation pond (V_{min}; m^3) is calculated as the larger of the following values (Fifield 1999):

$$V_{min} \geq 0.67 SA_{min}$$

or $V_{min} \geq$ runoff from two-year, 24-h storm up to $252\,\mathrm{m^3\,ha^{-1}}$ \hfill (10.2)

where SA_{min} is the minimum surface area of the pond, defined as:

$$SA_{min} = 120 \frac{Q_{out}}{v_s}$$ \hfill (10.3)

where Q_{out} = outflow from the pond (m^3 s^{-1}) and v_s is settling velocity of the particles (cm s^{-1}). The average depth of the pond must be ≥ 0.67 m with a minimum outlet depth of 0.61 m.

10.5

Stabilization structures

Stabilization structures are used to control erosion on steep slopes such as gully sidewalls, embankments and cuttings. Bioengineering techniques, like brush layering and wattling, use live cuttings of quick-rooting plant species, usually willow. They give an immediate reinforcement of the soil, as well as providing the basis for long-term slope stability as the cuttings take root and the vegetation grows.

With brush layering, live willow stakes are laid in lines across the slope on or close to the contour at 2 m intervals on slopes less than 30° and at 1 m interval on slopes of 30–45°; the method is not recommended for steeper slopes. The lines to be planted are marked out on the slope, with the first line 0.5 m from the bottom of the slope. Starting from the bottom, a small terrace, 400 mm wide, is cut with a 20 per cent fall into the back slope. Cuttings are placed on the terrace at 50 mm intervals, butt ends into the slope with at least one bud and up to one-third of the cutting protruding beyond the edge of the terrace. A 20 mm thick layer of soil is placed on the cuttings and a second layer of cuttings placed on this, staggered with the first layer. The terrace is then backfilled with the material excavated when forming the next terrace upslope and lightly compacted using foot pressure (Howell 1999).

Wattles, also termed fascines, are cigar-shaped bundles of six to eight live cuttings, each 200–250 mm in diameter, arranged with butt ends alternating, and tied at 300–400 mm intervals

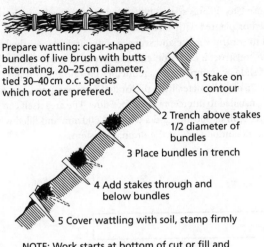

Prepare wattling: cigar-shaped
bundles of live brush with butts
alternating, 20–25 cm diameter,
tied 30–40 cm o.c. Species
which root are prefered.

1 Stake on
contour

2 Trench above stakes
1/2 diameter of
bundles

3 Place bundles in trench

4 Add stakes through and
below bundles

5 Cover wattling with soil, stamp firmly

NOTE: Work starts at bottom of cut or fill and
proceeds from step 1 through step 5

Fig. 10.3 Installation of wattles for slope stabilization (after Gray & Leiser 1982).

(Fig. 10.3). Species that root easily are used, such as *Salix*, *Leucaena*, *Baccharis* and *Tamarix*. The fascines are placed in shallow trenches on the contour, 0.3 m wide, up to 0.5 m deep and spaced at 4 m intervals on slopes less than 30° and at 2 m intervals on slopes of 30–45° (Howell 1999). The lines where the fascines will be installed are marked out on the slope. Then, starting from the bottom of the slope, 5 m lengths of trench, 100 mm deep and 200 mm wide, are dug at any one time. The fascines are placed in the trench, covered with soil so that about 10 per cent of the fascine is exposed and pegged to the slope at 0.5–0.8 m intervals using wooden stakes, 0.6 m long, driven vertically into the ground (Schiechtl & Stern 1996). Grasses and shrubs can be planted between the wattles.

The protection of steeper slopes usually requires the construction of a retaining wall at the base. This can be achieved using gabions. These are rectangular steel wire-mesh baskets, packed tightly with stones. They have the advantage over concrete structures of allowing seepage of water through the facing and of deforming by bending without loss of structural efficiency rather than by cracking. Gabions are supplied flat and then folded into their rectangular shape on site. They are placed in position and anchored before being filled with stones 125–200 mm in diameter. The gabion is filled to one-third of its depth, after which two connecting wires or braces are inserted front to back to prevent bulging of the wire basket on further filling. The bracing is repeated when the basket is two-thirds full. The gabion is slightly overfilled to allow for settlement, and the hinged lid is closed and wired to the sides. The simplest structure consists of one tier of gabions, 1 m high. A second tier can be positioned on top of the first, set back about 0.5 m. The addition of further tiers, however, requires bracing the structure against overturning and should not be attempted without civil engineering advice. Similar advice should be sought for other types of retaining walls. Provision of sound, dry foundations is important to the stability of the gabion wall and it may be necessary to install drainage from the lowest point of the foundations (Howell

1999). Soil can be used to fill the spaces between the stones within the gabion, allowing the structure to be seeded or planted with shrubs and trees.

Drainage is used to control surface and seepage water and prevent the build-up of soil water. The principles are to construct a diversion channel to run at grade across the hillslope, upslope of any area at risk of slide or slump, thereby reducing the amount of water coming into the area. Subsurface drains of 75 mm diameter PVC perforated pipe, wrapped in a filter cloth to prevent blockage, can also be installed to intercept subsurface flow. The area itself can be drained by rubble drains, excavated to a depth of 300 mm and a width of 500 mm and filled with rocks. Safe outlets must be provided for all components of the drainage system.

10.6

Geotextiles

Geotextile products commercially available for use in erosion control range from open-weave textile meshes, made from polypropylene, coir or jute, to blankets, containing natural or synthetic fibres that are woven, glued or structurally bound with nets or meshes. They are supplied in rolls, unrolled over the hillslope from the top and anchored with large pins. The natural fibre types are biodegradable and are designed to be laid over the surface of the slope to give temporary protection against erosion until a vegetation cover is established. The artificial fibre types, which include geowebs and geogrids as well as mats, are buried and designed to give permanent protection to a slope by reinforcing the soil; once a vegetation cover is established, the plant roots and the fibre act together to increase the cohesion of the soil and the fibre provides a back-up resistance should the vegetation fail.

Surface-laid mats of natural fibres are the most effective in controlling soil detachment by raindrop impact because they provide good surface cover, high water absorption, thick fibres to intercept splashed particles from their point of ejection and a rough surface in which water is ponded, thereby further inhibiting the splash action on the soil (Rickson 1995). In contrast, buried mats of artificial fibres do not effectively control the splash process. Their percentage cover and water-absorbing capacities are low and problems with backfilling them mean that they tend to be filled with highly erodible unconsolidated material. Despite the ability of natural fibres to hold water, no significant differences in runoff production were observed in laboratory experiments between an unprotected slope and slopes protected by natural or artificial fibres. However, erosion resulting from the runoff was significantly lower for slopes protected by surface-laid jute mats because of the higher roughness imparted to the flow. Although surface-laid mats of coir, wood-chip and artificial fibres also reduced soil loss by runoff, they were less effective than the jute because they did not adhere to the soil surface as well. The ability of the mat to drape naturally over the surface is important, because otherwise surface runoff will pass underneath it and erosion will occur beneath the mat.

Table 10.8 gives the C-factor values (section 6.2.1) for a range of erosion mats. It should be stressed that these relate to conditions immediately after installation of the geotextile materials. Over time, the natural fibre mats will become less effective as they biodegrade, whereas the performance of the artificial fibre mats is likely to remain constant. However, by the time degradation of the mat takes place, the vegetation cover should have established sufficiently to protect the slope, particularly since the mat modifies the soil's microclimate and improves the conditions for plant growth (Rickson 1995).

Table 10.8 *C-factor values for geotextile erosion mats*

Product	C-factor value
Woven jute mesh (500 g m⁻²)	0.01–0.07
Polypropylene mesh sewn together with wood wool (poplar, pine, aspen) (350–500 g m⁻²)	0.01–0.04
Mesh of nylon polyamide filaments (260 g m⁻²)	0.20–0.25
Multiple layered polyethylene mat (450 g m⁻²)	0.20–0.25
Polyethylene expanding cellular grid (1780 g m⁻²)	0.50–0.75
Woven polypropylene mat (88 g m⁻²)	0.06–0.10
Woven coir mat (400–900 g m⁻²)	0.02–0.12
Polymeric netting sewn together with straw and coconut fibres (270–550 g m⁻²)	0.01–0.08
Wood fibre hydromulch bonded to photodegradable netting (270 g m⁻²)	0.01
Non-woven UV-stabilized blanket and randomly oriented thermally welded PVC mono-filaments (1260 g m⁻²)	0.20–0.30

Values in g m⁻² denote mass per area of the material.
Source: after Fifield et al. (1989), Fifield and Malnor (1990), Cazzuffi et al. (1991), Krenistky and Carroll (1994), Rickson (1995), Sutherland and Ziegler (1996), Sutherland et al. (1997), Pazos and Gasca (1998).

10.7 Brush matting

Brush mats can be used as an alternative to geotextiles to provide an immediate cover to the slope and prevent surface erosion. Since they also provide the basis for the long-term vegetation cover, they are only suitable for sites where the species used is appropriate as the final land cover. With brush matting, cuttings of stems and branches of live willow are placed on the slope, butt ends downslope, at 20–50 stems per running metre, to give a minimum of 80 per cent ground cover (Schiechtl & Stern 1996). The thicker ends of the branches are covered with soil to aid rooting and then fixed with stones or pegs. The whole is covered with soil and the brush fixed to the slope using stakes or wires.

10.8 Gully control

Stabilization structures play an important role in gully reclamation and gully erosion control. Small dams, usually 0.4–2.0 m in height, made from locally available materials such as earth, wooden planks, brushwood or loose rock, are built across gullies to trap sediment and thereby reduce channel depth and slope. The structures have a high risk of failure but provide temporary stability and are therefore used in association with agronomic treatment of the surrounding land where grasses, trees and shrubs are planted. If the agronomic measures successfully hold the soil and reduce runoff, the dams can be allowed to fall into disrepair. Even though they are temporary, the dams have to be carefully designed. They must be provided with a spillway to deal with overtopping during high flows and installed at a spacing appropriate to the slope of the channel. Dam spacing should be based on the 'head-to-toe' rule, whereby the top of a downstream dam is level with the lowest elevation of the upstream dam.

The spacing of the dams can be determined from the formula of Heede (1976):

$$\text{spacing} = \frac{HE}{K \tan\theta \cos\theta} \tag{10.4}$$

where HE is the dam height, θ is the slope angle of the gully floor and K is a constant equal to 0.3 for $\tan\theta \leq 0.2$ and 0.5 for $\tan\theta > 0.2$. The dam height is measured from the crest of the spillway to the gully floor. Based on the costs of construction, loose-rock dams are only economical for heights up to 0.45 m; single-fence dams are the most economical for dam heights of 0.45–0.75 m; and double-fence dams for heights of 0.75–1.7 m (Heede & Mufich 1973). The gully depths for which these dam heights are recommended are less than 1.2, 1.2–1.5 and 1.5–2.1 m respectively.

Keying a check dam into the sides and floor of the gully is necessary for stability. This entails digging a trench, usually 0.6 m deep and wide, across the channel, but if the channel walls are deeply cracked and fissured, the trench should be increased in depth to 1.2 or even 1.8 m. Aprons must be installed on the gully floor downstream of the check dam to prevent flows from under-cutting the structure. Where the slope of the gully is less than 8.5°, the length of the apron should be 1.5 times the height of the structure; for steeper slopes it should be 1.75 times its height. At the downstream end of the apron, a loose rock sill about 0.15 m high should be built to create a pool to help to absorb the energy of the water falling over the spillway. The spillway should be designed to convey peak flows with a given return period, usually 25 years. It is recommended that the spillway be trapezoidal in cross-section, with a bottom length (L) that is equal to the bottom width of the gully. A longer spillway is not desirable because water flowing over the spill-way will strike the gully sides, where protection against erosion is less. The depth of the spillway (D) is given by the equation:

$$D = \left(\frac{Q}{1.65L}\right)^{2/3} \tag{10.5}$$

and, assuming the spillway sides are sloped at 1:1, the top length (L_t) is obtained from (Heede 1976):

$$L_t = L + D \tag{10.6}$$

This approach assumes that the spillways are approximate to a broad-crested weir.

Construction of a loose-rock dam (Fig. 10.4) begins by sloping back the tops of the banks. A trench is then dug across the floor of the gully and into the banks into which the large rocks are placed to form the toe of the structure. The dam is built upwards from the toe, using flatter rocks on the downstream face. Rocks smaller than 100 mm in diameter should not be used because they will be quickly washed out. A dam made of large rocks will leave large voids in the structure through which water jets may flow, weakening the dam. These jets will also carry sediment through the dam instead of allowing it to accumulate on the upstream side. To avoid these effects, the dam should be made with a graded rock structure. An effective composition is 25 per cent of

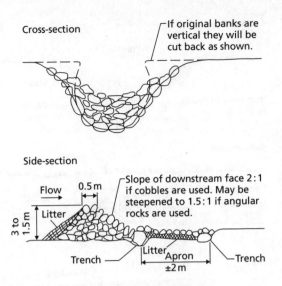

Fig. 10.4 Construction of a loose-rock dam (after Gray & Leiser 1982).

the rocks between 100 and 140 mm in diameter, 20 per cent between 150 and 190 mm, 25 per cent between 200 and 300 mm and 30 per cent between 310 and 450 mm (Heede 1976). The top of the dam should be shaped so that the central flow line is 150–450 mm lower than the sides. A second trench should be made to mark the downstream end of the apron and filled with heavy rocks. A 100 mm thick layer of litter, such as leaves, straw or fine twigs, is laid on the floor of the apron and covered with a solid pavement of rock. A thick layer of litter is also placed on the upstream face of the dam.

When a single-fence brush dam is being built, the gully banks are first sloped back and stout posts are then driven into the floor and banks of the gully to a depth of about 1 m below the surface and about 0.5 m apart (Fig. 10.5); willow is the recommended material. A 150 mm thick layer of litter is placed on the floor of the gully between the posts, extending upstream to the proposed base of the dam and downstream to the end of the apron. Green tree branches or brush are laid on the top of the litter, the longer ones at the bottom, with butt ends upstream. Usually the gully is filled with brush, which is trampled to compress it into a compact mass. Cross poles are fixed on the upstream side of the posts and the brush is tied to the structure with galvanized wire. A layer of litter is placed on the upstream face of the dam and packed into the openings between the butt ends of the brush.

For the double-fence brush dam, the gully banks are sloped back and two rows of stout posts are erected. A 150 mm thick litter layer is placed on the floor of the gully, again extending upstream to the proposed base of the dam and downstream to the end of the apron (Fig. 10.6). A 0.3 m layer of brush is positioned on the apron and attached to the lower row of posts. A row of stakes is driven through the middle of the apron into the gully floor and the brush tied to it to form a dense mat. The space between the two rows of posts is filled with brush laid across the gully; this is compressed tightly and held in position with wire. Litter is placed on the upstream face of the

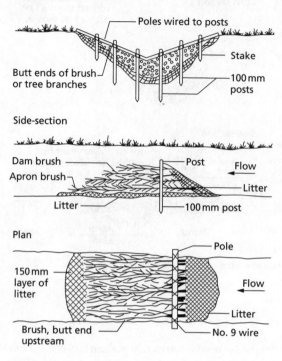

Fig. 10.5 Construction of a single-fence brush dam (after Gray & Leiser 1982).

dam. For steep slopes, more permanent material than brushwood should be used. Figure 10.7 shows a double-wicker fence dam designed by the author and Dr Michael Hann for installation in gullies along pipeline rights-of-way in Georgia.

More permanent structures are sometimes required on large gullies to control the overfall of water on the headwall. These are designed to deal safely with the peak runoff with a ten-year return period. They must therefore dissipate the energy of the flow in a manner that protects both the structure and the channel downstream. The structures comprise three components: an inlet, a conduit and an outlet. Various types of each component are outlined in Schwab et al. (1966). Where the drop is less than 3 m, the structure should incorporate a drop spillway. For drops between 3 and 6 m a chute is used and for greater drops a pipe spillway is required. These structures are expensive and, since they are built in adverse conditions with unstable soils subject to extreme fluctuations in moisture, their failure rate is also high. Their foundations may be undermined by animals and the structure may be circumvented if the gully cuts a new channel in the next major flood. Thus they cannot be generally recommended as an economic investment. If gully erosion is severe enough to require them, a cheaper alternative is to take the land out of use and allow it to revegetate naturally or by reseeding. Their greatest value is where agricultural land in flatter areas needs to be protected from channel erosion and where water needs to be conserved. Advice should be sought from civil engineers on the design and construction of the structures.

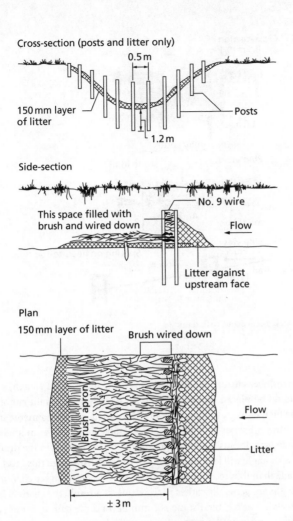

Cross-section (posts and litter only)

0.5 m

150 mm layer
of litter

Posts

1.2 m

Side-section

No. 9 wire

This space filled with
brush and wired down

Flow

Litter against
upstream face

Plan

150 mm layer of litter

Brush wired down

Brush apron

Flow

Litter

±3 m

Fig. 10.6 Construction of a double-fence brush dam (after Gray & Leiser 1982).

10.9

Footpaths

Stabilization structures are also used in the construction and maintenance of footpaths, especially in recreational areas on sloping land. The path itself must be constructed with either a camber or a cross-fall (Fig. 10.8) to shed runoff to a suitable outlet. On gently sloping land with porous soil on the slope above, the path can be outward-sloping and the runoff allowed to discharge on to the vegetated hillside below but, in other situations, it is preferable for the path to be inward-sloping with a fall of about 1:12 to a side drain that can collect runoff from both the path and the land above. The side drain should be graded across the slope to a safe outlet, either a vegetated waterway or a soak-away area. Culverts should be used to take water beneath the path. Where

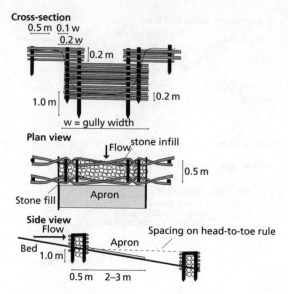

Fig. 10.7 Double wicker-fence and rock dam.

the path runs downslope, cut-off drains formed by making a small trench, reinforced with a wooden plank, should be placed at regular intervals to collect and drain runoff from the path. The plank must be high enough to divert the flow but small enough to merge into the profile of the path and not form a barrier to walking. The cut-offs should have an angle of 30–45° (Fig. 10.8). A shallower angle will cause water to pond, leading to siltation in the drain and the risk of overtopping. A steeper angle will lead to scouring of the path. Although this gives the cut-off drain a much steeper grade than that recommended above for diversion terrace channels, this is acceptable because footpaths are generally rather narrow and, as a result, the channel lengths are rather short. The cut-off should also extend some 300 mm beyond the path to avoid runoff by-passing the structure. Bends or turns in the path – for example, hair-pins on a steep slope – are vulnerable areas for damage and erosion. They should be made at zero slope by locally increasing the steepness of the path above and below the turn. Alternatively, but more expensively, the turn can be made using log or stone steps. Logs and stones may also be used to form revetments along the side of the path.

10.10
Windbreaks

Windbreaks are placed at right angles to erosive winds to reduce wind velocity and, by spacing them at regular intervals, break up the length of open wind blow. Windbreaks may be inert structures, such as stone walls, slat and brush fences and cloth screens, or living vegetation. Living windbreaks are known as shelterbelts. In addition to reducing wind speed, shelterbelts result in lower evapotranspiration, higher soil temperatures in winter and lower in summer, and higher soil moisture; in many instances, these effects can lead to increases in crop yield.

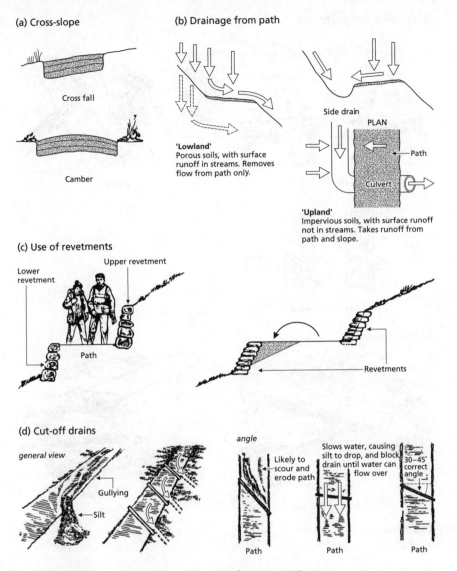

(a) Cross-slope

Cross fall

Camber

(b) Drainage from path

Side drain

PLAN

Path

Culvert

'Lowland'
Porous soils, with surface
runoff in streams. Removes
flow from path only.

'Upland'
Impervious soils, with surface runoff
not in streams. Takes runoff from
path and slope.

(c) Use of revetments

Upper revetment

Lower
revetment

Path

Revetments

(d) Cut-off drains

general view

Gullying

Silt

angle

Likely to
scour and
erode path

Slows water, causing
silt to drop, and block
drain until water can
flow over

30–45°
correct
angle

Path

Path

Path

Fig. 10.8 Measures for erosion control on footpaths (after Agate 1983).

A shelterbelt is designed so that it rises abruptly on the windward side and provides both a barrier and a filter to wind movement. A complete belt can vary from a single line of trees to one of two or three tree rows and up to three shrub rows, one of which is placed on the windward side. Belt widths vary from about 9 m for a two-row tree belt with associated shrubs to about 3 m for single-row hedge belts. These widths mean that belts can occupy about 3 per cent of the land they are protecting. The density of the belt should not be so great as to form an impermeable barrier nor so sparse that the belt is transparent. The correct density is equivalent to a porosity of 40–50 per cent. More open barriers do not reduce wind velocity sufficiently. Where only

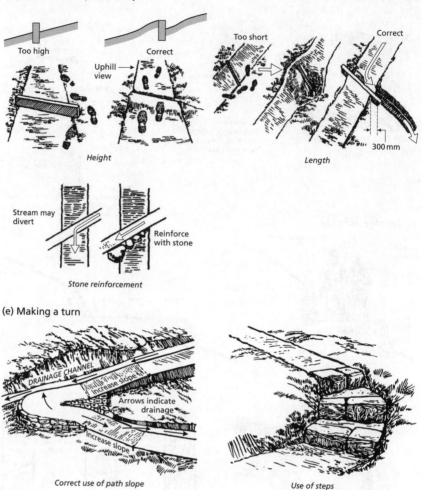

Fig. 10.8 *Continued*

a single row of trees is used, it is important that the branches and foliage extend to ground level to give the required level of porosity in the lower metre where most of the sediment movement by saltation takes place (section 2.8). With denser barriers there is a much greater reduction in wind speed initially but, since the velocity increases more rapidly with distance downwind than is the case for more porous barriers, they are effective for only short distances.

The reduction in wind velocity by a shelterbelt begins at a distance of about five times the height of the belt upwind and reaches a maximum of about 40 per cent of the original wind velocity at a distance of about three times the height of the belt downwind. Velocity then increases again, returning to the original wind speed at a distance of about 30 times the barrier height (Marshall 1967). Shelterbelts are designed to maintain the wind velocity at about 80 per cent of the open wind velocity. Wind tunnel studies by Woodruff and Zingg (1952) showed that tree belts at right angles to the wind afford this level of protection for distances up to 17 times their height for open wind velocities up to 44 km h^{-1}. Allowing for variations in wind speed and deviations in wind direction, they developed the following formula for determining shelterbelt spacing:

$$L = 17H(V_t/V)\cos\alpha \qquad (10.7)$$

where L is the spacing or distance apart of the belts (m), H is the height of the belt (m), V is the actual or design wind velocity measured at a height of 15 m above the ground surface (km h^{-1}), V_t is the threshold wind velocity for particle movement, taken as 34 km h^{-1}, and α is the angle of deviation of the prevailing wind from a line perpendicular to the belt. Effective protection in the field rarely reaches this theoretical level of 17H, however, being reduced in unstable air and by variable growth and poor maintenance of the trees. A distance of 10 or 12 times the height of the belt is more realistic. Where 5–7 m high hedges are used, the effective distance protected increases to about 30 times the barrier height but, because of their lower height, the absolute distance protected is much less and more frequent spacing is required.

Belt lengths should be a minimum of 12 times the belt height provided that the belt is at right-angles to the wind. To allow for deviations in wind direction, a longer length is desirable and a length of 24H is generally recommended (Bates 1924). For winds ranging from ±45° from the perpendicular, the effective area protected by the belt increases rapidly with belt length. Figure 10.9 (Olesen 1979) shows that for 2 m tall hedges, the area protected is 156 m^2 when the belt is 25 m long but increases to 2500 m^2 when the belt is 100 m long.

Where there is a dominant erosive wind from a single direction, the best protection is obtained by aligning the shelterbelts in parallel rows at right angles to it. Where erosive winds come from several directions, grid or herringbone layouts may be necessary. The requirement is to provide maximum protection averaged over all wind directions and all wind velocities above the threshold level. This may be achieved by a scheme in which complete protection is not obtained for any single wind direction. The effectiveness of shelterbelt layouts can be evaluated using eqn 3.8. This is first applied to obtain a measure of wind erosivity with no protection. A measure is then obtained for the shelterbelt layout by reducing the values of $\overline{V}_t$ by the ratio V_x/V_o, where V_x is the wind velocity at distance x from the belt, with x measured in units of barrier height, and V_o is the wind velocity in the open field. Values of V_x/V_o can be determined for a belt with 40 per cent porosity by the equation:

$$V_x/V_o = 0.85 - 4e^{-0.2H'} + e^{-0.3H'} + 0.0002H'^2 \qquad (10.8)$$

where $H' = x/\sin\beta$ when β is the acute angle of incident wind (Skidmore & Hagen 1977).

The greatest effect of shelterbelts is found where, as a result of farmer collaboration in a collective belt planting scheme, a regional framework exists of belts placed along property boundaries in a coordinated way so that they form part of a parallel series of main line barriers, 200–400 m apart. Within this framework, individual farmers are free to plant additional hedges. Collective shelterbelt schemes are encouraged by Hedeselskabet (Danish Land Development Service) to control erosion on the sandy soils (Olesen 1979).

The plant species selected for shelterbelts should be rapid growing, tolerant of wind and light and frost resistant where necessary. Their growth habit should give the required level of porosity at the time of year of greatest erosion risk and a conical or cylindrical shape, avoiding top-heavy crowns. The branches should be pliable so that they bend with the wind instead of breaking off. The root system should provide a firm anchorage to the soil. Preference should be given to local rather than imported species. The shelterbelt system developed by Hedeselskabet meets these

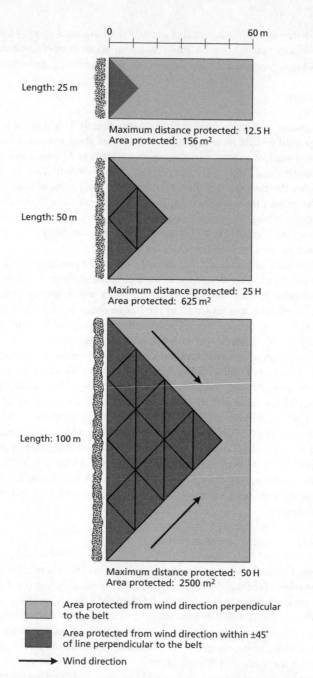

Fig. 10.9 Effect of shelterbelt length on size and shape of area protected by a 2-m tall hedge belt (after Olesen 1979).

requirements and, at the same time, provides for an ecological succession to give an effective barrier within three to four years and one with a life of 50–80 years. The belts are made up of three parallel rows, 1.25–1.5 m apart, each row comprising: nurse trees, such as alders and willows, which are fast-growing and provide the early protection; durable trees, such as oaks, sycamore, elm, maple and rowan, which take longer to grow but provide the long-term protection; and shade-tolerant bushes to provide undergrowth in the lower levels of the belt. The mixture of species provides a belt that is less vulnerable to attack by diseases and pests, visually more attractive and able to give a varied habitat for wildlife. The belt needs to be protected in its early years against damage by livestock and spray drift. After about five years and then at an interval of every three to four years, mechanical cutting of the sides of the belt is required to maintain the necessary shape, particularly the sharp rise from the ground on the windward side.

Windbreaks of brushwood or plastic meshes with about 50 per cent porosity are often used to help to stabilize mobile sand dunes and, thereby, provide a more suitable environment for vegetation growth. The windbreaks, sometimes termed sand fences, are placed at right angles to the wind (Savage & Woodhouse 1968). Deposition occurs windward of the barrier for a distance of 0.4–2.0 times the barrier height and in the lee of the barrier for a distance up to four times the height. Once the sand has accumulated and almost buried the fence, a second fence is built on top of the newly formed dune. Further fences are added until the dune is reformed by the wind into a streamlined shape so that air flows over it without loss of transport capacity. Where wind velocities exceed $18\,\mathrm{m\,s^{-1}}$, double or triple fence systems are used, spaced at intervals of four times the barrier height.

Box 10

Laying out terraces and waterways

1 Using aerial photographs, topographic maps and reconnaissance field surveys, determine the preliminary positions of the grass waterways. Locate the waterways in the natural depressions, hollows or drainage lines of the ground surface.

2 Locate the main breaks of slope and any badly eroded or gullied areas. Terrace banks should, wherever possible, be located to incorporate slope breaks and be positioned upslope of eroded lands.

3 Determine the spacing or vertical interval (VI) between the terraces. The computed spacings may be varied by 25–30 per cent to allow for adjustments in position of the terraces to conform with slope breaks and avoid eroded areas.

4 Determine terrace lengths. Terraces must be limited in length to avoid dangerous accumulations of runoff and large cross-sectional areas to the terrace channels.

5 Adjust the positions of the grass waterways if necessary to avoid excessive terrace lengths.

6 Locate paths and farm roads along the divides between separate terrace and waterway systems. The use of crest locations minimizes the catchment area contributing runoff to the road. Runoff may then be discharged into the surrounding land without the need for side drains. Crest locations also dispense with the need for bridges and culverts, avoid breaking up terraces to allow roads to cross them and keep vehicles away from grass waterways.

7 Using contour maps and aerial photographs, plan the layout of the system. Locate the grass waterways and diversion channels on the maps and photographs. Locate the top or key terrace and position the others in relation to it in accordance with the design spacings, lengths and gradients and in keeping with the location of

Continued

slope breaks and eroded areas. Note that terraces often begin at a ridge or high point on a spur and run away from it before turning approximately parallel to it and running across the slope.

8 Examine the layout to see if it is practical for farming. Check whether the smallest inter-terrace areas can be worked with the proposed machinery, assuming contour cultivation and allowing room for turning at each end; if they cannot, the terrace network may need adjustment or the land involved will have to be taken out of cultivation. The terrace spacing should be adjusted to the nearest multiple of the width of equipment to be used on the inter-terrace area.

9 On land of irregular topography, the terraces will converge in steeper sloping areas, resulting in many point rows during farming operations. This inconvenience can be minimized by the greater expense of installing a parallel terrace system. This involves levelling the land by removing material from the spurs and filling in the depressions. With parallel layouts, the recommended terrace spacing is 0.67 VI. Locate the top or key terrace and, using this new spacing, draw in successive terraces parallel to it. Check each terrace in turn to see that it always has a downhill grade, unless level terraces are being installed, and that the grade is nowhere excessive. When a terrace line is unsatisfactory because it has an uphill section or the downhill grade is too steep, remove this line and replace it with a new key line. Locate a new set of parallel terraces in relation to the second keyline.

10 Calculate the design dimensions of the grass waterways, terrace channels and diversion channels. It may be desirable to design the grass waterways in sections. This will allow adjustments to be made to the width and depth as slope steepness changes and avoid excessively wide channels at the top of the slope where runoff is

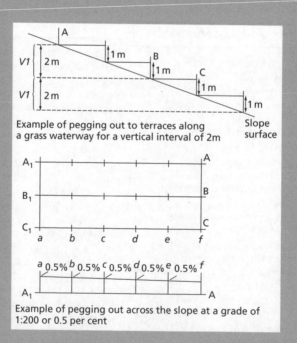

Example of pegging out to terraces along a grass waterway for a vertical interval of 2m

Example of pegging out across the slope at a grade of 1:200 or 0.5 per cent

Fig. B10.1 Laying out terraces in the field.

less. When the slope exceeds 11°, stone- or concrete-lined channels or drop structures should be used.

11 Stake out the waterway in the field and construct it by excavating soil from the centre and throwing it to each side to form the banks; check first, however, that the natural depression does not already meet the design depth and top width requirements. Seed the grass cover along with compost, fertilizer and mulch.

12 Stake out the position of the diversion channels and terrace channels along the outlet or grass waterway to the appropriate *VI* (Fig. B10.1). From these positions, the terraces are pegged out across the slope according to the selected grade. The vertical intervals should be checked regularly. Begin with the diversion channel, then the top terrace and work downslope. Minor adjustments to the position of the terrace channels can be made within a tolerance of ±50 mm *VI*. Mark the lines of the channels with a plough furrow.

13 Construct the diversion channel and then the terrace, beginning at the top and working downslope. This method of working is necessary because each terrace channel is designed to convey only the runoff from its inter-terrace area; it cannot cope with runoff from further upslope and so the protection works must be in place prior to construction. If the top soil is thin, it should be scraped off first, stored and then returned to the inter-terrace area after the terraces are complete.

After Schwab et al. (1966), Troeh et al. (1980), Hudson (1981) and Wenner (1981).

CHAPTER 11

Implementation

The ultimate objective of research on soil erosion is to solve erosion problems by adopting suitable protection measures. Suitability implies not just reducing erosion to an acceptable level; the measures must be capable of implementation. The aim of this chapter is to provide a background to some of the issues involved in implementing soil conservation proposals.

11.1

Socio-economic setting

As indicated in Chapter 6, soil erosion is frequently a response to the breakdown of farming systems. The three major periods in history associated with extensive soil erosion (Dregne 1982) all reflect the inability of existing farming systems to deal with population growth and the intensification of agriculture. These periods were: the expansion of agriculture into China, the Middle East and the Mediterranean some 1000–3000 years ago; the migration of Europeans to develop colonies some 50–150 years ago; and the expansion in the past 30–50 years of people on to marginal lands in Latin America, Africa and Asia. In all cases, the outcome was land degradation and the movement of population to new areas. The migrants took with them agricultural practices that had worked for hundreds of years in their home areas but that were generally unsuited to their new environments. Unfortunately, knowledge of this unsuitability was only available from experience.

Today, rural depopulation is causing erosion due to the lack of labour to maintain soil conservation works. Although most pronounced in Mediterranean Europe, it is a trend that will become increasingly important worldwide as all countries over the next century experience a decline in the proportion of their population employed in agriculture and a movement of people to the towns. The need to supply food for the urban market will also create an increasing strain on many existing farming systems. No longer is it appropriate to view soil erosion solely as a problem arising from population pressure. In many countries, a declining rural labour force and the need to increase agricultural production are leading to the amalgamation of holdings, increases in field sizes and wholesale levelling of land to make it easier to work with larger machinery, all trends that lead to increases in erosion.

Unfortunately, the overall take-up of soil conservation remains poor. After six decades of voluntary soil conservation programmes in the USA, erosion is still at an unacceptably high level (Swanson et al. 1986). Farmers are unlikely to adopt conservation measures if there is no immediate threat to the productivity of their land or if the main justification for their use is to prevent pollution and other off-site damage (Napier 1990), unless there are either incentives for their adoption or meaningful penalties for not using conservation production systems (Napier 1999). These issues bring into question the role that governments can play.

11.2 Political context

Although most governments have some form of soil protection policies that include erosion control, few are translated into effective action because of the lack of political will; soil conservation is not a vote-winning concern (Hudson 1981). Yet, under certain conditions, governments do respond quite rapidly. The United States Soil Conservation Service was created as a result of political pressure from all segments of society who were affected by severe wind erosion during the 1930s. Farmers lost their soil and often their farms, exacerbating the effects of the worldwide economic depression of the time; national food supplies were threatened; and off-site sedimentation created problems for the urban population and the non-agricultural population in rural areas (Rasmussen 1982). A much earlier but smaller-scale example occurred at the end of the nineteenth century in Iceland, where drifting sand seriously affected the livelihood of farmers in the southern and north-eastern parts of the country. Under pressure from the Agricultural Society of Iceland, the Icelandic Parliament (Althing) made available a small grant to bring in Danish specialists to investigate the problem and gave District Commissions the authority to take action. Without any financial wherewithal or knowledge of the best measures to take, however, this legislation was of little value. Political pressure from farmers continued and in 1907 the Act of Forestry and Prevention of Erosion of Land was passed, effectively marking the foundation of the State Soil Conservation Service of Iceland (Runólfsson 1978, 1987). These examples serve to indicate what can be achieved if the farmers in particular and society in general have sufficient political voice. Since this occurs only when disaster is imminent, the response of governments seems to be limited to crisis management.

Within the past 30 years many governments have endeavoured to set up conservation initiatives but have been constrained by lack of funds and the limited political value of being involved in conservation work. In Kenya, protection of the environment was formally endorsed by the President in 1977 and a permanent Presidential Commission on Soil Conservation and Afforestation was established in 1980 to develop strategy and policy, ensure coordination between interested bodies and monitor progress. This encouraged external financial and technical assistance such as that provided by the Swedish International Development Authority (SIDA) to the National Soil and Water Conservation Programme. In many countries, so many organizations are now involved in promoting soil conservation that new political problems have arisen, such as how the work might best be coordinated, how the responsibilities of each organization might be defined, and what should be the role of external organizations *vis-à-vis* those of the host country.

11.3 New approaches

The past three decades have seen considerable changes in the approaches used to promote and implement soil conservation. Perhaps the most fundamental has been the move from a top-down towards a bottom-up approach of participatory development in which the farming family, not just the head male, is involved in defining priorities. The job of the technical experts, including physical scientists and socio-economists, is to work with the family in producing a site-specific scheme. This approach aims to give farmers ownership of the proposals since they become major decision-makers throughout the soil conservation process. A key issue is how to provide that empowerment without reducing the influence and motivation of extension staff. In many cases, soil conservation is no longer promoted directly but is presented by stealth within a background of water conservation, improved soil fertility and overall promotion of wise land use and good land husbandry (Shaxson 1988). The new approach also depends on recognizing that many traditional agricultural systems involve soundly based soil protection practices (section 7.3.6) and that more acceptable conservation schemes can be developed by building on these.

Although the results of this new approach are promising, it is too early to know whether they can be sustained in the long term. Further, the implication that the top-down approach must always be unsuccessful cannot be supported. Experience from a number of projects in India, involving different farmers, shows that how farmers perceive the risk and the economic benefits is more important in the end. Projects where this is not recognized will fail no matter how much farmer participation is involved (Singh 1990).

A second area where a change of attitude is taking place is the move from soil conservation projects to soil conservation programmes. Projects have been favoured by aid agencies and other donors because money can be made available for a specific purpose over a finite period and the outcome monitored. The results can be seen in the number of kilometres of terraces and waterways constructed. Present-day food-for-work programmes follow a similar philosophy. Once the structures are in place, however, farmers are likely to remove them if they appear to have no benefit; for example, if they take up scarce land on a holding that is already too small (Holden & Shiferaw 1999). Since few of these projects have produced sustained conservation benefits, emphasis is now being placed on developing soil conservation programmes based on integrating research, extension, education and training. By having a longer life than the three to five years typical of soil conservation projects, soil conservation programmes should provide for greater continuity.

A soil conservation programme is based on a general mission statement and encompasses a range of activities targeted on individual farmers, communities and the population in general. Each activity has specific objectives, its performance is monitored and the programme is continually updated and modified. The Swedish Aid programme in Kenya worked with existing government structures to train staff and build the appropriate institutions to liaise with farmers. An important outcome was the training of soil conservation officers and technical assistants to maintain regular contact with and advise farmers through a Training and Visit System. Each technical assistant operated through a contact farmer and between four and eight follower farmers, visiting on a set day and time once a fortnight to discuss soil conservation, crops, mechanization, farm management and other issues. Where women's groups and other self-help groups exist, they were also used as contacts (Mbegera et al. 1992). The results were mixed because the contact farmers tended to be those who were already resource-rich and socially well connected, and were also men rather than women. Further, with its emphasis on crop production, very little attention was given to soil conservation. Since it was not meeting government objectives, the soil conservation programme was forced to develop a separate approach instead of working through the mainstream extension programme (Kiara et al. 1996).

Instead of working with contact farmers, the Department of Land Development in Thailand operates through land development villages, one in each administrative district of the country, where the farmers and government agencies work together to establish land development programmes that will prevent land degradation, restore already degraded land and adopt a land use pattern that will increase the income of the farmers (Attaviroj 1996). The emphasis of the programme is on soil and water conservation. The intention is that the techniques established at the development villages will be disseminated to neighbouring villages through field days with farmer-to-farmer contacts. At present, the Thai government is providing all the funding. The test of its success will be its sustainability once funding is reduced or withdrawn entirely.

A third area of reappraisal, arising from the above, is whether soil conservation should be approached at the level of the individual farmer, the village or some form of administrative district, or at a watershed. The watershed has the advantage of being the natural geomorphological unit for water erosion. The risk of erosion at any point within a watershed can be understood in relation to its topographic position and the effect this has on local hydrology and sediment production. The off-site effects of erosion are also more easily appreciated within a watershed than by the study of an individual field. Emphasis on a watershed, however, can result in too much reliance on mechanical measures aimed at runoff control and on a top-down approach to erosion control. Moreover, local communities do not always identify with a physical catchment. In large catchments, they identify with their village and local area so that farmers in the upper reaches may have little interest in the welfare of those at the lower end (Thomas 1996). The watershed approach can easily ignore the fact that the success or otherwise of soil conservation depends on the behaviour of the individual land user. At the most basic level, the effective watershed is a farmer's field, and with the increasing emphasis on agronomic methods of erosion control this seems the most appropriate unit at which to operate.

Administrative units are required through which contact can be made with farmers. In the USA, these take the form of Conservation Districts set up under the 1937 Standard State Soil Conservation Districts Enabling Law. The Districts provide the forum through which farmers can approach the various federal, state and local agencies that provide technical and financial assistance for conservation work. They bring together farmers, public-spirited citizens in business, industry and education, and officers of the Natural Resources Conservation Service, established in 1994 as the successor to the Soil Conservation Service, with a wider remit to cover soil, water, land and the natural environment. In Kenya, the Soil Conservation Districts are the administrative units for the planning and implementation of soil conservation programmes but the participation of farmers is organized within each District through catchment conservation committees responsible for watersheds, typically 200–500 ha in size. The committees comprise farmers, chiefs and assistant chiefs, as well as the soil conservation officer and technical assistant. All are involved in decision-making to develop a feasible and acceptable land management plan for each farm in the catchment (Kiara 2001). Similar approaches are being used in Burkina Faso (Eger & Bado 1992) and northern Thailand (Oberhauser & Limchoowong 1996).

Within the USA, individual counties and metropolitan areas have the responsibility for drawing up and managing Erosion and Sediment Control (ESC) programmes to comply with the requirements of the National Pollutants Discharge Elimination System for both point and non-point source pollution under the 1987 Clean Water Act. The ESC programmes must provide guidance on best management practices to prevent erosion and control sediment during construction. The most successful bring together environmental conservation groups, the building community, contractors, real estate people and the local government. This group of stakeholders has to decide on the measures that should be adopted to meet the required standards of water quality. These can range from ordinances to control land use and protect regional floodways and flood plains to the development and implementation of regulations with which construction firms must comply or face financial penalties. Education is an important activity of ESC programmes (Mitchell 1998). Training must be provided for the contractors in relation to best practices in soil management, soil handling, landscaping, operation of heavy machinery and other activities concerned with ground disturbance. Outreach campaigns educate elected officials, community residents and business leaders about the need for conservation systems with the aim of developing local community ownership of their rivers, lakes, wetlands and watersheds. Such ownership is important in order to secure sufficient funding to implement the programme, which comes from a mixture of state and local authority taxes, development permits, penalties for non-compliance with regulations and grants and donations from charities and private companies.

The advantages of involving the local community in erosion control programmes are not limited to urban areas and construction work. As shown by the Land Care movement in Australia, they also apply to groups of farmers and interested citizens in rural areas. The Land Care approach is essentially a bottom-up, community-led activity. The federal government of Australia provides some 21 per cent of the funding to Land Care community projects, the remainder coming from the farmers themselves, the local community and the state governments. The Land Care movement now directly influences many state government policies on environmental protection, the allocation of funds and the type of institutional support provided (Prior 1996). It plays an increasing role in: identifying problems of natural resource management; extension work, publicity and communication; networking; mobilizing local resources, including funding; and the management of natural resources. By taking over more responsibility in these areas from the state and federal governments, it is increasing local effectiveness and reducing government costs.

11.4

Responsible bodies

In some countries, soil conservation is the responsibility of a specific agency, which has a clearly defined mission and an administrative structure with a national core to define policy, and regional and local sub-divisions for implementation. The boundaries of its remit with respect to other agencies are also clear. This is generally the situation in countries with a formalized soil conservation service. Elsewhere, soil conservation is administered alongside other activities through advisory services to farmers. With the present focus on promoting soil conservation as an overall part of land husbandry, this might be considered a more appropriate model but, as seen above with respect to the US Natural Resources Conservation Service, many soil conservation services now have a wider environmental remit.

In many countries, the implementation of soil conservation projects and programmes is the responsibility of a multiplicity of organizations, including: national and local governments, often involving several ministries covering agriculture, forestry, public works and the environment; aid agencies, with several overseas governments and international organizations being involved; non-governmental organizations; and private companies. Often the result is ill-defined responsibilities, competition between bodies for scarce resources, particularly skilled labour, and confusion on the part of the farmer about whom to contact. In a review of the situation in Java, Indonesia, in the 1980s, McCauley (1988) found six government organizations that included aspects of soil conservation in their responsibilities working alongside FAO, UNDP, USAID, the Asian Development Bank, the World Bank and the Dutch government. There was little coordination of activities and often competition to work on different aspects within the same watershed project. It was to avoid this type of problem that the Kenyan government set up the Presidential Commission on Soil Conservation and Afforestation.

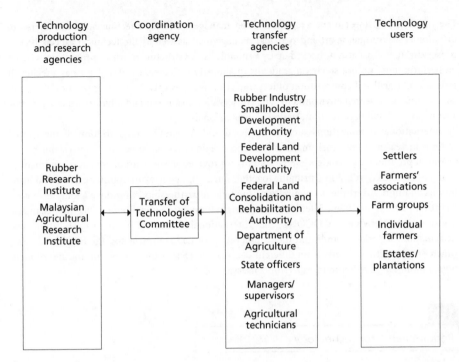

| Technology production and research agencies | Coordination agency | Technology transfer agencies | Technology users |

Fig. 11.1 Agencies involved in the transfer of technologies from research to practice in the rubber industry of Malaysia (after Shah et al. 1996).

In the rubber industry of Malaysia (Fig. 11.1), a key role is played by the Transfer of Technologies Committee, which vets technologies produced by the Rubber Research Institute of Malaysia before they are passed to the Extension and Development Department for on-farm testing; the successful ones are then released to the various implementing agencies through extension workers, publications, advisory visits and on-farm demonstrations (Shah et al. 1996). Although there are many implementing agencies, each one targets a different group of farmers. This example shows that as long as there is an overall coordinating agency through which responsibilities can be defined and collaboration achieved, there is no reason why all types of organizations cannot be usefully involved. Such coordination can also help to enlist the assistance of others, ranging from individuals working in research institutes and universities, to school teachers, farming cooperatives and women's groups. As shown by the work of the State Soil Conservation in Iceland, volunteers can play an important role. Examples are the pilots of Icelandic Air, who give their time free to fly planes for aerial seeding and spread of fertilizer in rangeland reclamation projects and school children who help in the collection of seed from the wild and with field days (Arnalds 1999).

In addition to ensuring that the roles of responsible bodies are clear, they must also have good leadership (Wenner 1988; Hudson 1991). The quality of leadership, from national down to village level, can influence whether or not a project succeeds. Indeed, a charismatic leader can often make a poorly designed project work, while a poor leader can mar the outcome of a well designed one.

The training of project managers, junior level managers and field technicians must therefore include organizational, team and personal management as well as the technical aspects of soil conservation. It must also be backed up by a suitable career structure in which promotion is based on performance and not solely on academic qualifications (Tejwani 1992). A post as a technical assistant or a field soil conservation officer must not in any way be an inferior position to one in an academic or research environment. The conservation extension and advisory service must be a truly professional organization, staffed by professionals.

International organizations such as the Food and Agriculture Organization of the United Nations increasingly view their major role as encouraging national governments to establish conservation programmes (Sanders 1988). Support is provided in the form of advice, secondment of trained staff and coordination, bringing together financing agencies, in-country experts and institutions and non-government organizations to work in partnership to produce a long-term conservation plan (Dent 1996). The FAO also strongly supports regional bodies, like the Asian Soil Conservation Network (ASOCON), the International Scheme for the Conservation and Reclamation of African Lands and the Conservation of Lands in Asia and the Pacific (CLASP), which bring together experts from different countries to share experiences with the aim of identifying more effective approaches to soil conservation.

11.5

Requirements for technology

Although, as stated earlier, the broad technology for controlling soil erosion is understood, it must satisfy a number of requirements in addition to being scientifically sound. These include (Hudson 1991):

- a high and quick financial return;
- a reduction in risk;
- no loss of existing benefits;
- accessibility to the farmer in terms of extra inputs of labour and capital;
- social acceptability, particularly in terms of gender issues;
- an extension or modification of an existing practice rather than something new.

The farmers must be convinced that the technology will work and there must be sufficient institutional support through the extension services. The technology should be effective on farms as well as on research stations. Trial farms thus form an important part of soil conservation programmes. In a scheme to introduce new technologies, including agroforestry and social forestry, to shifting cultivators in the uplands of Mindanao, Philippines, participatory programmes were introduced on selected farms in target areas. Within each target area, some ten farmers were selected as programme participants. Conservation measures were introduced on each chosen farm after discussion with the farmer family about their aspirations and the constraints of the present farming systems (Pava et al. 1990). Farmers trained through this programme were later asked to assist in promoting a similar programme run by the Mindanao Baptist Rural Life Center in the Pulangi watershed (Cruz 1996).

Legislative instruments

There is considerable disagreement among soil conservation workers on the importance of legislation. Many early soil and water conservation programmes relied on laws that required farmers to adopt certain techniques and desist from using others. Laws were also passed to restrict activities on certain types of land; for example, cultivation should not be permitted on slopes above a certain steepness, or forests in watershed areas should be protected. In some cases, legislation appears to work. Contour grass strips were introduced in Swaziland by royal decree and accepted by almost all farmers (section 8.3). In Iceland, the State Soil Conservation Service has the legal power to acquire degraded land from farmers. In reality, however, the Swazi farmers adopted the technology because it seemed to work. The Conservation Service in Iceland has never used its power because most farmers are only too willing to have the Service help them to restore their land. Generally, if legislation is effective it is because the principles behind it have already been accepted by the population (Hudson 1981).

If legislation is unworkable, the alternative is for farmers to adopt soil conservation voluntarily. The low extent of farmer uptake, however, suggests that this does not work either, particularly where, as in the USA and Europe, the benefits of conservation are acquired by the community at large rather than the individual farmer. In this situation, the role of government is to provide the necessary framework within which soil conservation can be promoted. In many cases, however, government frameworks have been detrimental to soil protection. The failure to contain erosion in the USA (Napier 1990) and the increasing erosion problem in parts of Europe (Chisci 1986; Boardman 1988) have been attributed to agricultural policies and have led to calls for more environmentally friendly farming. Within this context, the following options are available as ways of encouraging farmers to adopt more conservation-oriented measures.

11.6.1 Regulatory instruments

Examples of regulatory instruments include prohibiting arable farming on land classified as unsuitable for arable crops, setting a maximum acceptable rate of erosion and making land users liable to legal action if the rate is exceeded, and setting minimum requirements on farms for provision of wildlife habitats. As mentioned above, these measures are only effective if they can be monitored and enforced. Regulations based on rates of erosion would be almost impossible to apply because erosion is costly to measure and there is considerable uncertainty on the levels of accuracy that can be obtained. Proving in a court of law that soil loss from an area of land had exceeded a particular level would be extremely difficult. The regulations would need to be specific about the size of the area involved and the return period of the event. It would also be necessary to prove that the erosion was the result of mismanagement on the part of the land user and not a response to an extreme event.

11.6.2 Advisory work

An effective advisory service is the vital element behind all participatory approaches to soil conservation. An effective service, however, is costly, which means either a realistic charge must be made to farmers or the government must defray the expense. In many countries, the farmers are unable to pay and government finance is not available to run a staff of well trained and motivated professionals. Even where farmers can pay, they are less likely to seek advice than if the service were free. Whatever advice is given, it is clear from much of the evidence presented in sections 7.3 and 11.4 that it will only be taken up if economic benefits ensue to the farmer. Thus, it is often necessary to support advisory work with other financial incentives.

11.6.3 Financial support

Financial support should only be provided if adopting particular soil conservation practices results in farming below the economic optimum and there is a benefit beyond the individual farmer to the local community or to society in general. Incentives should be limited to stimulating the involvement of farmers and not aimed at buying their participation (IFAD 1992). The most suitable type of financial incentive is likely to be site specific. In some cases, it may consist of the provision of basic tools, such as pickaxes, wheelbarrows, seeds and cuttings; in others it may include the construction of terraces, the costs of training courses for farmers and the cost of land lost to cultivation by grassing valley floors or introducing contour grass strips or shelterbelts. Incentives may also be set within a broader environmental context; for example, the promotion of winter cover crops to control nitrate leaching. They can also take the form of policies that have an effect on the net income of farmers and land users, or that provide the necessary enabling conditions for the take-up of better conservation practices (Fig. 11.2; Enters 1999).

Cost-sharing programmes between farmer and government were the basis of the early work of the US Soil Conservation Service (Napier 1990). Their disadvantages are that they lead to complacency that soil conservation is being adequately supported and they require a long-term financial commitment. Moreover, once the incentive is withdrawn, the conservation ceases, a response that is typical of many schemes based on financial incentives worldwide (Huszar 1999). Programmes that rely on farmers taking land out of agricultural production suffer from the same short-term commitment. Under the 1985 Farm Bill, the United States government established three conservation schemes: the Conservation Reserve Program; Highly Erodible Land Protection, including the Sodbuster and Conservation Compliance; and Wetland Protection. Under the Conservation Reserve Program (CRP), farmers entered a contractual arrangement with the government to retire highly erodible land for ten years in return for a rent payment. Those who violated the contract within that time stood to lose federal farm benefits and were required to repay rent with interest. The Sodbuster Provision was designed to prevent new, highly erodible land from being brought into production. Anyone converting such land to crops without an approved conservation plan would lose all USDA programme benefits, such as price support, loans, disaster payments, federal crop insurance and CRP payments, on all the land they farmed. Under Conservation Compliance, an approved conservation programme was required on all currently cropped erodible land. The Wetland Protection operated through the Swampbuster programme, removing benefits to anyone who converted wetland into crop production. Between 1985 and 1990, nearly 13.7 million hectares of land were enrolled under CRP and conservation plans were approved for some 52.6 million hectares. Although a significant reduction in soil erosion resulted, the cost was extremely high because land owners could negotiate the price at which they were prepared to retire the land; in some locations the payments made were 200–300 per cent higher than local cash rents (Hoag 1999). Further, the eligibility rules, requiring fields to have erodible soils over at least two-thirds of their area and to have been cropped for two years between 1981 and 1985, meant that in Illinois, Iowa and Missouri some 25 per cent of the land classified as highly erodible was excluded from the scheme (Padgitt 1989). There were also concerns about the way the provisions in the programmes were implemented. One study concluded that some tens of millions of dollars in farm support went to farmers who should have been

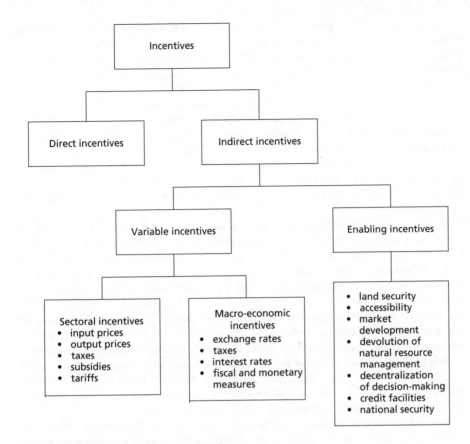

Fig. 11.2 Types of incentives (after Enters 1999).

ineligible because they did not comply with the conditions of Sodbuster, Swampbuster or Conservation Compliance (Hoag 1999). These concerns led to changes to both the scope of the programmes and the way they were implemented.

Since the Farm Bill of 1990, the emphasis has been on policies that reward farmers for adopting Codes of Good Agricultural Practice aimed at protection of the environment, particularly wildlife habitat, water quality and improvements in air quality by reducing wind erosion. In order to receive support under CRP, farmers had to demonstrate that their proposals would provide these additional environmental benefits, as well as reducing water or wind erosion. These changes led to more competitive bidding and to a more cost-effective programme. They also directed the programme away from areas of high erosion risk, like the Great Plains, to environmentally sensitive areas such as Chesapeake Bay, Long Island Sound and the Great Lakes (Hoag 1999). However, there was nothing to motivate land users to continue with conservation-based production systems should they choose to withdraw from the programme. If commodity prices continued to rise, farmers were more likely to leave the scheme and accept the loss of government benefits.

With the 1996 Farm Bill, many of the penalities for failing to comply with Conservation Compliance, Sodbuster and Swampbuster were removed under a programme to reduce costs by

Implementation

phasing out commodity support. The CRP remained, with the statutory upper limit being increased to 14 million hectares, but its focus changed to improving water quality rather than saving soil. As a result, a greater proportion of the funding was directed towards livestock farmers. The 2002 Farm Bill authorized the continuation of CRP through to 2007, with the upper limit for funding being increased to 15.9 million hectares. In addition, a new programme was introduced for soil erosion and sediment control in the Great Lakes Basin (Natural Resources Conservation Service 2002). To assist farmers in bidding for funds, the Natural Resources Conservation Service is promoting the CORE4 programme (Natural Resources Conservation Service 1999), established by the Conservation Technology Information Center, as a basis for planning, financing and implementing farm management systems. CORE4 has four aims – better soil, cleaner water, greater profit and brighter future – and four components: conservation tillage (section 9.2.3), crop nutrient management, weed and pest management and conservation buffers (section 8.3).

Experience in the United States and many other countries shows that financial support for soil conservation must be treated with care. Inappropriate support can either destroy the long-term prospects of a soil conservation programme or require a high level of long-term funding that cannot be sustained. Under certain circumstances, where damage from erosion can be clearly related to inadequate on-site management and to a specific land user, financial costs can be reduced through legislation based on the principle of the 'polluter pays'. The risk of prosecution is a major incentive to the land user to prevent erosion because of the associated embarrassment and inconvenience, in addition to the financial burden (Hannam 1999). In New South Wales, Australia, soil conservation offences are viewed as criminal offences but prosecutions are subject to strict rules of evidence that must show substantial environmental damage and wilfulness on the part of the land user. Although penalties can be up to A$110,000 for an offence and the court has the power to set conditions on land rehabilitation in perpetuity, the high costs involved in developing the case mean that prosecution is only used as a last resort.

Land users need to be persuaded that the cost of conservation is preferable to any court settlement that might be demanded. This is more likely to apply to land users responsible for preventing erosion on mining spoil, industrial land, pipeline corridors and construction sites than to farmers, where apportioning blame is more difficult. In the future, regular monitoring of the environment through remote sensing may make it easier to identify the sources of sediment delivered to rivers and lakes and it may be feasible to hold land users responsible where it can be shown that they have not complied with a Code of Good Agricultural Practice. The Ministries of Agriculture in many European countries have issued such codes and subsidies are provided for introducing riparian barriers, planting of trees and taking land out of production through set-aside programmes. Such support is best regarded as payments to land users for 'environmental services' for the public good (Giger et al. 1999; Enters 2001).

Box 11

Land Care

The Land Care Programme of Australia is now considered one of the most successful community-based soil conservation programmes in the world. It is a partnership between the federal government, the various state governments and the local community to combine the work of the extension services with local skills and energy to address local and regional problems of land management. It developed because the more traditional approach of using extension services to transfer technology to farmers and improve land management practices was viewed as a failure (Knowles-Jackson 1996).

There are now some 4000 Land Care or similar groups throughout Australia. Some deal only with single issues, such as controlling rabbits or planting trees. The majority, however, are concerned with land use, land degradation and land rehabilitation on a catchment basis. Land degradation covers soil erosion, salinity, loss of soil structure and quality, decline of native vegetation and loss of water quality. The groups are essentially 'self-starting, autonomous grassroots bodies' in which the members share aspirations, skills and experiences, seek advice from experts and identify funds from various state and federal government and private sources to implement their proposals (Roberts 1992). The membership includes representatives of the community, business and government, as well as farmers. The combination of peer pressure, camaraderie and sharing of work and information contributes to community ownership of the problem.

The Commonwealth Government of Australia provides funding without interfering with the autonomy of the Land Care Groups. Funds are provided to support:

- a regional and national facilitator to assist groups and encourage communication between groups and the government;
- Land Care Australia Ltd, a non-profit making public company that promotes the involvement of society in general and seeks sponsorship through donations and grants;
- partial funding of Land Care projects through the National Heritage Trust;
- the Australian Land Care Council, a community-based advisory body with a remit to further links between land care groups, the Commonwealth Government and the National Heritage Trust.

Land Care groups submit project proposals, prepared according to national guidelines, to the State Soil Conservation Service. Funding under the National Land Care programme is dependent on the benefits likely to accrue from the proposal to the community at large.

Although the support of the National Heritage Trust is effectively a financial incentive funded by taxation, the fact that it is partial funding, that the proposals come from local initiative and that long-term community involvement is anticipated makes Land Care a 'new generation' soil conservation programme, distinct from more traditional programmes based solely on government funding (Nabben 1999). Often the Land Care groups provide, through their own cash, materials and labour, more than twice the funding received from government sources. In addition, groups are able to secure sponsorship from business. For example, Alcoa, a mining company, is helping to support six groups, comprising 88 farming families, to implement farm and catchment plans in the Avon catchment, Western Australia, working with Agriculture Western Australia, which provides the technical expertise (Nabben 1999). Rather than expertise being presented as recipes by experts, in many Land Care programmes the group members work together with the extension agents to determine locally relevant topics and develop training programmes that, through discussion, lead to understanding of the causes of a problem and the planning of a long-term solution (Marston 1996).

The following are considered key characteristics of Land Care groups (Campbell 1994; Marston 1996):

- they are generally concerned with a broad range of multidisciplinary issues;
- they are based on neighbourhoods or catchments with contiguous boundaries, rather than merely groups of farmers with a common interest and narrow agenda;
- the impetus for establishing Land Care groups comes from the community;
- proposals for land use management and conservation are developed through a peer-group framework;
- the momentum and ownership of the programme is with the community.

An important feature of the Land Care programme is that the financial support comes mainly in the form of grants rather than cheap credit, loans or subsidies. These can be

Continued

targeted on programmes that encourage sustainable and profitable farming systems. They therefore encourage farmers to take the risk of developing such systems without commitment to long-term repayment or long-term government support. Studies of the Avon Landcare Programme show that the benefit–cost ratios (section 7.3.7) at the catchment level ranged from 1.7 to 2.3 and at the level of the individual farmer from 0.9 to 4.7 (Nabben 1999). In addition, the programme brought about positive sociological changes to the rural community.

Despite the apparent success of the Land Care programme, there is a concern that enthusiasm for it will decline. Bradsen (1994) questioned whether the programme was sufficiently funded or had the depth of technical rigour to address the issues, yet would be successful enough to prevent a more effective system of land conservation being put in place. However, the changes in attitude brought about by the programme, the fact that many projects appear to be economically sustainable and the high multiplier effect in terms of increased funding beyond that supplied by the government indicate that the Land Care programme seems to be beneficial overall and providing value for money.

CHAPTER 12

The way ahead

Soil is an important component of the global ecosystem. It is fundamental to life on earth. Yet, when compared to water and air, it is the poor relation in respect of policies designed to promote acceptable standards of quality and ensure its protection. As seen in Chapter 11, soil erosion has been recognized as a problem for centuries but the success rate in controlling it is poor. It has generally not proved economic for farmers to practise soil conservation and the political will to enforce erosion control has not been there. One reason for this is that from the viewpoint of agricultural production, it has not been globally required. Between 1945 and 1990, the rate of loss of agricultural land through erosion, at 0.1 per cent per year, was more than offset by annual increases in crop productivity of 1–2 per cent as a result of better farming practices and greater use of irrigation, pesticides and fertilizers. Against this background, Lomborg (2001) concludes that extra efforts to reduce erosion cannot be justified. However, this conclusion ignores the extent to which past and present soil conservation works may have helped to reduce the loss of land. Further, an important section of the world's population is still engaged in low-input agriculture and therefore effectively 'mining' the soil resource because that is the only way they can secure their present livelihood; through poverty, they do not have access to the resources needed to adopt more sustainable farming practices. There is also the issue of maintaining existing erosion control measures in areas where migration and disease are reducing the size of the rural population. A global analysis of the effects of erosion should also take account of resulting reductions in water quality and increases in flooding. These are likely to become the main drivers for implementing erosion control measures and the major justification why the costs should be borne by the community rather than the individual land user either through increased taxation or by paying higher prices for food, oil, gas, electricity and road and rail construction. Since, historically, erosion has increased whenever farmers have been unable to adjust their management practices to changing circumstances, there is the added uncertainty of how farmers worldwide will adapt to any changes in climate.

The reasons for failure of many soil conservation projects and programmes are listed in Table 12.1 (Hudson 1991). Whether or not existing approaches using participatory methods and working at the community level (section 11.3) will overcome these remains to be seen. Establishing an enthusiastic and professional extension service, encouraging the involvement of all the stakeholders in the community and improving the institutional framework for implementation will all undoubtedly help but, in the end, the success of the new approaches will depend on whether they help farmers and other land users to develop profitable and sustainable ways of managing the land. The willingness of the community to pay that proportion of the profit that relates to the environmental benefit may well be critical.

Table 12.1 Reasons for failure of soil conservation projects

Pre-project design errors	Overestimating effects of new practices
	Overestimating rate of adoption
	Overestimating ability of extension services to disseminate new ideas
	Underestimating time required to mobilize staff and materials
	Inadequate understanding of attitudes to risk
	Unreal estimate of economic benefits
	Underestimating problems of coordination between different ministries and departments
	Overestimating strength of national or local research base to find solutions
Weaknesses in implementation	Governments cannot afford the costs
	Insufficient knowledge of the cropping system
	Insufficient testing of new crop systems
	Optimistic assumptions on yields
	Underassessment of levels of farm labour required
	Proposals not attractive to farmers
	Unrealistically low prices for farm products
	Unhelpful marketing and pricing policies
	Inadequate size of extension staff
	Overloaded management with no clearly defined responsibilities
	Management divorced from implementing institutions
	Unstable government – uncertainty over long-term commitment

Source: after Hudson (1991).

Soil conservation is an interdisciplinary subject. It requires an understanding of geomorphological processes (Chapter 2), agricultural systems (Chapter 7) and the organizational structure of the society, as well as the ability to design sustainable farming systems (Chapters 8, 9 and 10), implement proposals (Chapter 11) and advise on the legislative framework to support them. With the current emphasis on socio-economic and political factors, there is a fear that biophysical scientists are being ignored and that, as a result, unsound practices are being proposed and promoted. There is a role for the physical scientist in adapting the principles and practices of erosion control to meet the specific requirements of a local community, taking account of the aspirations and constraints of farmers and their existing practices and knowledge. In addition, farmers should be encouraged to experiment with ideas and techniques and take part in on-farm research to supplement that carried out at research stations. Farmers are more likely to identify with measures that other farmers have shown to be effective.

Increasing use is being made of techniques such as Logical Framework Analysis (LFA) to specify the goals, objectives, outputs and activities of a project. Table 12.2 shows an LFA matrix. The left-hand column specifies the structure, namely the overall goal of the project, the immediate objectives, the outputs and the activities required to achieve the outputs. The second column describes how the success of the project will be evaluated using what are often termed 'objectively verifiable indicators'; these must be realistically achievable, scientifically appropriate, objective and measurable. The third column describes how the indicators will be measured. The fourth column lists the assumptions that were made when the project was drawn up and the risks that

Table 12.2 Matrix for logical framework analysis

Project strategy	Indicators of achievement	Means and sources of verification	Assumptions and risks
Goal What general purpose will the strategy achieve?	What are the relevant indicators (social, economic, environmental) to demonstrate that the goal has been achieved?	What data are required to verify that the goal has been achieved?	What external conditions or factors are required to sustain the project?
Objectives What are the immediate objectives that should be met? What benefits and costs are expected? What improvements are expected?	What are the relevant indicators to assess whether the objectives have been met, e.g. benefit–cost analysis, standards of service, environmental enhancement?	What data are required to verify that the objectives have been met? What are the methods of data collection, analysis and presentation?	What conditions are required to enable the objectives to meet the goals, e.g. project objectives related to erosion control may need to be consistent with sustainability?
Outputs What are the results of the project? Do they deliver the intended objectives?	What types, quality and quantity of output will be produced, by whom and by when?	What are the sources of information required to show that the outputs have been achieved? How will the data be collected, analysed and presented?	What conditions are required to enable the outputs to deliver the objectives?
Activities What activities must be undertaken to produce the intended outputs, by whom and by when?	What materials, equipment, labour and services are needed, over what period and at what cost?	What are the sources of information to show that the resources to implement the project have been met? How will the data be collected, analysed and presented?	What conditions must be met to ensure that the activities can be carried out as planned? What factors might influence the achievement of the outputs which are outside the control of the project management?

may impede its performance. The top two rows in the matrix help to set the criteria for project evaluation and the bottom two rows provide the framework for implementation. In many situations, it may be appropriate to construct matrices for all the stakeholder groups involved in a project, since each will have slightly different objectives and different roles to play. LFA places much greater emphasis on monitoring and post-project evaluation and encourages the use of programme audits to evaluate the success of the project and the performance of the various stakeholders.

Programme audits should be carried out by an independent organization and cover the performance of the people involved in the project, the project plans and the extent to which, for example, best management practices are being applied (Holbrook & Johns 2002). It is important to decide who and what to audit and then to establish the base-line conditions against which changes in conditions can be measured. The auditing team must decide what data to collect and what questions to ask, using the objectively verifiable indicators as a guide. Audits should be used to identify successes and deficiencies in a project and make recommendations for improvements in performance. The results of an audit can often lead to changes to design manuals describing best management practices and to improvements in administration and training programmes.

Arguably, the weakest part of soil conservation programmes to date has been the lack of an effective legal framework. Although many countries have a multiplicity of laws relating to land use planning, forest management, species and habitat protection and water management, many of which affect erosion indirectly, often they are not effective. Either they are not properly implemented or, in some cases, they enhance land degradation by encouraging inappropriate land use. The various global environmental treaties, conventions and strategies, such as Agenda 21 and the Conventions on Climatic Change, Biological Diversity and Combating Desertification, are encouraging some governments to review their legislation relating to land degradation (Hannam & Boer 2001). The various Directives related to water, air and the environment issued by the European Union are having a similar effect among the member states. As indicated in Chapter 11, most changes with respect to erosion and sedimentation relate to developing Codes of Best Management Practices, with some financial incentives to encourage their adoption. In contrast to legislation on water quality, there are no legal definitions of acceptable soil quality or acceptable levels of tolerable erosion rates. Partly, this is because these cannot be easily defined or cases of their violation cannot be supported by unequivocal evidence. Enforcement is either through cross-compliance, which means that land users can receive certain benefits only if they can demonstrate they have adopted specified management practices, or through penalties for environmental damage on the principle of the 'polluter pays'; although, as seen in section 11.6.3, the latter is rarely used. Generally, legislation is seen as providing an enabling role by establishing the institutional framework for implementing soil conservation and protection policies and encouraging the voluntary participation of land users. As yet little attention has been given to incorporating soil conservation practices into incentives for carbon sequestration and providing the legislation for its use in carbon credit trading.

According to Hannam (2001), an essential requirement for the future is to change public attitudes so that they will support the concept of 'natural rights for soil', which need to be protected. There are encouraging moves in this direction. Germany and Switzerland have specific Soil Protection Acts within their legislation, the USA has many state and local resource conservation laws that have evolved from the 1935 Soil Conservation Act and the 1937 Standard Soil Conservation District Law. Iceland has the 1965 Log um Landgræðslu to combat soil erosion in rangeland. The European Union is developing a Soils Directive and will expect each of its member states to support it by implementing soil strategies. As soil protection rises up the political agenda and national governments and communities become more aware of the substantial costs arising from environmental damage associated with erosion, the political will to do something will increase. The Land Care movement in Australia, its counterpart in New Zealand and the community-based Erosion and Sedimentation Control programmes in the USA indicate that local community groups, involving all the stakeholders with an interest in soil erosion and resource management, may be the most effective way of implementing policy. Since this takes responsibilities away from central governments, there is the question of whether governments will be prepared to provide the enabling financial support to community groups while allowing them local autonomy and the ability to function with minimum bureaucracy.

Soil erosion is an integral part of the natural and cultural environment; its rate and spatial and temporal distribution depend on the interaction of physical and human circumstances. Archaeological and historical studies show how the nature of this interaction has changed over time (section 1.2). Since the 1930s the need for erosion control has been driven largely by concerns over food production and security of food supply both for individual countries and for the increasing world population. Since the 1970s, there has been increasing concern about the environmental damage caused by erosion. Although much of this damage is associated with sediment derived from agricultural areas and the chemicals adsorbed to it, problems can also arise from erosion on road banks, urban land, pipeline corridors and recreational areas (Chapter 7). The people whose activities may contribute to erosion and who may be affected by any damage are now a much wider group than farmers. Erosion affects whole communities. The next few decades will reveal whether the spatial distribution of erosion will alter in relation to changes in climate and the resulting human responses, particularly changes in land cover. Areas that traditionally have not experienced erosion problems may well do so in the future.

References

Abrahams, A.D., Parsons, A.J. and Hirsch, P.J. 1992. Field and laboratory studies of resistance to interrill overland flow on semi-arid hillslopes, southern Arizona. In Parsons, A.J. and Abrahams, A.D. (eds), *Overland flow: hydraulics and erosion mechanics*. UCL Press, London: 1–23.

Abrahams, A.D., Parsons, A.J. and Luk, S.H. 1991. The effect of spatial variability in overland flow on the downslope pattern of soil loss on a semiarid hillslope, southern Arizona. *Catena* 18: 255–70.

Abrahim, Y.B. and Rickson, R.J. 1989. The effectiveness of stubble mulching in soil erosion control. In Schwertmann, U., Rickson, R.J. and Auerswald, K. (eds), *Soil erosion protection measures in Europe*. Soil Technology Series 1: 115–26.

Abtew, W., Gregory, J.M. and Borrelli, J. 1989. Wind profile: estimation of displacement height and aerodynamic roughness. *Transactions of the American Society of Agricultural Engineers* 32: 521–7.

Agate, E. 1983. *Footpaths*. British Trust for Conservation Volunteers, Wallingford.

Aina, P.O., Lal, R. and Taylor, G.S. 1977. Soil and crop management in relation to soil erosion in the rainforest of western Nigeria. In *Soil erosion: prediction and control*. Soil Conservation Society of America Special Publication 21: 75–82.

Aina, P.O., Lal, R. and Taylor, G.S. 1979. Effects of vegetal cover on soil erosion on an alfisol. In Lal, R. and Greenland, D.J. (eds), *Soil physical properties and crop production in the tropics*. Wiley, London: 501–8.

Al-Awadhi, J.M. and Willetts, B.B. 1999. Sand transport and deposition within arrays of non-erodible cylindrical elements. *Earth Surface Processes and Landforms* 24: 423–35.

Alberts, E.E., Moldenhauer, W.C. and Foster, G.R. 1980. Soil aggregates and primary particles transported in rill and interrill flow. *Soil Science Society of America Journal* 44: 590–5.

Alberts, E.E., Laflen, J.M., Rawls, W.J., Simanton, J.R. and Nearing, M.A. 1989. Soil component. In Lane, L.J. and Nearing, M.A. (eds), *USDA – Water Erosion Prediction Project: hillslope profile model documentation*. USDA-ARS National Soil Erosion Research Laboratory Report No. 2: 6.1–6.15.

Aldon, E.F. 1976. Soil ripping treatments for runoff and erosion control. In *Proceedings of the Federal Inter-Agency Sediment Conference, Denver, Colorado* 3: 2.24–2.29.

Al-Durrah, M.M. and Bradford, J.M. 1981. New methods of studying soil detachment due to water drop impact. *Soil Science Society of America Journal* 45: 949–53.

Al-Durrah, M.M. and Bradford, J.M. 1982. Parameters for describing soil detachment due to single water drop impact. *Soil Science Society of America Journal* 46: 836–40.

Alexander, E.B. 1988. Rates of soil formation: implications for soil loss tolerance. *Soil Science* 145: 37–45.

Alexander, M.J. 1990. Reclamation after tin mining on the Jos Plateau, Nigeria. *Geographical Journal* 156: 44–50.

Alfaro Moreno, J. 1988. Farmer income and soil conservation in the Peruvian Andes. In Rimwanich, S. (ed.), *Land conservation for future generations*. Department of Land Development, Bangkok: 711–26.

Amphlett, M.B. 1988. A nested catchment approach to sediment yield monitoring in the Magat catchment, central Luzon, the Philippines. In Rimwanich, S. (ed.), *Land conservation for future generations*. Land Development Department, Bangkok: 283–98.

Anderson, M.G. and Richards, K.S. 1987. *Slope stability: geotechnical engineering and geomorphology*. Wiley, Chichester.

André, J.E. and Anderson, H.W. 1961. Variation of soil erodibility with geology, geographic zone, elevation and vegetation type in northern California wildlands. *Journal of Geophysical Research* 66: 3351–8.

Armbrust, D.V. 1999. Effectiveness of polyacrylamide (PAM) for wind erosion control. *Journal of Soil and Water Conservation* 54: 557–9.

Armstrong, C.L. and Mitchell, J.K. 1987. Transformations of rainfall by plant canopy. *Transactions of the American Society of Agricultural Engineers* 30: 688–96.

Arnalds, A. 1987. Ecosystem disturbance in Iceland. *Arctic and Alpine Research* 19: 508–13.

Arnalds, A. 1999. Incentives for soil conservation in Iceland. In Sanders, D.W., Huszar, P.C., Sombatpanit, S. and Enters, T. (eds), *Incentives in soil conservation: from theory to practice*. Oxford and IBH Publishing, New Delhi: 135–50.

Arnalds, O. 2000. The Iceland 'rofabard' soil erosion features. *Earth Surface Processes and Landforms* 25: 17–28.

Arnalds, O., Aradóttir, A.L. and Thorsteinsson, I. 1987. The nature and restoration of denuded areas in Iceland. *Arctic and Alpine Research* 19: 518–25.

Arnalds, O., Thorarinsdottir, E.F., Metusalemsson, S., Jonsson, A., Gretarsson, E. and Arnason, A. 2001. *Soil erosion in Iceland*. The Soil Conservation Research Service and the Agricultural Research Institute, Reykjavik.

Arnoldus, H.M.J. 1980. An approximation of the rainfall factor in the Universal Soil Loss Equation. In De Boodt, M. and Gabriels, D. (eds), *Assessment of erosion*. Wiley, Chichester: 127–32.

Arulanandan, K. and Heinzen, R.T. 1977. Factors influencing erosion in dispersive clays and methods of identification. *International Association of Scientific Hydrology Publication* 122: 75–81.

Arulanandan, K., Loganathan, P. and Krone, R.B. 1975. Pore and eroding fluid influences on the surface erosion of a soil. *Journal of the Geotechnical Engineering Division ASCE* 101: 53–66.

Attaviroj, P. 1996. Land development villages and soil doctors: strategies towards better land husbandry in Thailand. In Sombatpanit, S., Zöbisch, M.A., Sanders, D.W. and Cook, M.G. (eds), *Soil conservation extension: from concepts to adoption*. Soil and Water Conservation Society of Thailand, Bangkok: 153–7.

Auerswald, K. and Schmidt, F. 1986. Atlas der Erosionsgefährdung in Bayern – Karten zum flächenhaften Bodenabtrung durch Regen. *GLA-Fachberichte* 1, Geologisches Landesamt, Munich.

Babaji, G.A. 1987. Some plant stem properties and overland flow hydraulics: a laboratory simulation. PhD Thesis, Cranfield Institute of Technology.

Bache, D.H. 1986. Momentum transfer to plant canopies: influence of structure and variable drag. *Atmospheric environment* 20: 1369–78.

Baffoe-Bonnie, E. and Quansah, C. 1975. The effect of tillage on soil and water loss. *Ghana Journal of Agricultural Science* 8: 191–5.

Bagnold, R.A. 1937. The transport of sand by wind. *Geographical Journal* 89: 409–38.

Bagnold, R.A. 1941. *The physics of blown sand and desert dunes*. Chapman and Hall, London.

Bagnold, R.A. 1951. The movement of a cohesionless granular bed by fluid flow over it. *British Journal of Applied Physics* 2: 29–34.

Bagnold, R.A. 1979. Sediment transport by wind and water. *Nordic Hydrology* 10: 309–22.

Baharuddin, K., Mokhtaruddin, A.M. and Majid, N.M. 1996. Effects of logging on soil physical properties in Peninsular Malaysia. *Land husbandry* 1(1/2): 33–41.

Baily, B., Collier, P., Farres, P., Inkpen, R. and Pearson, A. 2003. Comparative assessment of analytical and digital photogrammetric methods in the construction of DEMs of geomorphological forms. *Earth Surface Processes and Landforms* 28: 307–20.

Barahona, E., Quirantes, J., Guardiola, J.L. and Iriarte, A. 1990. Factors affecting the susceptibility of soils to interrill erosion in south-eastern Spain. In Rubio, J.L. and Rickson, R.J. (eds), *Strategies to combat desertification in Mediterranean Europe*. Commission of the European Communities Report EUR 11175 EN/ES: 216–27.

Barrow, E.G.C. 1989. The value of traditional knowledge in present-day soil-conservation practice: the example of West Pokot and Turkana. In Thomas, D.B., Biamah, E.K., Kilewe, A.M., Lundgren, L. and Mochoge, B.O. (eds), *Soil and water conservation in Kenya*. Department of Agricultural Engineering, University of Nairobi/Swedish International Development Authority: 471–85.

Barrow, G., Brotherton, D.I. and Maurice, O.C. 1973. Tarn Hows experimental restoration project. *Countryside Recreation News Supplement* 9: 13–18.

Bates, C.G. 1924. The windbreak as a farm asset. *USDA Farmers' Bulletin* 1405.

Bazzoffi, P., Torri, D. and Zanchi, C. 1980. Stima dell'erodibilità dei suoli mediante simulazione di pioggia in laboratorio. Nota 1: simulatore di pioggia. *Annali Istituto Sperimentale per lo Studio e la Difesa del Suolo* 11: 129–40.

Beasley, D.B., Huggins, L.F. and Monke, E.J. 1980. ANSWERS: a model for watershed planning. *Transactions of the American Society of Agricultural Engineers* 23: 938–44.

Begin, Z.B. and Schumm, S.A. 1979. Instability of alluvial valley floors: a method for its assessment. *Transactions of the American Society of Agricultural Engineers* 22: 347–50.

Bekele, M.W. and Thomas, D.B. 1992. The influence of surface residue on soil loss and runoff. In Hurni, H. and Tato, K. (eds), *Erosion, conservation and small-scale farming.* Geographica Bernensia, Bern: 439–52.

Bennett, H.H. 1939. *Soil conservation.* McGraw-Hill, New York.

Bennett, J.P. 1974. Concepts of mathematical modeling of sediment yield. *Water Resources Research* 10: 485–92.

Bensalem, B. 1977. Examples of soil and water conservation practices in North African countries. *FAO Soils Bulletin* 33: 151–60.

Berry, L. and Ruxton, B.P. 1960. The evolution of Hong Kong harbour basin. *Zeitschrift für Geomorphologie* 4: 97–115.

Beschta, R.L. 1978. Long-term patterns of sediment production following road construction and logging in the Oregon Coast Range. *Water Resources Research* 14: 1011–16.

Besler, H. 1987. Slope properties, slope processes and soil erosion risk in the tropical rain forest of Kalimantan Timur (Indonesian Borneo). *Earth Surface Processes and Landforms* 12: 195–204.

Betts, H.D., Trustrum, N.A. and DeRose, R.C. 2003. Geomorphic changes in a complex gully system measured from sequential digital elevation models, and implications for management. *Earth Surface Processes and Landforms* 28: 1043–58.

Beuselinck, L., Govers, G., Hairsine, P.B., Sander, G.C. and Breynaert M. 2002. The influence of rainfall on sediment transport by overland flow over areas of net deposition. *Journal of Hydrology* 257: 145–63.

Beven, K. and Binley, A. 1992. The future of distributed models: model calibration and uncertainty prediction. *Hydrological Processes* 6: 279–98.

Bharad, G.M. and Bathkal, B.C. 1991. Role of vetiver grass in soil and moisture conservation. *Vetiver Newsletter* 6: 15–17.

Bhardwaj, S.P., Khybri, M.L., Ram, S. and Prasad, S.N. 1985. Crop geometry – a nonmonetary input for reducing erosion in corn on four percent slope. In El-Swaify, S.A., Moldenhauer, W.C. and Lo, A. (eds), *Soil erosion and conservation.* Soil Conservation Society of America, Ankeny, IA: 644–8.

Bielders, C.L., Rajot, J.L. and Koala, S. 1998. Wind erosion research in Niger: the experience of ICRISAT and Advanced Research Organizations. In Sivakumar, M.V.K., Zöbisch, M.A., Koala, S. and Maukonen, T.

(eds), *Wind erosion in Africa and west Asia: problems and control strategies.* ICARDA, Aleppo: 95–123.

Biot, Y. 1990. THEPROM – an erosion productivity model. In Boardman, J., Foster, I.D.L. and Dearing, J.A. (eds), *Soil erosion on agricultural land.* Wiley, Chichester: 465–79.

Bisal, F. 1950. Calibration of splash cup for soil erosion studies. *Agricultural Engineering* 31: 621–2.

Bishop, D.M. and Stevens, M.E. 1964. Landslides on logged areas in southeast Alaska. *USDA Forest Research Service Paper* NOR-1.

Blong, R.J., Graham, O.P. and Veness, J.A. 1982. The role of sidewall processes in gully development: some NSW examples. *Earth Surface Processes and Landforms* 7: 381–5.

Boardman, J. 1988. Public policy and soil erosion in Britain. In Hooke, J.M. (ed), *Geomorphology in environmental planning.* Wiley, Chichester: 33–50.

Boardman, J. 1990. Soil erosion on the South Downs: a review. In Boardman, J., Foster, I.D.L. and Dearing, J.A. (eds), *Soil erosion on agricultural land.* Wiley, Chichester: 87–105.

Boardman, J. and Evans, R. 1994. Soil erosion in Britain: a review. In Rickson, R.J. (ed.), *Conserving soil resources: European perspectives.* CAB International, Wallingford: 3–12.

Bocharov, A.P. 1984. *A description of devices used in the study of wind erosion of soils.* Balkema, Rotterdam.

Boiffin, J. 1985. Stage and time-dependency of soil crusting *in situ.* In Callebaut, F., Gabriels, D. and De Boodt, M. (eds), *Assessment of surface crusting and sealing.* Flanders Research Centre for Soil Erosion and Conservation, State University of Gent: 91–8.

Boiffin, J. and Monnier, G. 1985. Infiltration rate as affected by soil surface crusting caused by rainfall. In Callebaut, F., Gabriels, D. and De Boodt, M. (eds), *Assessment of surface crusting and sealing.* Flanders Research Centre for Soil Erosion and Conservation, State University of Gent: 210–17.

Boix-Fayos, C., Calvo-Cases, A., Imeson, A.C. and Soriano-Soto, M.D. 2001. Influence of soil properties on the aggregation of some Mediterranean soils and the use of aggregate size and stability as land degradation indicators. *Catena* 44: 47–67.

Bojö, J. 1992. Cost–benefit analysis of soil and water conservation projects: a review of 20 empirical studies. In Tato, K. and Hurni, H. (eds), *Soil conservation for survival.* Soil and Water Conservation Society, Ankeny, IA: 195–205.

Bollinne, A. 1975. La mesure de l'intensité du splash sur sol limoneux. Mise au point d'une technique de terrain et premiers résultats. *Pédologie* 25: 199–210.

References

Bollinne, A. 1977. La vitesse de l'érosion sous culture en région limoneuse. *Pédologie* 27: 191–206.

Bollinne, A. 1978. Study of the importance of splash and wash on cultivated loamy soils of Hesbaye (Belgium). *Earth Surface Processes* 3: 71–84.

Bollinne, A., Laurant, A. and Boon, W. 1979. L'érosivité des précipitations à Florennes. Révision de la carte des isohyètes et de la carte d'érosivité de la Belgique. *Bulletin de la Société Géographique de Liège* 15: 77–99.

Bonell, M. and Gilmour, D.A. 1978. The development of overland flow in a tropical rain forest catchment. *Journal of Hydrology* 39: 365–82.

Bonsu, M. 1981. Assessment of erosion under different cultural practices on a savanna soil in the northern region of Ghana. In Morgan, R.P.C. (ed), *Soil conservation: problems and prospects*. Wiley, Chichester: 247–53.

Bonsu, M. 1985. Organic residues for less erosion and more grain in Ghana. In El-Swaify, S.A., Moldenhauer, W.C. and Lo, A. (eds), *Soil erosion and conservation*. Soil Conservation Society of America, Ankeny, IA: 615–20.

Bonsu, M. and Obeng, H.B. 1979. Effects of cultural practices on soil erosion and maize production in the semi-deciduous rain forest and forest-savanna transition zones of Ghana. In Lal, R. and Greenland, D.J. (eds), *Soil physical properties and crop production in the tropics*. Wiley, Chichester: 509–19.

Boon, W. and Savat, J. 1981. A nomogram for the prediction of rill erosion. In Morgan, R.P.C. (ed), *Soil conservation: problems and prospects*. Wiley, Chichester: 303–19.

Boonchee, S., Peukrai, S. and Chinabutr, N. 1988. Effects of land development and management on soil conservation in northern Thailand. In Rimwanich, S. (ed), *Land conservation for future generations*. Department of Land Development, Bangkok: 1101–12.

Borghi, C.E., Giannoni, S.M. and Martinez-Rica, J.P. 1990. Soil removed by voles of the genus *Pitymys* in the Spanish Pyrenees. *Pirineos* 136: 3–17.

Borst, H.L. and Woodburn, R. 1942. The effect of mulching and methods of cultivation on runoff and erosion from Muskingum silt loam. *Agricultural Engineering* 23: 19–22.

Bork, H.R. 1989. The history of soil erosion in southern Lower Saxony. *Landschaftsgenese und Landschaftsökologie* 16: 135–63.

Bork, H.R. and Rohdenburg, H. 1981. Rainfall simulation in south-east Spain. In Morgan, R.P.C. (ed), *Soil conservation: problems and prospects*. Wiley, Chichester: 293–302.

Bork, H.R., Bork, H., Dalchow, C., Faust, B., Priorr, H.P. and Schatz, T. 1998. *Landschaftsentwicklung in Mit-*

teleuropa. Wirkungen des Menschen auf die Landschaften. Klett-Perthes, Gotha.

Boubakari, M. and Morgan, R.P.C. 1999. Contour grass strips for soil erosion control on steep lands: a laboratory simulation. *Soil Use and Management* 15: 21–6.

Bouma, N.A. and Imeson, A.C. 2000. Investigation of relationships between measured field indicators and erosion processes on badland surfaces at Petrer, Spain. *Catena* 40: 147–71.

Bouyoucos, G.J. 1935. The clay ratio as a criterion of susceptibility of soils to erosion. *Journal of the American Society of Agronomy* 27: 738–51.

Bowyer-Bower, T.A.S. 1993. Effects of rainfall intensity and antecedent moisture on the steady-state infiltration rate in a semi-arid region. *Soil Use and Management* 9: 69–76.

Brabben, T., Bird, J. and Bolton, P. 1988. Improving the survey and computation of reservoir sedimentation. *ODU Bulletin* 12: 4–7.

Bradford, J.M. and Piest, R.F. 1980. Erosional development of valley-bottom gullies in the Upper Mid-Western United States. In Coates, D.R. and Vitek, J.D. (eds), *Thresholds in geomorphology*. Allen and Unwin, London: 75–101.

Bradford, J.M., Truman, C.C. and Huang, C. 1992. Comparisons of three measures of resistance of soil surface seals to raindrop splash. *Soil Technology* 5: 47–56.

Bradsen, J. 1994. Natural resource conservation in Australia: some fundamental issues. In Napier, T.L., Camboni, S.M. and El-Swaify, S.A. (eds), *Adopting conservation on the farm: an international perspective on the socioeconomics of soil and water conservation*. Soil and Water Conservation Society, Ankeny, IA: 435–60.

Brandt, C.J. 1989. The size distributions of throughfall drops under vegetation canopies. *Catena* 16: 507–24.

Brandt, C.J. 1990. Simulation of the size distribution and erosivity of raindrops and throughfall drops. *Earth Surface Processes and Landforms* 15: 687–98.

Brazier, R.E., Beven, K.J., Freer, J. and Rowan, J.S. 2000. Equifinality and uncertainty in physically based soil erosion models: application of the GLUE methodology to WEPP. *Earth Surface Processes and Landforms* 25: 825–45.

Bridges, E.M. and Oldeman, L.R. 2001. Food production and environmental degradation. In Bridges, E.M., Hannam, I.D., Oldeman, L.R., Penning de Vries, F.W.T., Scherr, S.J. and Sombatpanit, S. (eds), *Response to land degradation*. Science Publishers, Enfield, NH: 36–43.

Brierley, J.S. 1976. Kitchen gardens in the West Indies with a contemporary study from Grenada. *Journal of Tropical Geography* 43: 30–40.

Brown, L.C. and Foster, G.R. 1987. Storm erosivity using idealized intensity distributions. *Transactions of the American Society of Agricultural Engineers* 30: 379–86.

Browning, G.M., Norton, R.A., McCall, A.G. and Bell, F.G. 1948. *Investigation in erosion control and the reclamation of eroded land at the Missouri Valley Loess Conservation Experiment Station, Clarinda, Iowa.* USDA Technical Bulletin 959.

Bruce-Okine, E. and Lal, R. 1975. Soil erodibility as determined by raindrop technique. *Soil Science* 119: 149–57.

Brunori, F., Penzo, M.C. and Torri, D. 1989. Soil shear strength: its measurement and soil detachability. *Catena* 16: 59–71.

Brunsden, D. and Prior, D.B. 1984. *Slope instability.* Wiley, Chichester.

Bryan, R.B. 1968. The development, use and efficiency of indices of soil erodibility. *Geoderma* 2: 5–26.

Bryan, R.B. 1979. The influence of slope angle on soil entrainment by sheetwash and rainsplash. *Earth Surface Processes* 4: 43–58.

Bryan, R.B. and Poesen, J.W.A. 1989. Laboratory experiments on the influence of slope length on runoff, percolation and rill development. *Earth Surface Processes and Landforms* 14: 211–31.

Bryan, R.B. and Yair, A. 1982. Perspectives on studies of badland geomorphology. In Bryan, R. and Yair, A. (eds), *Badland geomorphology and piping.* Geo Books, Norwich: 1–12.

Bubenzer, G.D. 1979. Rainfall characteristics important for simulation. In *Proceedings, Rainfall simulator workshop, Tucson, Arizona.* USDA-SEA Agricultural Reviews and Manuals ARM-W-10: 22–34.

Bubenzer, G.D. and Jones, B.A. 1971. Drop size and impact velocity effects on the detachment of soil under simulated rainfall. *Transactions of the American Society of Agricultural Engineers* 14: 625–8.

Bunte, K. and Poesen, J. 1994. Effects of rock fragment size and cover on overland flow hydraulics, local turbulence and sediment yield on an erodible soil surface. *Earth Surface Processes and Landforms* 19: 115–35.

Buol, S.W., Hole, F.D. and McCracken, P.J. 1973. *Soil genesis and classification.* Iowa State University Press, Ames, IA.

Buttner, G. and Csillag, F. 1989. Comparative study of crop and soil mapping using multitemporal and multispectral SPOT and Landsat Thematic Mapper data. *Remote Sensing of the Environment* 29: 241–9.

Cammeraat, L.H. 2002. A review of two strongly contrasting geomorphological systems within the context of scale. *Earth Surface Processes and Landforms* 27: 1201–22.

Campbell, A. 1994. *Landcare: community shaping the land and the future.* Allen and Unwin, Sydney.

Carson, M.A. and Kirkby, M.J. 1972. *Hillslope form and process.* Cambridge University Press, Cambridge.

Carte, A. 1971. Raindrop spectra in Pretoria. *South African Geographical Journal* 53: 100–3.

Carter, C.E., Greer, J.D., Braud, H.J. and Floyd, J.M. 1974. Raindrop characteristics in south central United States. *Transactions of the American Society of Agricultural Engineers* 17: 1033–7.

Carver, M. and Schreier, H. 1995. Sediment and nutrient budgets over four spatial scales in the Jhikhu Khola watershed: implications for land use management. In Schreier, H., Shah, P.B. and Brown, S. (eds), *Challenges in mountain resource management in Nepal: processes, trends and dynamics in middle mountain watersheds.* ICIMOD/IDRC/UBC, Kathmandu: 163–70.

Cazzuffi, D., Monti, R. and Rimoldi, P. 1991. Geosynthetics subjected to different conditions of rain and runoff in erosion control application: a laboratory investigation. In *Erosion control: a global perspective.* International Erosion Control Association, Steamboat Springs, CO: 191–208.

Cerdà, A., Ibáñez, S. and Calvo, A. 1997. Design and operation of a small and portable rainfall simulator for rugged terrain. *Soil Technology* 11: 163–70.

Cerdan, O., Le Bissonnais, Y., Couturier, A., Bourennane, H. and Souchère, V. 2002a. Rill erosion on cultivated hillslopes during two extreme rainfall events in Normandy, France. *Soil and Tillage Research* 67: 99–108.

Cerdan, O., Le Bissonnais, Y., Martin, P. and Lecomte, V. 2002b. Sediment concentration in interrill flow: interactions between soil surface conditions, vegetation and rainfall. *Earth Surface Processes and Landforms* 27: 193–205.

Cerro, C., Bech, J., Codina, B. and Lorente, J. 1998. Modeling rain erosivity using disdrometric techniques. *Soil Science Society of America Journal* 62: 731–5.

Cervera, M., Clotet, N. and Sala, M. 1990. Runoff and sediment production on small badland basins in the Upper Llobregat catchment (submediterranean environment). Internal Report, LUCDEME Project.

Chamen, T., Alakukku, L., Pires, S., Sommer, C., Spoor, G., Tijink, F. and Weisskopf, P. 2003. Prevention strategies for field traffic-induced subsoil compaction: a review. Part 2. Equipment and field practices. *Soil and Tillage Research* 73: 161–74.

Chan, C.C. 1981a. Evaluation of soil loss factors on cultivated slopelands of Taiwan. *Food and Fertilizer Technology Center Taipei, Technical Bulletin* 55.

References

Chan, C.C. 1981b. Conservation measures in the cultivated slopes of Taiwan. *Food and Fertilizer Technology Center Taipei, Extension Bulletin* 137.

Chapman, D.M. 1990. Aeolian sand transport – an optimized model. *Earth Surface Processes and Landforms* 15: 751–60.

Chapman, G. 1948. Size of raindrops and their striking force at the soil surface in a red pine plantation. *Transactions of the American Geophysical Union* 29: 664–70.

Chappell, A., McTainsh, G., Leys, J. and Strong, G. 2003. Simulations to optimize sampling of aeolian sediment transport in space and time for mapping. *Earth Surface Processes and Landforms* 28: 1223–41.

Charman, P.E.V. 1978. Soils of New South Wales: their characterisation, classification and conservation. *NSW Soil Conservation Service Technical Handbook* 1.

Cheatle, R.J. and Njoroge, S.N.J. 1993. Smallholder adoption of some land husbandry practices in Kenya. In Hudson, N. and Cheatle, R.J. (eds), *Working with farmers for better land husbandry*. IT Publications, Southampton: 130–41.

Chepil, W.S. 1945. Dynamics of wind erosion. III. Transport capacity of the wind. *Soil Science* 60: 475–80.

Chepil, W.S. 1946. Dynamics of wind erosion. VI. Sorting of soil material by wind. *Soil Science* 61: 331–40.

Chepil, W.S. 1950. Properties of soil which influence wind erosion. II. Dry aggregate structure as an index of erodibility. *Soil Science* 69: 403–14.

Chepil, W.S. 1960. Conversion of relative field erodibility to annual soil loss by wind. *Soil Science Society of America Proceedings* 24: 143–5.

Chepil, W.S., Siddoway, F.H. and Armbrust, D.V. 1964. Wind erodibility of knolly terrain. *Journal of Soil and Water Conservation* 19: 179–81.

Chepil, W.S. and Woodruff, N.P. 1963. The physics of wind erosion and its control. *Advances in Agronomy* 15: 211–302.

Chepil, W.S., Woodruff, N.P. and Zingg, A.W. 1955. Field study of wind erosion in western Texas. *USDA Soil Conservation Service Technical Paper* SCS-TP-125.

Chisci, G. 1986. Influence of change in land use and management on the acceleration of land degradation phenomena in the Apennines hilly areas. In Chisci, G. and Morgan, R.P.C. (eds), *Soil erosion in the European Community: impact of changing agriculture*. Balkema, Rotterdam: 3–18.

Chisci, G., Bazzoffi, P. and Mbagwu, J.S.C. 1989. Comparison of aggregate stability indices for soil classification and assessment of soil management practices. *Soil Technology* 2: 113–33.

Chisci, G. and Zanchi, C. 1981. The influence of different tillage systems and different crops on soil losses on hilly silty-clayey soil. In Morgan, R.P.C. (ed.), *Soil conservation: problems and prospects*. Wiley, Chichester: 211–17.

Chisci, G., Zanchi, C. and Biagi, B. 1978. Drenaggio profondo in un suolo limo-argilloso su sedimenti plioceni marini in funzione di differenti colture. *Annali Istituto Sperimentale per lo Studio e la Difesa del Suolo* 9: 257–72.

Chorley, R.J. 1959. The geomorphic significance of some Oxford soils. *American Journal of Science* 257: 503–15.

Clark, R., Durón, G., Quispe, G. and Stocking, M.A. 1999. Boundary bunds or piles of stones? Using farmers' practices in Bolivia to aid soil conservation. *Mountain Research and Development* 19: 235–40.

Clarke, M.A., Walsh, R.P.D., Sinun, W., Larenus, J. and Hanapi, J. 2002. Soil erosion and surface roughness dynamics in relation to slope angle in primary and logged tropical rainforest, eastern Sabah. In Rubio, J.L., Morgan, R.P.C., Asins, S. and Andreu, V. (eds), *Soil at the third millennium*. Geoforma Ediciones, Logroño: 1471–84.

Cogo, N.P., Moldenhauer, W.C. and Foster, G.R. 1984. Soil loss reductions from conservation tillage practices. *Soil Science Society of America Journal* 48: 368–73.

Colclough, J.D. 1965. Soil conservation and soil erosion control in Tasmania: tunnel erosion. *Tasmanian Journal of Agriculture* 36: 7–12.

Coleman, R. 1981. Footpath erosion in the English Lake District. *Applied Geography* 1: 121–31.

Collins, A.L. and Walling, D.E. 2002. Selecting finger-print properties for discriminating sediment sources in river basins. *Journal of Hydrology* 261: 218–44.

Combeau, A. and Monnier, G. 1961. A method for the study of structural stability: application to tropical soils. *African Soils* 6: 33–52.

Cooke, R.U. and Doornkamp, J.C. 1974. *Geomorphology in environmental management*. Oxford University Press, Oxford.

Coote, D.R., Malcolm-McGovern, C.A., Wall, G.J., Dickinson, W.T. and Rudra, R.P. 1988. Seasonal variation of erodibility indices based on shear strength and aggregate stability in some Ontario soils. *Canadian Journal of Soil Science* 68: 405–16.

Coppin, N.J. and Richards, I.G. 1990. *Use of vegetation in civil engineering*. CIRIA/Butterworths, London.

Coutinho, M.A. and Tomás, P.P. 1995. Characterisation of raindrop size distributions at the Vale Formoso Experimental Erosion Center. *Catena* 25: 187–97.

Critchley, W.R.S. and Bruijnzeel, L.A. 1995. Terrace risers: erosion control or sediment source? In Singh, R.B. and Haigh, M.J. (eds), *Sustainable reconstruction of highland*

and headwater regions. Oxford and IBH Publishing, New Delhi: 529–41.

Critchley, W.R.S., Reij, C. and Willcocks, T.J. 1994. Indigenous soil and water conservation: a review of the state of knowledge and prospects of building on traditions. *Land Degradation and Rehabilitation* 5: 293–314.

Critchley, W.R.S., Sombatpanit, S. and Medina, S.M. 2001. Uncertain steps? Terraces in the tropics. In Bridges, E.M., Hannam, I.D., Oldeman, L.R., Penning de Vries, F.W.T., Scherr, S.J. and Sombatpanit, S. (eds), *Response to land degradation.* Science Publishers, Enfield, NH: 325–38.

Crosson, P. 1995. Soil erosion estimates and costs. *Science* 269: 461–4.

Crouch, R.J. 1976. Field tunnel erosion – a review. *Journal of the Soil Conservation Service NSW* 32: 98–111.

Crouch, R.J. 1978. Variation in the structural stability of soil in a tunnel-eroding area. In Emerson, W.W., Bond, R.D. and Dexter, A.R. (eds), *Modification of soil structure.* Wiley, Chichester: 267–74.

Crouch, R.J. 1990a. Erosion processes and rates for gullies in granitic soils, Bathurst, New South Wales, Australia. *Earth Surface Processes and Landforms* 15: 169–73.

Crouch, R.J. 1990b. Rates and mechanisms of discontinuous gully erosion a red-brown earth catchment, New South Wales, Australia. *Earth Surface Processes and Landforms* 15: 277–82.

Crouch, R.J., McGarity, J.W. and Storrier, R.R. 1986. Tunnel formation processes in the Riverina area of NSW, Australia. *Earth Surface Processes and Landforms* 11: 157–68.

Crowley, C.M. 2003. Ten rules of thumb for culvert crossings. *Erosion Control* 10(6): 72–5.

Cruse, R.M. and Larson, W.E. 1977. Effect of soil shear strength on soil detachment due to raindrop impact. *Soil Science Society of America Journal* 41: 777–81.

Cruz, E.B. 1996. The adoption of hedgerows as a soil conservation measure in the Philippines. In Sombatpanit, S., Zöbisch, M.A., Sanders, D.W. and Cook, M.G. (eds), *Soil conservation extension: from concepts to adoption.* Soil and Water Conservation Society of Thailand, Bangkok: 447–55.

Currence, H.D. and Lovely, W.G. 1970. The analysis of surface roughness. *Transactions of the American Society of Agricultural Engineers* 13: 710–14.

Daddow, R.L. and Warrington, G.E. 1983. Growth-limiting soil bulk densities as influenced by soil texture. *WDG Report,* WSDG-TN-00005, USDA Forest Service.

Davidson, J.L. and Quirk, J.P. 1961. The influence of dissolved gypsum on pasture establishment on irrigated sodic clays. *Australian Journal of Agricultural Research* 12: 100–10.

De Bano, L.F., Mann, L.D. and Hamilton, D.A. 1970. Translocation of hydrophobic substances into soil by burning organic litter. *Soil Science Society of America Proceedings* 34: 130–3.

Dedkov, A.P. and Mozzherin, V.I. 1996. Erosion and sediment yield on the Earth. *International Association of Scientific Hydrology Publication* 236: 29–33.

De Graaf, J. 2001. The economic appraisal of soil and water conservation measures. In Bridges, E.M., Hannam, I.D., Oldeman, L.R., Penning de Vries, F.W.T., Scherr, S.J. and Sombatpanit, S. (eds), *Response to land degradation.* Science Publishers, Enfield, NH: 274–83.

De Jong, E., Begg, C.B.M. and Kachanoski, R.G. 1983. Estimates of soil erosion and deposition for some Saskatchewan soils. *Canadian Journal of Soil Science* 63: 607–17.

De Jong, S.M. 1994. Applications of reflective remote sensing for land degradation studies in a Mediterranean environment. *Nederlandse Geografische Studies* 177.

De Jong, S.M., Paracchini, M.L., Bertolo, F., Folving, S., Megier, J. and De Roo, A.P.J. 1999. Regional assessment of soil erosion using the distributed model SEMMED and remotely sensed data. *Catena* 37: 291–308.

De Kesel, M. and De Vleeschauwer, D. 1981. Sand dune fixation in Tunisia by means of polyurea polyalkylene oxide (Uresol). In Lal, R. and Russell, E.W. (eds), *Tropical agricultural hydrology.* Wiley, Chichester: 273–81.

De Meis, M.R.M. and De Moura, J.R.S. 1984. Upper Quaternary sedimentation and hillslope evolution: southeastern Brazilian plateau. *American Journal of Science* 284: 241–54.

Dent, F.J. 1996. The FAO land conservation and rehabilitation scheme. In Sombatpanit, S., Zöbisch, M.A., Sanders, D.W. and Cook, M.G. (eds), *Soil conservation extension: from concepts to adoption.* Soil and Water Conservation Society of Thailand, Bangkok: 301–14.

De Ploey, J. 1981. The ambivalent effects of some factors of erosion. *Mémoirs, Institute de Géologie, l'Université de Louvain* 31: 171–81.

De Ploey, J. 1982. A stemflow equation for grasses and similar vegetation. *Catena* 9: 139–52.

De Ploey, J. 1989a. A model for headcut retreat in rills and gullies. *Catena Supplement* 14: 81–6.

De Ploey, J. 1989b. Erosional systems and perspectives for erosion control in European loess areas. In Schwertmann, U., Rickson, R.J. and Auerswald, K. (eds), *Soil erosion protection measures in Europe.* Soil Technology Series 1: 93–102.

De Ploey, J. and Cruz, O. 1979. Landslides in the Serra do Mar, Brazil. *Catena* 6: 111–22.

De Ploey, J. and Gabriels, D. 1980. Measuring soil loss and experimental studies. In Kirkby, M.J. and Morgan, R.P.C. (eds), *Soil erosion*. Wiley, Chichester: 63–108.

De Ploey, J., Savat, J. and Moeyersons, J. 1976. The differential impact of some soil factors on flow, runoff creep and rainwash. *Earth Surface Processes* 1: 151–61.

De Roo, A.P.J. 1996. Validation problems of hydrologic and soil-erosion catchment models: examples from a Dutch erosion project. In Anderson, M.G. and Brooks, S.M. (eds), *Advances in hillslope processes*. Wiley, Chichester: 669–83.

DeRose, R.C., Gomez, B., Marden, M. and Trustrum, N.A. 1998. Gully erosion in Mangatu Forest, New Zealand, estimated from digital elevation models. *Earth Surface Processes and Landforms* 23: 1045–53.

Dickinson, W.T., Rudra, R.P. and Wall, G.J. 1986. Identification of soil erosion and fluvial sediment problems. *Hydrological Processes* 1: 111–24.

Dimitrakopoulos, A.P., Martin, R.E. and Papamichos, N.T. 1994. A simulation model of soil heating during wildland fires. In Sala, M. and Rubio, J.L. (eds), *Soil erosion and degradation as a consequence of forest fires*. Geoforma Ediciones, Logroño: 199–206.

Dimitrakopoulos, A.P. and Seilopoulos, D. 2002. Effects of rainfall and burning intensity on early post-fire soil erosion in a Mediterranean forest of Greece. In Rubio, J.L., Morgan, R.P.C., Asins, S. and Andreu, V. (eds), *Soil at the third millennium*. Geoforma Ediciones, Logroño: 1351–7.

Disrud, L.A. and Krauss, R.K. 1971. Examining the process of soil detachment from clods exposed to wind-driven simulated rainfall. *Transactions of the American Society of Agricultural Engineers* 14: 90–2.

Dissmeyer, G.E. and Foster, G.R. 1981. Estimating the cover-management factor (C) in the Universal Soil Loss Equation for forest conditions. *Journal of Soil and Water Conservation* 36: 235–40.

Dlamini, P.M. and Maro, P.S. 1988. The economic behaviour of the farming communities on conservation programmes: Swaziland and the SADCC region. Paper presented to Second Workshop, Economics of Conservation (Swaziland).

Doerr, S.H., Leighton-Boyce, G., Shakesby, R.A. and Walsh, R.P.D. 2002. The role of hydrophobicity in soil erosion processes: evidence from Portuguese forest soils. In Rubio, J.L., Morgan, R.P.C., Asins, S. and Andreu, V. (eds), *Soil at the third milllennium*. Geoforma Ediciones, Logroño: 1311–22.

Doerr, S.H., Shakesby, R.A. and Walsh, R.P.D. 2000. Soil water repellency: its causes, characteristics and hydrogeomorphological significance. *Earth-Science Reviews* 51: 33–65.

Dolgilevich, M.I., Sofronova, A.A. and Mayevskaya, L.L. 1973. Klassifikatsia pochv zapadnoy Sibiri, severnogo Kazakhstana po stepeniy podatlivosti k vetrovoy eroziy. *Bulletin VNIIA* No. 12.

Dong, Z., Liu, X., Li, F., Wang, H. and Zhao, A. 2002. Impact-entrainment relationship in a saltating cloud. *Earth Surface Processes and Landforms* 27: 641–58.

Dotterweich, M., Schmitt, A., Schmidtchen, G. and Bork, H.R. 2003. Quantifying historical gully erosion in northern Bavaria. *Catena* 50: 135–50.

Douglas, I. 1978. The impact of urbanisation on fluvial geomorphology in the humid tropics. *Géo-Eco-Trop* 2: 239–42.

Douglas, J.T. and Goss, M.J. 1982. Stability and organic matter content of surface soil aggregates under different methods of cultivation and in grassland. *Soil and Tillage Research* 2: 155–76.

Downes, R.G. 1946. Tunnelling erosion in north-eastern Victoria. *Journal of the Council of Scientific and Industrial Research* 19: 283–92.

Dregne, H.E. 1982. Desertification: man's abuse of the land. *Journal of Soil and Water Conservation* 33: 11–14.

D'Souza, V.P.C. and Morgan, R.P.C. 1976. A laboratory study of the effect of slope steepness and curvature on soil erosion. *Journal of Agricultural Engineering Research* 21: 21–31.

Duley, F.L. and Russel, J.C. 1943. Effect of stubble mulching on soil erosion and runoff. *Soil Science Society of America Proceedings* 7: 77–81.

Dunne, T. and Aubry, B.F. 1986. Evaluation of Horton's theory of sheetwash and rill erosion on the basis of field experiments. In Abrahams, A.D. (eds), *Hillslope processes*. Allen and Unwin, London: 31–53.

Dunne, T. and Black, R.D. 1970. Partial area contributions to storm runoff in a small New England watershed. *Water Resources Research* 6: 1296–311.

Dyrness, C.T. 1975. Grass-legume mixtures for erosion control along forest roads in western Oregon. *Journal of Soil and Water Conservation* 30: 169–73.

Edwards, K. 1993. Soil erosion and conservation in Australia. In Pimental, D. (ed), *World soil erosion and conservation*. Cambridge University Press, Cambridge: 147–69.

Edwards, W.M. and Owens, L.B. 1991. Large storm effects on total soil loss. *Journal of Soil and Water Conservation* 46: 75–8.

Eger, H. and Bado, J. 1992. Village-level land management in the central plateau of Burkina Faso. In Hurni, H. and Tato, K. (eds), *Erosion, conservation and small-scale farming*. Geographica Bernensia, Bern: 531–40.

Ekwue, E.I. 1990. Effect of organic matter on splash

detachment and the processes involved. *Earth Surface Processes and Landforms* 15: 175–81.

Ekwue, E.I. 1992. Effect of organic and fertiliser treatments on soil physical properties and erodibility. *Soil and Tillage Research* 22: 199–209.

Ekwue, E.I., Ohu, J.O. and Wakawa, I.H. 1993. Effects of incorporating two organic materials at varying levels on splash detachment of some soils from Borno State, Nigeria. *Earth Surface Processes and Landforms* 18: 399–406.

Ellis-Jones, J. and Mason, T. 1999. Livelihood strategies and assets of small farmers in the evaluation of soil and water management practices in the temperate Inter-Andean valleys of Bolivia. *Mountain Research and Development* 19: 221–34.

Ellison, W.D. 1944. Two devices for measuring soil erosion. *Agricultural Engineering* 25: 53–5.

El-Swaify, S.A. 1994. State of the art for assessing soil and water conservation needs and technologies. In Napier, T.L., Camboni, S.M. and El-Swaify, S.A. (eds), *Adopting conservation on the farm. An international perspective on the socioeconomics of soil and water conservation.* Soil and Water Conservation Society, Ankeny, IA: 13–27.

El-Swaify, S.A., Dangler, E.W. and Armstrong, C.L. 1982. *Soil erosion by water in the tropics.* College of Tropical Agriculture and Human Resources, University of Hawaii.

El-Swaify, S.A., Pathak, P., Rego, T.J. and Singh, S. 1985. Soil management for optimized productivity under rainfed conditions in the semi-arid tropics. *Advances in Soil Science* 1: 1–64.

Elwell, H.A. 1978. Modelling soil losses in Southern Africa. *Journal of Agricultural Engineering Research* 23: 111–27.

Elwell, H.A. 1981. A soil loss estimation technique for southern Africa. In Morgan, R.P.C. (ed.), *Soil conservation: problems and prospects.* Wiley, Chichester: 281–92.

Elwell, H.A. and Stocking, M.A. 1976. Vegetal cover to estimate soil erosion hazard in Rhodesia. *Geoderma* 15: 61–70.

Elwell, H.A. and Wendelaar, F.E. 1977. To initiate a vegetal cover data bank for soil loss estimation. *Department of Conservation and Extension, Research Bulletin* 23.

Emama Ligdi, E. and Morgan, R.P.C. 1995. Contour grass strips: a laboratory simulation of their role in soil erosion control. *Soil Technology* 8: 109–17.

Emmett, W.W. 1965. The vigil network: methods of measurement and a sample of data collected. *International Association of Scientific Hydrology Publication* 66: 89–106.

Emmett, W.W. 1970. The hydraulics of overland flow on hillslopes. *USGS Professional Paper* 662-A.

Engman, E.T. 1986. Roughness coefficients for routing surface runoff. *Journal of the Irrigation and Drainage Division ASCE* 112: 39–53.

Enters, T. 1999. Incentives as policy instruments – key concepts and definitions. In Sanders, D.W., Huszar, P.C., Sombatpanit, S. and Enters, T. (eds), *Incentives in soil conservation: from theory to practice.* Oxford and IBH Publishing, New Delhi: 25–40.

Enters, T. 2001. Incentives for soil conservation. In Bridges, E.M., Hannam, I.D., Oldeman, L.R., Penning de Vries, F.W.T., Scherr, S.J. and Sombatpanit, S. (eds), *Response to land degradation.* Science Publishers, Enfield, NH: 351–60.

Environment Agency, 2002. *Agriculture and natural resources: benefits, costs and potential solutions.* Environment Agency, Bristol.

Eppink, L.A.A.J. and Spaan, W.P. 1989. Agricultural wind erosion control measures in The Netherlands. In Schwertmann, U., Rickson, R.J. and Auerswald, K. (eds), *Soil erosion protection measures in Europe.* Soil Technology Series 1: 1–13.

Ervin, D.E. and Washburn, R. 1981. Profitability of soil conservation practices in Missouri. *Journal of Soil and Water Conservation* 36: 107–11.

Escarameia, M. 1998. *River and channel revetments – a design manual.* Thomas Telford, London.

Evans, A.C. 1948. Studies on the relationships between earthworms and soil fertility. II. Some effects of earthworms and soil structure. *Applied Biology* 35: 1–13.

Evans, R. 1977. Overgrazing and soil erosion on hill pastures with particular reference to the Peak District. *Journal of the British Grassland Society* 32: 65–76.

Evans, R. 1980. Mechanics of water erosion and their spatial and temporal controls: an empirical viewpoint. In Kirkby, M.J. and Morgan, R.P.C. (eds), *Soil erosion.* Wiley, Chichester: 109–28.

Evans, R. and Nortcliff, S. 1978. Soil erosion in north Norfolk. *Journal of Agricultural Science Cambridge* 90: 185–92.

Everaert, W. 1991. Empirical relations for the sediment transport capacity of interrill flow. *Earth Surface Processes and Landforms* 16: 513–32.

Eyles, R.J. 1967. Laterite at Kerdau, Pahang, Malaya. *Journal of Tropical Geography* 25: 18–23.

Fang, Z.S., Zhou, P.H., Liu, Q.D., Liu, B.H., Ren, L.T. and Zhang, H.X. 1981. Terraces in the loess plateau of China. In Morgan, R.P.C. (ed.), *Soil conservation: problems and prospects.* Wiley, Chichester: 481–513.

FAO 1965. *Soil erosion by water.* FAO, Rome.

FAO 1976. A framework for land evaluation. *FAO Soils Bulletin* 32.

Farres, P. 1978. The role of time and aggregate size in the crusting process. *Earth Surface Processes* 3: 243–54.

Fenster, C.R. and McCalla, T.M. 1970. Tillage practices in western Nebraska with a wheat-fallow rotation. *Nebraska Agriculture Station, Lincoln, Bulletin* 597.

Fenton, E.W. 1937. The influence of sheep on the vegetation of hill grazings in Scotland. *Journal of Ecology* 25: 424–30.

Fifield, J.S. 1999. Effective sediment and erosion control for construction sites. *Erosion Control* 5(6): 36–40.

Fifield, J.S. and Malnor, L.K. 1990. Erosion control materials vs a semiarid environment. What has been learned from three years of testing? In *Erosion control: technology in transition*. International Erosion Control Association, Steamboat Springs, CO: 233–48.

Fifield, J.S., Malnor, L.K. and Dezman, L.E. 1989. Effectiveness of erosion control products on steep slopes to control sediment and to establish dryland grasses. In *Erosion knows no boundaries*. International Erosion Control Association, Steamboat Springs, CO: 45–55.

Finney, H.J. 1984. The effect of crop covers on rainfall characteristics and splash detachment. *Journal of Agricultural Engineering Research* 29: 337–43.

Flach, K.W., Barnwell Jr, T.O. and Crosson, P. 1997. Impacts of agriculture on atmospheric carbon dioxide. In Paul, E.A., Paustian, K., Elliott, E.T. and Cole, C.V. (eds), *Soil organic matter in temperate agroecosystems: long-term experiments in North America*. CRC Press, Boca Raton, FL: 3–13.

Folly, A., Quinton, J.N. and Smith, R.E. 1999. Evaluation of the EUROSEM model using data from the Catsop watershed, The Netherlands. *Catena* 37: 507–19.

Foster, G.R. 1982. Modeling the erosion process. In Haan, C.T., Johnson, H.P. and Brakensiek, D.L. (eds), *Hydrologic modeling of small watersheds*. American Society of Agricultural Engineers Monograph 5: 297–380.

Foster, G.R., Johnson, C.B. and Moldenhauer, W.C. 1982. Hydraulics of failure of unanchored corn-stalk and wheat straw mulches for erosion control. *Transactions of the American Society of Agricultural Engineers* 25: 940–7.

Foster, G.R., Lane, L.J., Nowlin, J.D., Laflen, J.M. and Young, R.A. 1981. Estimating erosion and sediment yield on field-sized areas. *Transactions of the American Society of Agricultural Engineers* 24: 1253–63.

Foster, G.R. and Meyer, L.D. 1972. A closed-form soil erosion equation for upland areas. In Shen, H.W. (ed.), *Sedimentation*. Department of Civil Engineer-ing, Colorado State University, Fort Collins: 12.1–12.19.

Foster, R.L. and Martin, G.L. 1969. Effect of unit weight and slope on erosion. *Journal of the Irrigation and Drainage Division ASCE* 95: 551–61.

Fournier, F. 1960. *Climat et érosion: la relation entre l'érosion du sol par l'eau et les précipitations atmosphériques*. Presses Universitaires de France, Paris.

Fournier, F. 1972. *Soil conservation*. Nature and Environment Series, Council of Europe.

Fox, D. and Bryan, R.B. 1992. Influence of a polyacrylamide soil conditioner on runoff generation and soil erosion: field tests in Baringo District, Kenya. *Soil Technology* 5: 101–19.

Fox, F.A., Flanagan, D.C., Wagner, L.E. and Deer-Ascough, L. 2001. WEPS and WEPP science commonality project. In Ascough II, J.C. and Flanagan, D.C. (eds), *Proceedings of the ASAE symposium on soil erosion research for the 21st century*: 376–79.

Francis, I.S. and Taylor, J.A. 1989. The effect of forestry drainage operations on upland sediment yields: a study of two peat-covered catchments. *Earth Surface Processes and Landforms* 14: 73–83.

Franke, R. and Chasin, B.H. 1981. Peasants, peanuts, profits and pastoralists. *The Ecologist* 11: 156–68.

Fredén, C. and Furuholm, L. 1978. The Säterberget Gully at Brattforsheden, Värmland. *Geologiska Föreningens i Stockholm Förhandlingar* 100: 231–5.

Free, G.R. 1960. Erosion characteristics of rainfall. *Agricultural Engineering* 41: 447–9, 455.

Freer-Hewish, R.J. 1991. Erosion of road shoulders from rainfall and runoff. In *Erosion control: a global perspective*. International Erosion Control Association, Steamboat Springs, CO: 263–73.

Froehlich, W. and Starkel, L. 1993. The effects of deforestation on slope and channel evolution in the tectonically active Darjeeling Himalaya. *Earth Surface Processes and Landforms* 18: 285–90.

Fryrear, D.W. 1984. Soil ridges-clods and wind erosion. *Transactions of the American Society of Agricultural Engineers* 27: 445–8.

Fryrear, D.W. 1986. A field dust sampler. *Journal of Soil and Water Conservation* 41: 117–20.

Fryrear, D.W., Saleh, A., Bilbro, J.D., Schomberg, H.M., Stout, J.E. and Zobeck, T.M. 1998. *Revised Wind Erosion Equation (RWEQ)*. Wind Erosion and Water Conservation Research Unit, USDA-ARS, Southern Plains Area Cropping Systems Research Laboratory, Technical Bulletin 1.

Fryrear, D.W., Stout, J.E., Hagen, L.J. and Vories, E.D. 1991. Wind erosion: field measurement and analysis. *Transactions of the American Society of Agricultural Engineers* 34: 155–60.

Fullen, M.A., Tye, A.M. and Cookson, K.E. 1995. Effects of 'Agri-SC' soil conditioner on soil structure and erodibility: some further observations. *Soil Use and Management* 11: 183–5.

Gabriels, D. 2002. Rain erosivity in Europe. In Rubio, J.L., Morgan, R.P.C., Asins, S. and Andreu, V. (eds), *Man and soil at the third millennium*. Geoforma Ediciones, Logroño: 99–108.

Gabriels, D., Cornelis, W., Pollet, I., van Coillie, T. and Ouessar, M. 1997. The ICE wind tunnel for wind and water erosion studies. *Soil Technology* 10: 1–8.

Gabriels, D. and De Boodt, M. 1978. Evaluation of soil conditioners for water erosion control and sand stabilization. In Emerson, W.W., Bond, R.D. and Dexter, A.R. (eds), *Modification of soil structure*. Wiley, Chichester: 341–8.

Gabriels, D., Maene, L., Lenvain, J. and De Boodt, M. 1977. Possibilities of using soil conditioners for soil erosion control. In Greenland, D.J. and Lal, R. (eds), *Soil conservation and management in the humid tropics*. Wiley, London: 99–108.

Gabriels, D., Pauwels, J.M. and De Boodt, M. 1975. The slope gradient as it affects the amounts and size distribution of soil loss material from runoff on silt loam aggregates. *Mededelingen Fakulteit Landbouwwetenschappen, Rijksuniversiteit Gent* 40: 1333–8.

Gale, W.J., McColl, R.W. and Xie, F. 1993. Sandy fields traditional farming water conservation in China. *Journal of Soil and Water Conservation* 48: 474–7.

Gantzer, C.J., Anderson, S.H., Thompson, A.L. and Brown, J.R. 1990. Estimating soil erosion after 100 years of cropping on Sanborn field. *Journal of Soil and Water Conservation* 45: 641–4.

George, H. and Jarvis, M.G. 1979. Land for camping and caravan sites, picnic sites and footpaths. In Jarvis, M.G. and Mackney, D. (eds), *Soil survey applications*. Soil Survey of England and Wales Technical Monograph 13: 166–83.

Gerlach, T. 1966. Wspólczesby rozwóy stoków w dorzeczu górnego Grajcarka (Beskid Wysoki-Karpaty Zachodnie). *Prace Geograf. IG PAN* 52 (with French summary).

Gerontidis, S., Kosmas, C., Detsis, B., Marathianou, M., Zafiriou, Th. and Tsara, M. 2001. The effect of mouldboard plough on tillage erosion along a hillslope. *Journal of Soil and Water Conservation* 56: 147–52.

Ghadiri, H. and Payne D. 1979. Raindrop impact and soil splash. In Lal, R. and Greenland, D.J. (eds), *Soil physical properties and crop production in the tropics*. Wiley, Chichester: 95–104.

Ghadiri, H., Rose, C.W. and Hogarth, W.L. 2001. The influence of grass and porous barrier strips on runoff hydrology and sediment transport. *Transactions of the American Society of Agricultural Engineers* 44: 259–68.

Gichangi, E.M., Jones, R.K., Njarui, D.M., Simpson, J.R., Mututho, J.M.N. and Kitheka, S.K. 1992. Pitting practices for rehabilitating eroded grazing land in the semi-arid tropics of eastern Kenya: a progress report. In Hurni, H. and Tato, K. (eds), *Erosion, conservation and small-scale farming*. Geographica Bernensia, Bern: 313–27.

Gichungwa, J.K. 1970. *Soil conservation in Central Province (Kenya)*. Ministry of Agriculture, Nairobi.

Gifford, G.F. 1976. Applicability of some infiltration formulae to rangeland infiltrometer data. *Journal of Hydrology* 28: 1–11.

Giger, M., Liniger, H.P. and Critchley, W. 1999. Use of direct incentives and profitability of soil and water conservation in eastern and southern Africa. In Sanders, D.W., Huszar, P.C., Sombatpanit, S. and Enters, T. (eds), *Incentives in soil conservation: from theory to practice*. Oxford and IBH Publishing, New Delhi: 247–74.

Gillette, D.A. and Stockton, P.H. 1989. The effect of nonerodible particles on wind erosion on erodible surfaces. *Journal of Geophysical Research* 94: 12885–93.

Gilley, J.E., Woolhiser, D.A. and McWhorter, D.B. 1985. Interrill soil erosion. Part II. Testing and use of model equations. *Transactions of the American Society of Agricultural Engineers* 28: 154–9.

Giovannini, G. 1994. The effect of fire on soil quality. In Sala, M. and Rubio, J.L. (eds), *Soil erosion and degradation as a consequence of forest fires*. Geoforma Ediciones, Logroño: 15–27.

Giovannini, G., Lucchesi, S. and Giachetti, M. 1988. Effects of heating on some physical and chemical parameters related to soil aggregation and erodibility. *Soil Science* 146: 255–61.

Glasstetter, M. and Prasuhn, V. 1992. The influence of earthworm activity on erodibility and soil losses in central European agricultural soils. In Hurni, H. and Tato, K. (eds), *Erosion, conservation and small-scale farming*. Geographica Bernensia, Bern: 285–98.

Godwin, R.J. and Spoor, G. 1977. Soil failure with narrow tines. *Journal of Agricultural Engineering Research* 22: 213–28.

Golosov, V. 2003. Application of Chernobyl-derived ^{137}Cs for the assessment of soil redistribution within a cultivated field. *Soil and Tillage Research* 69: 85–98.

Gong, S.Y. and Jiang, D.Q. 1977. Soil erosion and its control in small gully watersheds in the rolling loess area on the middle reaches of the Yellow River. *Paris symposium on erosion and soil matter transport in inland waters*, preprint.

Goosens, D., Offer, Z. and London, G. 2000. Wind tunnel and field calibration of five aeolian sand traps. *Geomorphology* 35: 233–52.

References

Gosh, R.L., Kaul, O.N. and Subba-Rao, B.K. 1978. Some aspects of water relations and nutrition in *Eucalyptus* plantations. *Indian Forester* 104: 517–24.

Govers, G. 1985. Selectivity and transport capacity of thin flows in relation to rill erosion. *Catena* 12: 35–49.

Govers, G. 1987. Initiation of motion in overland flow. *Sedimentology* 34: 1157–64.

Govers, G. 1989. Grain velocities in overland flow: a laboratory study. *Earth Surface Processes and Landforms* 14: 481–98.

Govers, G. 1990. Empirical relationships for the transporting capacity of overland flow. *International Association of Hydrological Sciences Publication* 189: 45–63.

Govers, G. 1991. Time-dependency of runoff velocity and erosion: the effect of the initial soil moisture profile. *Earth Surface Processes and Landforms* 16: 713–29.

Govers, G. 1992. Relationship between discharge, velocity and flow area for rills eroding loose, non-layered materials. *Earth Surface Processes and Landforms* 17: 515–28.

Govers, G., Everaert, W., Poesen, J., Rauws, G. and De Ploey, J. 1987. Susceptibilité d'un sol limoneux à l'érosion par rigoles: essais dans le grand canal de Caen. *Bulletin du Centre de Géomorphologie CNRS Caen* 33: 83–106.

Govers, G. and Poesen, J. 1985. A field-scale study of surface sealing and compaction on loam and sandy loam soils. Part I. Spatial variability of soil surface sealing and crusting. In Callebaut, F., Gabriels, D. and De Boodt, M. (eds), *Assessment of soil surface crusting and sealing*. Flanders Research Centre for Soil Erosion and Conservation, State University of Gent: 171–82.

Govers, G. and Poesen, J. 1988. Assessment of the interrill and rill contributions to total soil loss from an upland field plot. *Geomorphology* 1: 343–54.

Govers, G. and Rauws, G. 1986. Transporting capacity of overland flow on plane and on irregular beds. *Earth Surface Processes and Landforms* 11: 515–24.

Govers, G., Vandaele, K., Desmet, P., Poesen, J. and Bunte, K. 1994. The role of tillage in soil redistribution on hillslopes. *European Journal of Soil Science* 45: 469–78.

Gray, D.H. and Leiser, A.T. 1982. *Biotechnical slope protection and erosion control*. Van Nostrand Reinhold, New York.

Green, W.H. and Ampt, G.A. 1911. Studies on soil physics. I: The flow of air and water through soils. *Journal of Agricultural Science* 4: 1–24.

Greenway, D.R. 1987. Vegetation and slope stability. In Anderson, M.G. and Richards, K.S. (eds), *Slope stability: geotechnical engineering and geomorphology*. Wiley, Chichester: 187–230.

Gregory, K.J. and Walling, D.E. 1973. *Drainage basin form and process*. Edward Arnold, London.

Grierson, I.T. 1978. Gypsum and red-brown earths. In Emerson, W.W., Bond, R.D. and Dexter, A.R. (eds), *Modification of soil structure*. Wiley, Chichester: 315–24.

Gril, J.J., Canler, J.P. and Carsoulle, J. 1989. The benefits of permanent grass and mulching for limiting runoff and erosion in vineyards. Experimentations using rainfall simulations in the Beaujolais. In Schwertmann, U., Rickson, R.J. and Auerswald, K. (eds), *Soil erosion protection measures in Europe*. Soil Technology Series 1: 157–66.

Grissinger, E.H. 1966. Resistance of selected clay systems to erosion by water. *Water Resources Research* 2: 131–8.

Grissinger, E.H. and Asmussen, L.E. 1963. Discussion of channel stability in undisturbed cohesive soils by E.M. Flaxman. *Journal of the Hydraulics Division ASCE* 89: 529–64.

Gupta, A. 1982. Observations on the effects of urbanization on runoff and sediment production in Singapore. *Singapore Journal of Tropical Geography* 3: 137–46.

Gupta, J.P. 1979. Some observations on the periodic variations in moisture in stabilized and unstabilized dunes of the Indian desert. *Journal of Hydrology* 41: 153–6.

Guy, B.T. and Dickinson, W.T. 1990. Inception of sediment transport in shallow overland flow. *Catena Supplement* 17: 91–109.

Guy, B.T., Dickinson, W.T., Rudra, R.P. and Wall, G.J. 1990. Hydraulics of sediment-laden sheetflow and the influence of simulated rainfall. *Earth Surface Processes and Landforms* 15: 101–18.

Gyssels, G., Poesen, J., Nachtergaele, J. and Govers, G. 2002. The impact of sowing density of small grains on rill and ephemeral gully erosion in concentrated flow zones. *Soil and Tillage Research* 64: 189–201.

Haas, H.J., Evans, C.E. and Miles, E.F. 1957. Nitrogen and carbon changes in Great Plains soils as influenced by cropping and soil treatments. *USDA Technical Bulletin* 1164.

Hagen, L.J. 1991. A wind erosion prediction system to meet user needs. *Journal of Soil and Water Conservation* 46: 106–11.

Hall, G.F., Daniels, R.B. and Foss, J.E. 1979. Soil formation and renewal rates in the US. In *Symposium on determinants of soil loss tolerance*. Soil Science Society of America Annual Meeting, Fort Collins, CO.

Hall, M.J. 1970. A critique of methods of simulating rainfall. *Water Resources Research* 6: 1104–14.

Hall, R.L. and Calder, I.R. 1993. Drop size modification by forest canopies: measurements using a disdrometer. *Journal of Geophysical Research* 98: 18465–70.

Hallsworth, E.G. 1987. *Anatomy, physiology and psychology of erosion*. Wiley, Chichester.

References

Hann, M.J. and Morgan, R.P.C. 2003. Prevention of biorestoration failures along pipeline rights-of-way. In Moore, H.M., Fox, H.R. and Elliott, S. (eds), *Land reclamation: extending the boundaries*. Balkema, Rotterdam: 227–33.

Hannam, I. 1999. Soil conservation incentives in New South Wales, Australia. In Sanders, D.W., Huszar, P.C., Sombatpanit, S. and Enters, T. (eds), *Incentives in soil conservation: from theory to practice*. Oxford and IBH Publishing, New Delhi: 183–96.

Hannam, I.D. 2001. A global view of the law and policy to manage land degradation. In Bridges, E.M., Hannam, I.D., Oldeman, L.R., Penning de Vries, F.W.T., Scherr, S.J. and Sombatpanit, S. (eds), *Response to land degradation*. Science Publishers, Enfield, NH: 385–94.

Hannam, I.D. and Boer, B.W. 2001. Land degradation and international environmental law. In Bridges, E.M., Hannam, I.D., Oldeman, L.R., Penning de Vries, F.W.T., Scherr, S.J. and Sombatpanit, S. (eds), *Response to land degradation*. Science Publishers, Enfield, NH: 429–38.

Hannam, I.D. and Hicks, R.W. 1980. Soil conservation and urban land use planning. *Journal of the Soil Conservation Service NSW* 36: 134–45.

Hansen, B. 1990. Afstrømning og stoftransport i Rabis og Syvbaek. *NPo-forskning fra Miljøstyrelsen* B9.

Hardin, G. 1968. The tragedy of the commons. *Science* 162: 1243–8.

Harper, D.E. and El-Swaify, S.A. 1988. Sustainable agricultural development in north Thailand: conservation as a component of success in assistance projects. In Moldenhauer, W.C. and Hudson, N.W. (eds), *Conservation farming on steep lands*. Soil and Water Conservation Society, Ankeny, IA: 77–92.

Hashim, G.M., Ciesiolka, C.A.A., Yusoff, W.A., Nafis, A.W., Mispan, M.R., Rose, C.W. and Coughlan, K.J. 1995. Soil erosion processes on sloping land in the east coast of Peninsular Malaysia. *Soil Technology* 8: 215–33.

Hashim, G.M. and Wong, N.C. 1988. Erosion from steep land under various plant covers and terrains. In Tay, T.H., Mokhtaruddin, A.M. and Zahari, A.B. (eds), *Steepland agriculture in the humid tropics*. MARDI/Malaysian Society of Soil Science, Kuala Lumpur: 424–61.

Hasholt, B. 1991. Influence of erosion on the transport of suspended sediment and phosphorus. *International Association of Hydrological Sciences Publication* 203: 329–38.

Hazelhoff, L., van Hoof, P., Imeson, A.C. and Kwaad, F.J.P.M. 1981. The exposure of forest soil to erosion by earthworms. *Earth Surface Processes and Landforms* 6: 235–50.

He, Q. and Walling, D.E. 1996. Interpreting particle size effects in the adsorption of ^{137}Cs and unsupported ^{210}Pb by mineral soils and sediments. *Journal of Environmental Radioactivity* 30: 117–37.

Hedfors, L. 1983. Evaluation and economic appraisal of soil conservation in a pilot area: a summarised report. In Thomas, D.B. and Senga, W.M. (eds), *Soil and water conservation in Kenya*. Institute of Development Studies and Faculty of Agriculture, University of Nairobi, Occasional Paper 42: 257–62.

Heede, B.H. 1975a. Watershed indicators of landform development. In *Hydrology and water resources in Arizona and the Southwest, Volume 5*. Proceedings, 1975 Meeting, Arizona Section, American Water Resources Association and Hydrology Section of the Arizona Academy of Science: 43–6.

Heede, B.H. 1975b. Stages of development of gullies in the west. In *Present and prospective technology for predicting sediment yields and sources*. USDA-ARS Publication ARS-S-40: 155–61.

Heede, B.H. 1976. Gully development and control: the status of our knowledge. *USDA Forest Service Research Paper* RM-169.

Heede, B.H. and Mufich, J.G. 1973. Functional relationships and a computer program for structural gully control. *Journal of Environmental Management* 1: 321–44.

Hénin, S.G., Monnier, G. and Combeau, A. 1958. Méthode pour l'étude de la stabilité structurale des sols. *Annales Agronomie* 9: 73–92.

Herwitz, S.R. 1986. Infiltration-excess caused by stemflow in a cyclone-prone tropical rain forest. *Earth Surface Processes and Landforms* 11: 401–12.

Hesp, P. 1979. Sand trapping ability of culms of marram grass (*Ammophila arenaria*). *Journal of the Soil Conservation Service NSW* 35: 156–60.

Hewlett, H.W.M., Boorman, L.A. and Bramley, M.E. 1987. *Design of reinforced grass waterways*. CIRIA Report 116.

Hijkoop, J., van der Poel, P. and Kaya, B. 1991. *Une lutte de longue haleine. Aménagements anti-érosifs et gestion de terroir*. IER Bamako / KIT Amsterdam.

Hills, R.C. 1970. The determination of the infiltration capacity of field soils using the cylinder infiltrometer. *British Geomorphological Research Group Technical Bulletin* 5.

Hjulström, F. 1935. Studies of the morphological activity of rivers as illustrated by the River Fyries. *Bulletin of the Geological Institute, University of Uppsala* 25: 221–527.

Hoag, D. 1999. Soil conservation incentives in the 1985–1996 US Farm Bills. In Sanders, D.W., Huszar,

P.C., Sombatpanit, S. and Enters, T. (eds), *Incentives in soil conservation: from theory to practice*. Oxford and IBH Publishing, New Delhi: 165–82.

Holbrook, F. and Johns, J.P. 2002. EPSC goes under the microscope: an in-depth approach to program audits. *Erosion Control* 9(6): 42–9.

Holden, S.T. and Shiferaw, B. 1999. Incentives for sustainable land management in peasant agriculture in the Ethiopian Highlands. In Sanders, D.W., Huszar, P.C., Sombatpanit, S. and Enters, T. (eds), *Incentives in soil conservation: from theory to practice*. Oxford and IBH Publishing, New Delhi: 275–94.

Hoogerbrugge, I.D. and Fresco, L.O. 1993. *Homegarden systems: agricultural characteristics and challenges*. International Institute for Environment and Development Gatekeeper Series 39.

Hoogmoed, W.B. and Stroosnijder, L. 1984. Crust formation on sandy soils in the Sahel. I. Rainfall and infiltration. *Soil and Tillage Research* 4: 5–24.

Horton, R.E. 1945. Erosional development of streams and their drainage basins: a hydrophysical approach to quantitative morphology. *Bulletin of the Geological Society of America* 56: 275–370.

Horváth, V. and Erődi, B. 1962. Determination of natural slope category limits by functional identity of erosion intensity. *International Association of Scientific Hydrology Publication* 59: 131–43.

Houze, R.A., Hobbs, P.V., Parsons, D.B. and Herzeg, P.H. 1979. Size distribution of precipitation particles in frontal clouds. *Journal of Atmospheric Science* 36: 156–62.

Howell, J. 1999. *Roadside bio-engineering*. Department of Roads, HM Government of Nepal, Kathmandu.

Howell, J.H., Clark, J.E., Lawrance, C.J. and Sunwar, I. 1991. *Vegetation structures for stabilising highway slopes. A manual for Nepal*. Department of Roads, HM Government of Nepal, Kathmandu.

Huang, C., Bradford, J.M. and Cushman, J.H. 1982. A numerical study of raindrop impact phenomena: the rigid case. *Soil Science Society of America Journal* 46: 14–19.

Hudson, N.W. 1957. The design of field experiments on soil erosion. *Journal of Agricultural Engineering Research* 2: 56–65.

Hudson, N.W. 1963. Raindrop size distribution in high intensity storms. *Rhodesian Journal of Agricultural Research* 1: 6–11.

Hudson, N.W. 1964. Field measurements of accelerated soil erosion in localized areas. *Rhodesian Agricultural Journal* 31: 46–8.

Hudson, N.W. 1965. The influence of rainfall on the mechanics of soil erosion with particular reference to Southern Rhodesia. MSc Thesis, University of Cape Town.

Hudson, N.W. 1981. *Soil conservation*, 2nd edn. Batsford, London.

Hudson, N.W. 1987. Soil and water conservation in semi-arid areas. *FAO Soils Bulletin* 57.

Hudson, N.W. 1988. Soil conservation strategies for the future. In Rimwanich, S. (ed.), *Land conservation for future generations*. Department of Land Development, Bangkok: 117–30.

Hudson, N.W. 1991. Reasons for success and failure of soil conservation projects. *FAO Soils Bulletin* 64.

Hudson, N.W. 1993a. Field measurement of soil erosion and runoff. *FAO Soils Bulletin* 68.

Hudson, N.W. 1993b. *Land husbandry*. Batsford, London.

Hudson, N.W. and Cheatle, R.J. 1993. *Working with farmers for better land husbandry*. IT Publications, Southampton.

Hudson, N.W. and Jackson, D.C. 1959. Results achieved in the measurement of erosion and runoff in Southern Rhodesia. *Proceedings of the Third Inter-African Soils Conference*. Dalaba: 575–83.

Huggins, D.R., Buyanovsky, G.A., Wagner, G.H., Brown, J.R., Darmody, R.G., Peck, T.R., Lesoing, G.W., Vanotti, M.B. and Bundy, L.G. 1998. Soil organic C in the tallgrass prairie-derived region of the corn belt: effects of long-term crop management. *Soil and Tillage Research* 47: 219–34.

Hulugalle, N.R. 1988. Properties of tied ridges in the Sudan savannah of the West African semi-arid tropics. In Rimwanich, S. (ed.), *Land conservation for future generations*. Department of Land Development, Bangkok: 693–709.

Hulugalle, N.R., Lal, R. and Terkuile, C.H.H. 1984. Soil physical changes and crop root growth following different methods of land clearing in western Nigeria. *Soil Science* 138: 172–9.

Humbert-Droz, B. 1996. Integrating indigenous knowledge in soil conservation in a watershed development programme. In Sombatpanit, S., Zöbisch, M.A., Sanders, D.W. and Cook, M.G. (eds), *Soil conservation extension: from concepts to adoption*. Soil and Water Conservation Society of Thailand, Bangkok: 67–75.

Hurni, H. 1984. *Soil conservation research project Ethiopia. Volume 4. Third progress report (Year 1983)*. University of Bern/United Nations University.

Hurni, H. 1986. *Guidelines for development agents on soil conservation in Ethiopia*. Ministry of Agriculture, Addis Ababa.

Hurni, H. 1987. Erosion–productivity–conservation systems in Ethiopia. In Pla Sentis, I. (ed.), *Soil conser-*

vation and productivity. Sociedad Venezolana de la Ciencia del Suelo, Maracay: 654–74.

Hurni, H. 1993. Land degradation, famine and land resource scenarios in Ethiopia. In Pimental, D. (ed.), *World soil erosion and conservation*. Cambridge University Press, Cambridge: 27–61.

Husse, B. 1991. Soil conservation in vineyards. *Universität Trier Forschungstelle Bodenerosion* 10: 89–96.

Hussein, M.H. and Laflen, J.M. 1982. Effects of crop canopy and residue on rill and interrill soil erosion. *Transactions of the American Society of Agricultural Engineers* 25: 1310–15.

Huszar, P.C. 1999. Justification for using soil conservation incentives. In Sanders, D.W., Huszar, P.C., Sombatpanit, S. and Enters, T. (eds), *Incentives in soil conservation: from theory to practice*. Oxford and IBH Publishing, New Delhi: 57–68.

IFAD 1992. *Soil and water conservation in sub-Saharan Africa. Towards sustainable development by the rural poor*. Rome.

Imeson, A.C. 1976. Some effects of burrowing animals on slope processes in the Luxembourg Ardennes. Part I. The excavation of animal mounds in experimental plots. *Geografiska Annaler* 58-A: 115–25.

Imeson, A.C. 1977. A simple field-portable rainfall simulator for difficult terrain. *Earth Surface Processes* 2: 431–6.

Ireland, H.A., Sharpe, C.F.S. and Eargle, D.H. 1939. Principles of gully erosion in the piedmont of South Carolina. *USDA Technical Bulletin* 633, Washington, DC.

Iversen, J.D. 1985. Aeolian threshold: effect of density ratio. In Barndorff-Nielsen, O.E., Møller, J.T., Rasmussen, K.R. and Willetts, B.B. (eds), *Proceedings of international workshop on the physics of blown sand*. Memoir No. 8, Department of Theoretical Statistics, University of Aarhus: 67–81.

Jacinthe, P.A., Lal, R. and Kimble, J.M. 2002. Carbon dioxide evolution in runoff from simulated rainfall on long-term no-till and plowed soils in southwestern Ohio. *Soil and Tillage Research* 66: 23–33.

Jackson, S.J. 1984. The role of slope form on soil erosion at different scales as a possible link between soil erosion surveys and soil erosion models. PhD Thesis, Cranfield Institute of Technology.

Janssen, W. and Tetzlaff, G. 1991. Constructon and calibration of a registering sediment trap. In Karácsony, J. and Szalai, G. (eds), *Proceedings of the international wind erosion workshop of CIGR*. University of Agricultural Sciences, Gödöllő, Hungary.

Janzen, H.H., Campbell, C.A., Izaurralde, R.C., Ellert, B.H., Juma, N., McGill, W.B. and Zentner, R.P. 1998. Management effects on soil C storage on the Canadian prairies. *Soil and Tillage Research* 47: 181–95.

Jayawardena, A.W. and Rezaur, R.B. 2000. Drop size distribution and kinetic energy load of rainstorms in Hong Kong. *Hydrological Processes* 14: 1069–82.

Jeffers, J.N.R. 1986. The role of ecosystem theory in upland land use and management. In Chisci, G. and Morgan, R.P.C. (eds), *Soil erosion in the European Community: impact of changing agriculture*. Balkema, Rotterdam: 51–65.

Jensen, J.L. and Sørensen, M. 1986. Estimation of some aeolian saltation transport parameters: a reanalysis of Williams' data. *Sedimentology* 33: 547–58.

Jiang, D., Qi, L. and Tan, J. 1981. Soil erosion and conservation in the Wuding River Valley, China. In Morgan, R.P.C. (ed.), *Soil conservation: problems and prospects*. Wiley, Chichester: 461–79.

Johnson, C.B. and Moldenhauer, W.C. 1979. Effect of chisel versus moldboard plowing on soil erosion by water. *Soil Science Society of America Journal* 43: 177–9.

Jones, M.J. 1971. The maintenance of soil organic matter under continuous cultivation at Samaru, Nigeria. *Journal of Agricultural Science Cambridge* 77: 473–82.

Jones, R.G.B. and Keech, M.A. 1966. Identifying and assessing problem areas in soil erosion surveys using aerial photographs. *Photogrammetric Record* 5(27): 189–97.

Jusoff, K. and Majid, N.M. 1988. Effects of site preparation methods on soil physical characteristics. In Rimwanich, S. (ed.), *Land conservation for future generations*. Department of Land Development, Bangkok: 745–54.

Kamalu, C. 1991. Soil erosion on road shoulders: the effects of rainfall-runoff interaction. In *Erosion control: a global perspective*. International Erosion Control Association, Steamboat Springs, CO: 249–62.

Kamalu, C. 1993. Soil erosion on road shoulders. PhD Thesis, Cranfield University.

Kampen, J., Hari Krishna, J. and Pathak, P. 1981. Rainy season cropping on deep vertisols in the semi-arid tropics: effects on hydrology and soil erosion. In Lal, R. and Russell, E.W. (eds), *Tropical agricultural hydrology*. Wiley, Chichester: 257–71.

Kamphorst, A. 1987. A small rainfall simulator for the determination of soil erodibility. *Netherlands Journal of Agricultural Science* 35: 407–15.

Keech, M.A. 1969. Mondaro Tribal Trust Land. Determination of trend using air photo analysis. *Rhodesian Agricultural Journal* 66: 3–10.

Keech, M.A. 1992. The photogrammetrical evaluation of areal and volumetric change in a gully in Zimbabwe. In Hurni, H. and Tato, K (eds), *Erosion, conservation and small-scale farming.* Geographica Bernensia, Bern: 51–62.

Kellman, M.C. 1969. Some environmental components of shifting cultivation in upland Mindanao. *Journal of Tropical Geography* 28: 40–56.

Kemper, B. and Derpsch, R. 1981. Results of studies made in 1978 and 1979 to control erosion by cover crops and no-tillage techniques in Paraná, Brazil. *Soil and Tillage Research* 1: 253–68.

Kermack, K.A. and Haldane, J.B.S. 1950. Organic correlation and allometry. *Biometrika* 37: 30–41.

Khybri, M.L. 1989. Mulch effects on soil and water loss in maize in India. In Moldenhauer, W.C., Hudson, N.W., Sheng, T.C. and Lee, S.W. (eds), *Development of conservation farming on hillslopes.* Soil and Water Conservation Society, Ankeny, IA: 195–8.

Kiara, J.K. 2001. The catchment approach to soil and water conservation in Kenya. In Bridges, E.M., Hannam, I.D., Oldeman, L.R., Penning de Vries, F.W.T., Scherr, S.J. and Sombatpanit, S. (eds), *Response to land degradation.* Science Publishers, Enfield, NH: 370–1.

Kiara, J.K., Skoglund, E. and Eriksson, A. 1996. Development of soil conservation extension in Kenya. In Sombatpanit, S., Zöbisch, M.A., Sanders, D.W. and Cook, M.G. (eds), *Soil conservation extension: from concepts to adoption.* Soil and Water Conservation Society of Thailand, Bangkok: 35–45.

Kiepe, P. 1995. No runoff, no soil loss: soil and water conservation in hedgerow barrier systems. *Tropical Resource Management Papers* 10, Wageningen.

Kinnell, P.I.A. 1974. Splash erosion: some observations on the splash-cup technique. *Soil Science Society of America Proceedings* 38: 657–60.

Kinnell, P.I.A. 1981. Rainfall intensity–kinetic energy relationships for soil loss prediction. *Soil Science Society of America Journal* 45: 153–5.

Kinnell, P.I.A. 1987. Rainfall energy in eastern Australia: intensity–kinetic energy relationships for Canberra, ACT. *Australian Journal of Soil Research* 25: 547–53.

Kinnell, P.I.A. 1990. The mechanics of raindrop-induced flow transport. *Australian Journal of Soil Research* 28: 497–516.

Kirkby, M.J. 1969a. Infiltration, throughflow and overland flow. In Chorley, R.J. (ed.), *Water, earth and man.* Methuen, London: 215–27.

Kirkby, M.J. 1969b. Erosion by water on hillslopes. In Chorley, R.J. (ed.), *Water, earth and man.* Methuen, London: 229–38.

Kirkby, M.J. 1971. Hillslope process–response models based on the continuity equation. In Brunsden, D. (ed.), *Slopes: form and process.* Institute of British Geographers Special Publication 3: 15–30.

Kirkby, M.J. 1980a. The problem. In Kirkby, M.J. and Morgan, R.P.C. (eds), *Soil erosion.* Wiley, Chichester: 1–16.

Kirkby, M.J. 1980b. Modelling water erosion processes. In Kirkby, M.J. and Morgan, R.P.C. (eds), *Soil erosion.* Wiley, Chichester: 183–216.

Kleene, P., Sanogo, B. and Vierstra, G. 1989. *A partir de Fonsébougou. Présentation, objectifs et méthodologie du 'Volet Fonsébougou' (1977–1987).* IER, Bamako/IRT, Amsterdam.

Klingebiel, A.A. and Montgomery, P.H. 1966. Land capability classification. *USDA Soil Conservation Service Agricultural Handbook* 210.

Knisel, W.G. 1980. CREAMS: a field scale model for chemicals, runoff and erosion from agricultural management systems. *USDA Conservation Research Report* 26.

Knowles-Jackson, C. 1996. The Landcare movement: its growth in a soil conservation district. In Sombatpanit, S., Zöbisch, M.A., Sanders, D.W. and Cook, M.G. (eds), *Soil conservation extension: from concepts to adoption.* Soil and Water Conservation Society of Thailand, Bangkok: 293–9.

Kolenbrander, G.J. 1974. Efficiency of organic manure in increasing soil organic matter content. *Transactions, 10th International Congress of Soil Science* 2: 129–36.

König, D. 1992. The potential of agroforestry methods for erosion control in Rwanda. *Soil Technology* 5: 167–76.

Konuche, P.K.A. 1983. Effects of forest management practices on soil and water conservation in Kenya forests. In Thomas, D.B. and Senga, W.M. (eds), *Soil and water conservation in Kenya.* Institute of Development Studies and Faculty of Agriculture, University of Nairobi, Occasional Paper 42: 350–9.

Kowal, J.M. and Kassam, A.H. 1976. Energy load and instantaneous intensity of rainstorms at Samaru, northern Nigeria. *Tropical Agriculture* 53: 185–97.

Krenitsky, E.C. and Carroll, M.J. 1994. Use of erosion control materials to establish turf. In *Sustaining environmental quality: the erosion control challenge.* International Erosion Control Association, Steamboat Springs, CO: 79–90.

Kronvang, B. 1990. Sediment-associated phosphorus transport from two intensively farmed catchment areas. In Boardman, J., Foster, I.D.L. and Dearing, J.A. (eds), *Soil erosion on agricultural land.* Wiley, Chichester: 313–30.

Kronvang, B., Laubel, A.R., Larsen, S.E. and Iversen, S.L. 2000. Soil erosion and sediment delivery through buffer zones in Danish slope units. *International Association of Scientific Hydrology Publication* 263: 67–73.

Kumar, A., Samra, J.S., Singh, N.T. and Batra, L. 1990. Tree canopy–intercrop relationships in an agroforestry system. In *Proceedings, International symposium on water erosion, sedimentation and resource conservation.* Central Soil and Water Conservation Research and Training Institute, Dehra Dun: 446–52.

Kutiel, P. 1994. Fire and ecosystem heterogeneity: a Mediterranean case study. *Earth Surface Processes and Landforms* 19: 187–94.

Kutílek, M., Krejča, M., Haverkamp, R., Rendón, L.P. and Parlange, J.Y. 1988. On extrapolation of algebraic infiltration equations. *Soil Technology* 1: 47–61.

Kwaad, F.J.P.M. 1977. Measurements of rainsplash erosion and the formation of colluvium beneath deciduous woodland in the Luxembourg Ardennes. *Earth Surface Processes* 2: 161–73.

Kwaad, F.J.P.M. 1991. Summer and winter regimes of runoff generation and soil erosion on cultivated loess soils (The Netherlands). *Earth Surface Processes and Landforms* 16: 653–62.

Laflen, J.M. and Colvin, T.S. 1981. Effect of crop residue on soil loss from continuous row cropping. *Transactions of the American Society of Agricultural Engineers* 24: 605–9.

Laflen, J.M., Johnson, H.P. and Reeve, R.C. 1972. Soil loss from tile-outlet terraces. *Journal of Soil and Water Conservation* 27: 74–7.

Laflen, J.M. and Moldenhauer, W.C. 1979. Soil and water losses from corn-soybean rotations. *Soil Science Society of America Journal* 43: 1213–15.

Laing, D. 1992. From the International Center for Tropical Agriculture (CIAT), Cali, Colombia. *Vetiver Newsletter* 8: 13–15.

Laing, I.A.F. 1978. Soil surface treatment for runoff inducement. In Emerson, W.W., Bond, R.D. and Dexter, A.R. (eds), *Modification of soil structure.* Wiley, Chichester: 249–56.

Lakew, D.T. and Morgan, R.P.C. 1996. Contour grass strips: a laboratory simulation of their role in erosion control using live grasses. *Soil Technology* 9: 83–9.

Lakshmipathy, B.M. and Narayanswamy, S. 1956. Bench terracing in the Nilgiris. *Journal of Soil and Water Conservation in India* 4: 161–8.

Lal, R. 1976. *Soil erosion problems on an alfisol in western Nigeria and their control.* IITA Monograph 1.

Lal, R. 1977a. Soil-conserving versus soil-degrading crops and soil management for erosion control. In Greenland, D.J. and Lal, R. (eds), *Soil conservation and management in the humid tropics.* Wiley, London: 81–6.

Lal, R. 1977b. Soil management systems and erosion control. In Greenland, D.J. and Lal, R. (eds), *Soil conservation and management in the humid tropics.* Wiley, Chichester: 93–7.

Lal, R. 1981. Deforestation of tropical rainforest and hydrological problems. In Lal, R. and Russell, E.W. (eds), *Tropical agricultural hydrology.* Wiley, Chichester: 131–40.

Lal, R. 1982. Effect of slope length and terracing on runoff and erosion on a tropical soil. *International Association of Hydrological Sciences Publication* 137: 23–31.

Lal, R. 1987. Research achievements towards soil and water conservation in the tropics: potential and priorities. In Pla Sentis, I. (ed.), *Soil conservation and productivity.* Sociedad Venezolana de la Ciencia del Suelo, Maracay: 755–87.

Lal, R. 1988. Soil erosion control with alley cropping. In Rimwanich, S. (ed.), *Land conservation for future generations.* Land Development Department, Bangkok: 237–45.

Lal, R. 1995. Global soil erosion by water and carbon dynamics. In Lal, R., Kimble, J.M., Levine, E. and Stewart, B.A. (eds), *Soils and global change.* CRC/Lewis, Boca Raton, FL: 131–41.

Lal, R. 2002. Soil conservation and restoration to sequester carbon and mitigate the greenhouse effect. In Rubio, J.L., Morgan, R.P.C., Asins, S. and Andreu, V. (eds), *Man and soil at the third millennium.* Geoforma Ediciones, Logroño: 37–51.

Lal, R. and Cummings, D.J. 1979. Clearing a tropical forest. I. Effects on soil and microclimate. *Field Crops Research* 2: 91–107.

Lang, R.D. and McCaffrey, L.A.H. 1984. Ground cover: its effects on soil loss from grazed runoff plots, Gunnedah. *Journal of the Soil Conservation Service NSW* 40: 56–61.

Langdale, G.W., Mills, W.C. and Thomas, A.W. 1992. Conservation tillage development for soil erosion control in the southern Piedmont. In Hurni, H. and Kebedo Tato (eds), *Erosion, conservation and small-scale farming.* Geographica Bernensia, Bern: 453–8.

Larson, W.E., Pierce, F.J. and Dowdy, R.H. 1985. Loss in long-term productivity from soil erosion in the United States. In El-Swaify, S.A., Moldenhauer, W.C. and Lo, A. (eds), *Soil erosion and conservation.* Soil Conservation Society of America, Ankeny, IA: 262–71.

Laurent, A. and Bollinne, A. 1978. Caractérisation des pluies en Belgique du point de vue de leur intensité et de leur erosivité. *Pédologie* 28: 214–32.

Lawrance, C.J., Rickson, R.J. and Clark, J.E. 1996. The effect of grass roots on the shear strength of colluvial soils in Nepal. In Anderson, M.G. and Brooks, S.M. (eds), *Advances in hillslope processes.* Wiley, Chichester: 857–68.

Laws, J.O. and Parsons, D.A. 1943. The relationship of raindrop size to intensity. *Transactions of the American Geophysical Union* 24: 452–60.

References

Le Bissonnais, Y. 1990. Experimental study and modelling of soil surface crusting processes. *Catena Supplement* 17: 13–28.

Le Bissonnais, Y., Montier, C., Jamagne, M., Daroussin, J. and King, D. 2002. Mapping erosion risk for cultivated soils in France. *Catena* 46: 207–20.

Leigh, C. 1982. Sediment transport by surface wash and throughflow at the Pasoh Forest Reserve, Negri Sembilan, Peninsular Malaysia. *Geografiska Annaler* 64-A: 171–80.

Lenvain, J.S., Sakala, W.K. and Pauwelyn, P.L.L. 1988. Iso-erodent map of Zambia. Part II. Erosivity prediction and mapping. *Soil Technology* 1: 251–62.

Leopold, L.B., Wolman, M.G. and Miller, J.P. 1964. *Fluvial processes in geomorphology*. Freeman, San Francisco.

Lettau, K. and Lettau, H.H. 1978. Experimental and micrometeorological studies of dune migration. In Lettau, H.H. and Lettau, K. (eds), *Exploring the world's driest climates*. Institute of Environmental Science Report No. 101, Center for Climatic Research, University of Wisconsin: 110–47.

Li, G., Rozelle, S. and Brandt, L. 1998. Tenure, land rights and farmer investment incentives in China. *Agricultural Economics* 19: 61–71.

Li, X.Y. and Liu, L.Y. 2003. Effect of gravel mulch on aeolian dust accumulation in the semiarid region of northwest China. *Soil and Tillage Research* 70: 73–81.

Li, X.Y., Liu, L.Y. and Gong, J.D. 2001. Influence of pebble mulch on soil erosion by wind and trapping capacity for windblown sediment. *Soil and Tillage Research* 59: 137–42.

Liao, M.C. and Wu, H.L. 1987. *Soil conservation on steeplands in Taiwan*. Chinese Soil and Water Conservation Society, Taipei.

Liddle, M.J. 1973. The effects of trampling and vehicles on natural vegetation. PhD Thesis, University College of North Wales, Bangor.

Lim, K.H. 1988. A study on soil erosion control under mature oil palm in Malaysia. In Rimwanich, S. (ed.), *Land conservation for future generations*. Department of Land Development, Bangkok: 783–95.

Lindstrom, M.J., Nelson, W.W. and Schumacher, T.E. 1992. Quantifying tillage erosion rates due to mouldboard plowing. *Soil and Tillage Research* 24: 243–55.

Liu, B.Y., Nearing, M.A., Baffaut, C. and Ascough II, J.C. 1997. The WEPP watershed model. III. Comparisons to measured data from small watersheds. *Transactions of the American Society of Agricultural Engineers* 40: 945–51.

Livingstone, I. and Warren, A. 1996. *Aeolian geomorphology*. Addison Wesley Longman, Harlow.

Lloyd, S.D., Bishop, P. and Reinfelds, I. 1998. Shoreline erosion: a cautionary note in using small farm dams to determine catchment erosion rates. *Earth Surface Processes and Landforms* 23: 905–12.

Lobb, D.A., Kachanoski, R.G. and Miller, M.H. 1999. Tillage translocation and tillage erosion in the complex landscapes of south-western Ontario, Canada. *Soil and Tillage Research* 51: 189–209.

Logie, M. 1982. Influence of roughness elements and soil moisture on the resistance of sand to wind erosion. *Catena Supplement* 1: 161–73.

Lomborg, B. 2001. *The skeptical environmentalist. Measuring the real state of the world*. Cambridge University Press, Cambridge.

López-Bermúdez, F. and Romero-Diaz, M.A. 1989. Piping erosion and badland development in south-east Spain. *Catena Supplement* 14: 59–73.

Low, F.K. 1967. Estimating potential erosion in developing countries. *Journal of Soil and Water Conservation* 22: 147–8.

Lugo-Lopez, M.A. 1969. Prediction of the erosiveness of Puerto Rican soils on a basis of the percentage of particles of silt and clay when aggregated. *Journal of Agriculture, University of Puerto Rico* 53: 187–90.

Luk, S.H. 1981. Variability of rainwash erosion within small sample areas. In *Proceedings, Twelfth Binghampton Geomorphology Symposium*. Allen and Unwin, London: 243–68.

Lundgren, B. and Nair, P.K.R. 1985. Agroforestry for soil conservation. In El-Swaify, S.A., Moldenhauer, W.C. and Lo, A. (eds), *Soil erosion and conservation*. Soil Conservation Society of America, Ankeny, IA: 703–11.

Lundgren, L. 1978. Studies of soil and vegetation development on fresh landslide scars in the Mgeta Valley, western Uluguru Mountains, Tanzania. *Geografiska Annaler* 60-A: 91–127.

Lundgren, L. and Rapp, A. 1974. A complex landslide with destructive effects on the water supply of Morogoro town, Tanzania. *Geografiska Annaler* 56-A: 251–60.

Lvovich, M.I., Karasik, G.Ya., Bratseva, N.L., Medvedeva, G.P. and Maleshko, A.V. 1991. *Contemporary intensity of the world land intracontinental erosion*. USSR Academy of Sciences, Moscow.

Lyles, L. and Allison, B.E. 1981. Equivalent wind erosion protection from selected crop residues. *Transactions of the American Society of Agricultural Engineers* 24: 405–9.

Lyles, L., Dickerson, J.D. and Schmeidler, M.F. 1974a. Soil detachment from clods by rainfall: effects of wind, mulch cover and initial soil moisture. *Transactions of the American Society of Agricultural Engineers* 17: 697–700.

Lyles, L., Schrandt, R.L. and Schmeidler, M.F. 1974b. How aerodynamic roughness elements control sand movement. *Transactions of the American Society of Agricultural Engineers* 17: 134–9.

McCauley, D.S. 1988. Overcoming institutional and organizational constraints to watershed management for the densely populated island of Java. In Rimwanich, S. (ed.), *Land conservation for future generations*. Department of Land Development, Bangkok: 1039–60.

McConkey, B.G., Liang, B.C., Campbell, C.A., Curtin, D., Moulin, A., Brandt, S.A. and Lafond, G.P. 2003. Crop rotation and tillage impact on carbon sequestration in Canadian prairie soils. *Soil and Tillage Research* 74: 81–90.

McCormack, D.E. and Young, K.K. 1981. Technical and societal implications of soil loss tolerance. In Morgan, R.P.C. (ed.), *Soil conservation: problems and prospects*. Wiley, Chichester: 365–76.

McCormack, R.J. 1971. The Canada Land Use Inventory: a basis for land use planning. *Journal of Soil and Water Conservation* 26: 141–6.

McCuen, R.H. 1973. The role of sensitivity analysis in hydrologic modelling. *Journal of Hydrology* 18: 37–53.

McGregor, J.C. and Mutchler, C.K. 1978. The effect of crop canopy on raindrop size distribution. *USDA Sedimentation Laboratory Annual Report*. Oxford MS.

McHugh, M., Harrod, T. and Morgan, R. 2002. The extent of soil erosion in upland England and Wales. *Earth Surface Processes and Landforms* 27: 99–107.

McIsaac, G.F. 1990. Apparent geographic and atmospheric influences on raindrop sizes and rainfall kinetic energy. *Journal of Soil and Water Conservation* 45: 663–6.

Maene, L.M., Thong, K.C., Ong, T.S. and Mokhtaruddin, A.M. 1979. Surface wash under mature oil palm. In Pushparajah, E. (ed.), *Proceedings, Symposium on water in Malaysian agriculture*. Malaysian Society of Soil Science, Kuala Lumpur: 203–16.

Magrath, W.B. and Arens, P. 1989. The cost of soil erosion on Java: a natural resource accounting approach. *Environment Department Working Paper* 18. World Bank Policy Planning and Research Staff, World Bank, Washington, DC.

Marshall, J.K. 1967. The effect of shelter on the productivity of grasslands and field crops. *Field Crops Abstracts* 20: 1–14.

Marshall, J.S. and Palmer, W.M. 1948. Relation of rain drop size to intensity. *Journal of Meteorology* 5: 165–6.

Marshall, R.G. and Ruban, T.F. 1983. Geotechnical aspects of pipeline construction in Alberta. *Canadian Geotechnical Journal* 20: 1–10.

Marston, D. 1996. Extending soil conservation from concept to action. In Sombatpanit, S., Zöbisch, M.A., Sanders, D.W. and Cook, M.G. (eds), *Soil conservation extension: from concepts to adoption*. Soil and Water Conservation Society of Thailand, Bangkok: 27–34.

Marston D. and Hird, C. 1978. Effect of stubble management on the structure of black cracking clays. In Emerson, W.W., Bond, R.D. and Dexter, A.R. (eds), *Modification of soil structure*. Wiley, Chichester: 411–17.

Marston, D. and Perrens, S.J. 1981. Effect of stubble on erosion of a black earth soil. In Tingsanchali, T. and Eggers, H. (eds), *Southeast Asian regional symposium on problems of soil erosion and sedimentation*. Asian Institute of Technology, Bangkok: 289–300.

Martel, Y.A. and Paul, E.A. 1974. Effects of cultivation on the organic matter of grassland soils as determined by fractionation and radiocarbon dating. *Canadian Journal of Soil Science* 54: 419–26.

Martin, L. and Morgan, R.P.C. 1980. Soil erosion in mid-Bedfordshire. In Doornkamp, J.C., Gregory, K.J. and Burn, A.S. (eds), *Atlas of drought in Britain 1975–76*. Institute of British Geographers, London: 47.

Martin, P. 1999. Reducing flood risk from sediment-laden agricultural runoff using intercrop management techniques in northern France. *Soil and Tillage Research* 52: 233–245.

Martin, P., Le Bissonnais, Y., Benkhadra, H., Ligneau, L. and Ouvry, J.F. 1997. Mesures du ruissellement et de l'érosion diffuse engendrés par les pratiques culturales en Pays de Caux (Normandie). *Géomorphologie: Relief, Processus, Environment* 2: 143–54.

Martínez-Mena, M., Williams, A.G., Ternan, J.L. and Fitzjohn, C. 1998. Role of antecedent soil water content on aggregates stability in a semi-arid environment. *Soil and Tillage Research* 48: 71-80.

Mason, B.J. and Andrews, J.B. 1960. Drop size distributions from various types of rain. *Quarterly Journal of the Royal Meteorological Society* 86: 346–53.

Mason, B.J. and Ramandham, R. 1953. A photoelectric spectrometer. *Quarterly Journal of the Royal Meteorological Society* 79: 490–5.

Mathieu, R., King, C. and Le Bissonnais, Y. 1997. Contribution of multi-temporal SPOT data to the mapping of a soil erosion index. The case of the loamy plateaux of northern France. *Soil Technology* 10: 99–110.

Mati, B.M. 1999. Erosion hazard assessment in the Upper Ewaso Ng'iro basin of Kenya: application of GIS, USLE and EUROSEM. PhD Thesis, Cranfield University.

Mazurak, A.P. and Mosher, P.N. 1968. Detachment of soil particles in simulated rainfall. *Soil Science Society of America Proceedings* 32: 716–19.

References

Mbegera, M., Eriksson, A. and Njoroge, S.N.J. 1992. Soil and water conservation training and extension: the Kenyan experience. In Tato, K and Hurni, H. (eds), *Soil conservation for survival*. Soil and Water Conservation Society, Ankeny, IA: 284–95.

Mead, R. and Curnow, R.N. 1983. *Statistical methods in agriculture and experimental biology*. Chapman and Hall, London.

Mech, S.J. and Free, G.R. 1942. Movement of soil during tillage operations. *Agricultural Engineering* 23: 379–82.

Megahan, W.F. and Molitor, D.C. 1975. Erosional effects of wildfire and logging in Idaho. *Proceedings, Symposium on watershed management*. American Society of Civil Engineers, St Giles, MI: 423–44.

Mein, R.G. and Larson, C.L. 1973. Modeling infiltration during a steady rain. *Water Resources Research* 9: 384–94.

Melville, N. and Morgan, R.P.C. 2001. The influence of grass density on effectiveness of contour grass strips for control of soil erosion on low-angle slopes. *Soil Use and Management* 17: 278–81.

Merritt, E. 1984. The identification of four stages during microrill development. *Earth Surface Processes and Landforms* 9: 493–6.

Merten, G.H., Nearing, M.A. and Borges, A.L.O. 2001. Effect of sediment load on soil detachment and deposition in rills. *Soil Science Society of America Journal* 65: 861–8.

Merzouk, A. and Blake, G.R. 1991. Indices for the estimation of interrill erodibility of Moroccan soils. *Catena* 18: 537–50.

Meyer, L.D. 1965. Mathematical relationships governing soil erosion by water. *Journal of Soil and Water Conservation* 20: 149–50.

Meyer, L.D. 1981. How rain intensity affects interrill erosion. *Transactions of the American Society of Agricultural Engineers* 24: 1472–5.

Meyer, L.D., Foster, G.R. and Römkens, M.J.M. 1975. Source of soil eroded by water from upland slopes. In *Present and prospective technology for predicting sediment yields and sources*. USDA-ARS Publication ARS-S-40: 177–89.

Meyer, L.D. and Harmon, W.C. 1979. Multiple-intensity rainfall simulator for erosion research on row sideslopes. *Transactions of the American Society of Agricultural Engineers* 22: 100–3.

Meyer, L.D. and Monke, E.J. 1965. Mechanics of soil erosion by rainfall and overland flow. *Transactions of the American Society of Agricultural Engineers* 8: 572–7.

Meyer, L.D. and Wischmeier, W.H. 1969. Mathematical simulation of the process of soil erosion by water. *Transactions of the American Society of Agricultural Engineers* 12: 754–8, 762.

Middleton, H.E. 1930. Properties of soils which influence soil erosion. *USDA Technical Bulletin* 178.

Mihara, Y. 1951. Raindrops and soil erosion. *Bulletin of the National Institute of Agricultural Science*, Series A, 1 (in Japanese).

Miller, W.P. and Sumner, M.E. 1988. Dispersion processes affecting runoff and erosion in highly weathered soils. In Rimwanich, S. (ed.), *Land conservation for future generations*. Department of Land Development, Bangkok: 419–27.

Millington, A.C. 1984. Indigenous soil conservation studies in Sierra Leone. *International Association of Hydrological Sciences Publication* 144.

Millington, A. 1987. Local farmer perceptions of soil erosion hazards and indigenous soil conservation strategies in Sierra Leone, West Africa. In Pla Sentis, I. (ed.), *Soil conservation and productivity*. Sociedad Venezolana de la Ciencia del Suelo, Maracay: 675–90.

Milner, C. and Douglas, M.G. 1989. *Problems of land degradation in Commonwealth Africa*. Commonwealth Secretariat, London.

Mirtskhoulava, Ts.E. 2001. *Soil erosion: forecasting, risk, conservation*. Georgian Institute of Water Management and Engineering Ecology, Tbilisi.

Mitchell, J.K., Bubenzer, G.D., McHenry, J.R. and Ritchie, J.C. 1980. Soil loss estimation from the fallout caesium-137 measurements. In De Boodt, M. and Gabriels, D. (eds), *Assessment of erosion*. Wiley, Chichester: 393–401.

Mitchell, M. 1998. Evaluating the effectiveness of erosion control programs. *Erosion Control* 5(6): 30–40.

Moeyersons, J. 1983. Measurements of splash-saltation fluxes under oblique rain. *Catena Supplement* 4: 19–31.

Moeyersons, J. and De Ploey, J. 1976. Quantitative data on splash erosion simulated on unvegetated slopes. *Zeitschrift für Geomorphologie Supplementband* 25: 120–131.

Moldenhauer, W.C. 1965. Procedure for studying soil characteristics using disturbed samples and simulated rainfall. *Transactions of the American Society of Agricultural Engineers* 8: 74–5.

Moldenhauer, W.C. 1985. A comparison of conservation tillage systems for reducing erosion. In D'Itri, F.M. (ed.), *A systems approach to conservation tillage*. Lewis Publishers, Chelsea, MI: 111–20.

Moldenhauer, W.C. and Onstad, C.A. 1975. Achieving specified soil loss levels. *Journal of Soil and Water Conservation* 30: 166–8.

Monnier, G. and Boiffin, J. 1986. Effect of the agricultural use of soils on water erosion: the case of cropping systems in western Europe. In Chisci, G. and Morgan, R.P.C. (eds), *Soil erosion in the European Community: impact of changing agriculture*. Balkema, Rotterdam: 17–32.

Montgomery, D.R. and Dietrich, W.E. 1994. Landscape dissection and drainage area-slope thresholds. In Kirkby, M.J. (ed.), *Process models and theoretical geomorphology*. Wiley, Chichester: 221–45.

Moore, I.D., Burch, G.J. and Mackenzie, D.H. 1988. Topographic effects on the distribution of surface soil water and the location of ephemeral gullies. *Transactions of the American Society of Agricultural Engineers* 34: 1098–107.

Morel-Seytoux, H.J. and Khanji, J. 1974. Derivation of an equation of infiltration. *Water Resources Research* 10: 794–800.

Moretti, S. and Rodolfi, G. 2000. A typical 'calanchi' landscape on the eastern Appenine margin (Atri, central Italy): geomorphological features and evolution. *Catena* 40: 217–28.

Morgan, R.P.C. 1974. Estimating regional variations in soil erosion hazard in Peninsular Malaysia. *Malayan Nature Journal* 28: 94–106.

Morgan, R.P.C. 1976. The role of climate in the denudation system: a case study from Peninsular Malaysia. In Derbyshire, E. (ed.), *Climate and geomorphology*. Wiley, London: 317–43.

Morgan, R.P.C. 1980a. Field studies of sediment transport by overland flow. *Earth Surface Processes* 3: 307–16.

Morgan, R.P.C. 1980b. Soil erosion and conservation in Britain. *Progress in Physical Geography* 4: 24–47.

Morgan, R.P.C. 1981. Field measurement of splash erosion. *International Association of Scientific Hydrology Publication* 133: 373–82.

Morgan, R.P.C. 1985a. Assessment of soil erosion risk in England and Wales. *Soil Use and Management* 1: 127–31.

Morgan, R.P.C. 1985b. Effect of corn and soybean canopy on soil detachment by rainfall. *Transactions of the American Society of Agricultural Engineers* 28: 1135–40.

Morgan, R.P.C. 1989. Design of in-field shelter systems for wind erosion control. In Schwertmann, U., Rickson, R.J. and Auerswald, K. (eds), *Soil erosion protection measures in Europe*. Soil Technology Series 1: 15–23.

Morgan, R.P.C. 1996. Verification of the European Soil Erosion Model (EUROSEM) for varying slope and vegetation conditions. In Anderson, M.G. and Brooks, S.M. (eds), *Advances in hillslope processes*. Wiley, Chichester: 657–68.

Morgan, R.P.C. 2001. A simple approach to soil loss prediction: a revised Morgan–Morgan–Finney model. *Catena* 44: 305–22.

Morgan, R.P.C. and Finney, H.J. 1987. Drag coefficients of single crop rows and their implications for wind erosion control. In Gardiner, V. (ed.), *International geomorphology 1986. Part II*. Wiley, Chichester: 449–58.

Morgan, R.P.C. and Hann, M.J. 2003. Design of diverter berms for erosion control and biorestoration along pipeline rights-of-way. Poster paper presented to Land Reclamation 2003, Runcorn.

Morgan, R.P.C., Hann, M.J., Shilston, D., Lee, E.M., Mirtskhoulava, Ts.E., Nadirashvili, V., Topuria, L., Clarke, J. and Sweeney, M. 2004. Use of terrain analysis as a basis for erosion risk assessment: a case study from pipeline rights-of-way in Georgia. Paper presented to International Conference on Terrain and geohazard challenges facing onshore oil and gas pipelines. BP/ICE, London.

Morgan, R.P.C., McIntyre, K., Vickers, A.W., Quinton, J.N. and Rickson, R.J. 1997a. A rainfall simulation study of soil erosion on rangeland in Swaziland. *Soil Technology* 11: 291–9.

Morgan, R.P.C., Martin, L. and Noble, C.A. 1986. *Soil erosion in the United Kingdom: a case study from mid-Bedfordshire*. Silsoe College, Occasional Paper 14.

Morgan, R.P.C., Mirtskhoulava, Ts.E., Nadirashvili, V., Hann, M.J. and Gasca, A.H. 2003. Spacing of berms for erosion control along pipeline rights-of-way. *Biosystems Engineering* 85: 249–59.

Morgan, R.P.C. and Mngomezulu, D. 2003. Threshold conditions for initiation of valley-side gullies in the Middle Veld of Swaziland. *Catena* 50: 401–14.

Morgan, R.P.C., Morgan, D.D.V. and Finney, H.J. 1982. Stability of agricultural ecosystems: documentation of a simple model for soil erosion assessment. *International Institute for Applied Systems Analysis Collaborative Paper* CP-82-50.

Morgan, R.P.C., Morgan, D.D.V. and Finney, H.J. 1984. A predictive model for the assessment of soil erosion risk. *Journal of Agricultural Engineering Research* 30: 245–53.

Morgan, R.P.C., Quinton, J.N., Smith, R.E., Govers, G., Poesen, J.W.A., Auerswald, K., Chischi, G., Torri, D. and Styczen, M.E. 1998. The European Soil Erosion Model (EUROSEM): a dynamic approach for predicting sediment transport from fields and small catchments. *Earth Surface Processes and Landforms* 23: 527–44.

Morgan, R.P.C., Rickson, R.J., McIntyre, K., Brewer, T.R. and Altshul, H.J. 1997b. Soil erosion survey of the central part of the Swaziland Middleveld. *Soil Technology* 11: 263–89.

Morin, J., Goldberg, D. and Seginer, I. 1967. A rainfall simulator with a rotating disc. *Transactions of the American Society of Agricultural Engineers* 10: 74–7, 79.

Morse, G. 1992. Revegetation at Heavenly Ski Resort: past, present and future. In *The environment is our future*.

References

International Erosion Control Association, Steamboat Springs, CO: 37–42.

Mosley, M.P. 1973. Rainsplash and the convexity of badland divides. *Zeitschrift für Geomorphologie Supplementband* 18: 10–25.

Mosley, M.P. 1982. The effect of a New Zealand beech forest canopy on the kinetic energy of water drops and on surface erosion. *Earth Surface Processes and Landforms* 7: 103–7.

Moss, A.J., Green, P. and Hutka, J. 1982. Small channels: their formation, nature and significance. *Earth Surface Processes and Landforms* 7: 401–15.

Moss, A.J. and Walker, P.H. 1978. Particle transport by continental water flows in relation to erosion, deposition, soils and human activities. *Sedimentary Geology* 20: 81–139.

Mrabet, R., Saber, N., El-Brahli, A., Lahlou, S. and Bessam, F. 2001. Total particulate organic matter and structural stability of a Calcixeroll soil under different wheat rotations and tillage systems in a semiarid area of Morocco. *Soil Tillage Research* 57: 225–35.

Mueller, D.H., Klemme, R.M. and Daniel, T.C. 1985. Short- and long-term cost comparisons of conventional and conservation tillage systems in corn production. *Journal of Soil and Water Conservation* 40: 466–70.

Mulligan, K.R. 1988. Velocity profiles measured on the windward slope of a transverse dune. *Earth Surface Processes and Landforms* 13: 573–82.

Murdoch, G., Webster, R. and Lawrance, C.J. 1971. *A land system atlas of Swaziland*. Military Vehicle Experimental Establishment, Christchurch.

Murgatroyd, A.L. and Ternan, J.L. 1983. The impact of afforestation on stream bank erosion and channel form. *Earth Surface Processes and Landforms* 8: 357–69.

Musgrave, G.W. 1947. The quantitative evaluation of factors in water erosion: a first approximation. *Journal of Soil and Water Conservation* 2: 133–8.

Musick, H.B. and Gillette, D.A. 1990. Field evaluation of relationships between a vegetation structural parameter and sheltering against wind erosion. *Land Degradation and Rehabilitation* 2: 87–94.

Mutchler, C.K. and Young, R.A. 1975. Soil detachment by raindrops. In *Present and prospective technology for predicting sediment yields and sources*. USDA-ARS Publication ARS-S-40: 113–17.

Mututho, J.M.N. 1989. Some aspects of soil conservation on grazing lands. In Thomas, D.B., Biamah, E.K., Kilewe, A.M., Lundgren, L. and Mochoge, B.O. (eds), *Soil and water conservation in Kenya*. Department of Agricultural Engineering, University of Nairobi/SIDA: 315–22.

Mwichabe, S. 1992. The effect of range management on vegetation cover in the Narok area, Kenya. In Hurni, H. and Tato, K. (eds), *Erosion, conservation and small-scale farming*. Geographica Bernensia, Bern: 407–13.

Nabben, T. 1999. Funding to community Landcare groups in Western Australia: a powerful 'new generation' of soil conservation incentives. In Sanders, D.W., Huszar, P.C., Sombatpanit, S. and Enters, T. (eds), *Incentives in soil conservation: from theory to practice*. Oxford and IBH Publishing, New Delhi: 197–214.

Nachtergaele, J. 2001. A spatial and temporal analysis of the characteristics, importance and prediction of ephemeral gully erosion. PhD Thesis, Katholieke Universiteit Leuven.

Nachtergaele, J., Poesen, J., Oostwoud Wijdenes, D. and Vandekerckhove, L. 2002. Medium-term evolution of a gully developed in a loess-derived soil. *Geomorphology* 46: 223–39.

Napier, T.L. 1988. Socio-economic factors influencing the adoption of soil erosion control practices in the United States. In Morgan, R.P.C. and Rickson, R.J. (eds), *Erosion assessment and modelling*. Commission of the European Communities Report EUR 10860 EN: 299–327.

Napier, T.L. 1990. The evolution of US soil-conservation policy: from voluntary adoption to coercion. In Boardman, J., Foster, I.D.L. and Dearing, J.A. (eds), *Soil erosion on agricultural land*. Wiley, Chichester: 627–44.

Napier, T.L. 1999. Inadequacies of voluntary soil and water conservation incentives threaten adoption of regulatory approaches. In Sanders, D.W., Huszar, P.C., Sombatpanit, S. and Enters, T. (eds), *Incentives in soil conservation: from theory into practice*. Oxford and IBH Publishing, New Delhi: 151–64.

Nash, J.E. and Sutcliffe, J.V. 1970. River flow forecasting through conceptual models. I. Discussion of principles. *Journal of Hydrology* 10: 282–90.

Nassif, S.H. and Wilson, E.M. 1975. The influence of slope and rain intensity on runoff and infiltration. *Hydrological Sciences Bulletin* 20: 539–53.

Natural Resources Conservation Service 1999. *CORE4 conservation practices training guide. The common sense approach to natural resources conservation*. USDA Natural Resources Conservation Service (http://www.nrcs.usda.gov/technical/ECS/agronomy/core4.pdf).

Natural Resources Conservation Service 2002. *Farm Bill 2002: conservation provisions overview*. USDA Natural Resources Conservation Service (http://www.nrcs.usda.gov/programs/farmbill/2002).

Naveh, Z. 1975. The evolutionary sequence of fire in the Mediterranean region. *Vegetation* 9: 199–06.

Nearing, M.A. 1998. Why soil erosion models overpredict small soil losses and underpredict large soil losses. *Catena* 32: 15–22.

Nearing, M.A. 2000. Evaluating soil erosion models using measured plot data: accounting for variability in the data. *Earth Surface Processes and Landforms* 25: 1035–43.

Nearing, M.A., Deer-Ascough, L. and Chaves, H.M.L. 1989a. WEPP model sensitivity analysis. In Lane, L.J. and Nearing, M.A. (eds), *USDA Water Erosion Prediction Project: hillslope model documentation*. USDA-ARS NSERL Report 2: 14.1–14.33.

Nearing, M.A., Foster, G.R., Lane, L.J. and Finckner, S.C. 1989b. A process-based soil erosion model for USDA-Water Erosion Prediction Project technology. *Transactions of the American Society of Agricultural Engineers* 32: 1587–93.

Newson, M. 1980. The erosion of drainage ditches and its effect on bed-load yields in mid-Wales: reconnaissance case studies. *Earth Surface Processes* 3: 275–90.

Nickling, W.G. and McKenna Neuman, C.K. 1995. Development of deflation lag surfaces. *Sedimentology* 42: 403–14.

Nicks, A.D., Lane, L.J. and Gander, G.A. 1995. Weather generator. In Flanagan, D.C. and Nearing, M.A. (eds), *USDA-Water Erosion Prediction Project: hillslope profile and watershed model documentation*. USDA-ARS NSERL Report 10.

Noble, C.A. and Morgan, R.P.C. 1983. Rainfall interception and splash detachment with a Brussels sprouts plant: a laboratory simulation. *Earth Surface Processes and Landforms* 8: 569–77.

Norton, L.D., Cogo, N.P. and Moldenhauer, W.C. 1985. Effectiveness of mulch in controlling erosion. In El-Swaify, S.A., Moldenhauer, W.C. and Lo, A. (eds), *Soil erosion and conservation*. Soil Conservation Society of America, Ankeny, IA: 598–606.

Nossin, J.J. 1964. Geomorphology of the surroundings of Kuantan (Eastern Malaya). *Geologie en Mijnbouw* 43: 157–82.

Nsibandze, S.M. 1987. The history of soil conservation in Swaziland. Paper presented at SADCC Seminar, Maputo.

Nyamulinda, V. and Ngiruwonsanga, V. 1992. Lutte anti-érosif et strategies paysannes dan les montagnes du Rwanda. *Réseau Erosion Bulletin* 12: 71–82.

Nyssen, J., Poesen, J., Haile, M., Moeyersons, J. and Deckers, J. 2000. Tillage erosion on slopes with soil conservation structures in the Ethiopian highlands. *Soil and Tillage Research* 57: 115–27.

Nyssen, J., Poesen, J., Moeyersons, J., Luyten, E., Veyret-Picot, M., Deckers, J., Haile, M. and Govers, G. 2002. Impact of road building on gully erosion risk: a case study from the northern Ethiopian Highlands. *Earth Surface Processes and Landforms* 27: 1267–83.

Oberhauser, U. and Limchoowong, S. 1996. Soil conservation by local communities. In Sombatpanit, S., Zöbisch, M.A., Sanders, D.W. and Cook, M.G. (eds), *Soil conservation extension: from concepts to adoption*. Soil and Water Conservation Society of Thailand, Bangkok: 59–66.

Odemerho, F.O. 1986. Variation in erosion – slope relationship on cut slopes along a tropical highway. *Singapore Journal of Tropical Geography* 7: 98–107.

Odemerho, F.O. and Avwunudiogba, A. 1993. The effects of changing cassava management practices on soil loss: a Nigerian example. *Geographical Journal* 159: 63–9.

Okigbo, B.N. 1977. Farming systems and soil erosion in West Africa. In Greenland, D.J. and Lal, R. (eds), *Soil conservation and management in the humid tropics*. Wiley, London: 151–63.

Okigbo, B.N. and Lal, R. 1977. Role of cover crops in soil and water conservation. *FAO Soils Bulletin* 33: 97–108.

Oldeman, L.R. 1994. An international methodology for an assessment of soil degradation, land georeferenced soils and terrain data base. In *The collection and analysis of land degradation data*. FAO-RAPA Publication 1994/3, Bangkok: 35–60.

Olesen, F. 1979. *Læplantning*. Det Danske Hedeselskab, Viborg.

O'Loughlin, C.L. 1974. A study of tree root strength deterioration following clearfelling. *Canadian Journal of Forest Research* 4: 107–13.

O'Loughlin, C.L. and Watson, A. 1979. Root-wood strength deterioration in radiata pine after clearfelling. *New Zealand Journal of Forestry Science* 9: 284–93.

Onaga, K., Shirai, K. and Yoshinaga, A. 1988. Rainfall erosion and how to control its effects on farmland in Okinawa. In Rimwanich, S. (ed.), *Land conservation for future generations*. Department of Land Development, Bangkok: 627–39.

Osuji, G.E. 1989. Raindrop characteristics in the humid tropics. *Journal of Environmental Management* 28: 227–33.

Osuji, G.E., Babalola, O. and Aboaba, F.O. 1980. Rainfall erosivity and tillage practices affecting soil and water loss on a tropical soil in Nigeria. *Journal of Environmental Management* 10: 207–17.

Othieno, C.O. 1978. An assessment of soil erosion on a field of young tea under different soil management practices. In *Soil and water conservation in Kenya*. Institute of Development Studies, University of Nairobi, Occasional Paper 27: 62–73.

References

Øygarden, L. 1995. Runoff and erosion in small catchments. *European Society for Soil Conservation Newsletter* 1/2: 17–20.

Øygarden, L. 2003. Rill and gully development during an extreme winter runoff event in Norway. *Catena Supplement* 50: 217–42.

Padgitt, M. 1989. Soil diversity and the effects of field eligibility rules in implementing soil conservation programs targeted to highly erodible land. *Journal of Soil and Water Conservation* 44: 91–5.

Palmer, R.S. 1964. The influence of a thin water layer on water-drop impact forces. *International Association of Scientific Hydrology Publication* 65: 141–8.

Paningbatan, E.P., Ciesiolka, C.A., Coughlan, K.J. and Rose, C.W. 1995. Alley cropping for managing soil erosion of hilly lands in The Philippines. *Soil Technology* 8: 193–204.

Paracchini, M.L., Minacapilli, M., Bertolo, F. and Folving, S. 1997. Soil erosion modelling and coastal dynamics: a case study from Sicily. In *Remote sensing: observations and interactions*. Remote Sensing Society, Nottingham: 334–9.

Parsons, A.J., Abrahams, A.D. and Luk, S.H. 1990. Hydraulics of interrill overland flow on a semi-arid hillslope, southern Arizona. *Journal of Hydrology* 117: 255–73.

Partap, T. and Watson, H.R. 1994. Sloping agricultural land technology (SALT). A regenerative option for sustainable mountain farming. *ICIMOD Occasional Paper* 23.

Pathak, P., Miranda, S.M. and El-Swaify, S.A. 1985. Improved rainfed farming for semiarid tropics: implications for soil and water conservation. In El-Swaify, S.A., Moldenhauer, W.C. and Lo, A. (eds), *Soil erosion and conservation*. Soil and Water Conservation Society of America, Ankeny, IA: 338–54.

Pava, H.M., Arances, J.B., Magallanes, J.M., Mugot, I.O., Manubag, J.M. and Sealza, I.S. 1990. Participatory processes to upland development: the MUSUAN model. *MUSUAN Program Monograph Series* 2, Central Mindanao University, Bukidnon.

Pazos, A. and Gasca, A.H. 1998. Prueba piloto para control de erosión. Linea de Flujo Cupiagua XC-TIE IN. Paper presented to 7th Congress, Sociedad Colombiana de Geotecnía.

Pearce, A.J. 1976. Magnitude and frequency of erosion by Hortonian overland flow. *Journal of Geology* 84: 65–80.

Pease, M. and Truong, P. 2002. Vetiver grass technology: a tool against environmental degradation in southern Europe. In Rubio, J.L., Morgan, R.P.C., Asins, S. and Andreu, V. (eds), *Soil at the third millennium*. Geoforma Ediciones, Logroño: 2167–78.

Peden, D.G. 1987. Livestock and wildlife population distributions in relation to aridity and human population in Kenya. *Journal of Range Management* 40: 67–71.

Perrens, S.J. and Trustrum, N.A. 1984. *Assessment and evaluation for soil conservation policy*. East – West Environment and Policy Institute, Workshop Report, Honolulu, HI.

Petryk, S. and Bosmajian, G. 1975. Analysis of flow through vegetation. *Journal of the Hydraulics Division ASCE* 101: 871–4.

Phien, T. and Tu Siem, N. 1996. Soil conservation extension in agricultural development in Vietnam. In Sombatpanit, S., Zöbish, M.A., Sanders, D.W. and Cook, M.G. (eds), *Soil conservation extension: from concepts to adoption*. Soil and Water Conservation Society of Thailand, Bangkok: 359–63.

Philip, J.R. 1957. The theory of infiltration. I. The infiltration equation and its solution. *Soil Science* 83: 345–57.

Phillips, J.M., Webb, B.W., Walling, D.E. and Leeks, G.J.L. 1999. Estimating the suspended sediment loads of rivers in the LOIS study area using infrequent samples. *Hydrological Processes* 13: 1035–50.

Pidgeon, J.D. and Soane, B.D. 1978. Soil structure and strength relations following tillage, zero tillage and wheel traffic in Scotland. In Emerson, W.W., Bond, R.D. and Dexter, A.R. (eds), *Modification of soil structure*. Wiley, Chichester: 371–8.

Pihan, J. 1979. Risques climatiques d'érosion hydrique des sols en France. In Vogt, H. and Vogt, Th. (eds), *Colloque sur l'érosion agricole des sols en milieu tempéré non Mediterranéen*. L'Université Louis Pasteur, Strasbourg: 13–18.

Pilgrim, D.H. and Huff, D.D. 1983. Suspended sediment in rapid subsurface stormflow on a large field plot. *Earth Surface Processes and Landforms* 8: 451–63.

Pimental, D., Allen, J., Beers, A., Guinand, L., Hawkins, A., Linder, R., McLaughlin, P., Meer, B., Musonda, D., Perdue, D., Poisson, S., Salazar, R., Siebert, S. and Stoner, K. 1993. Soil erosion and agricultural productivity. In Pimental, D. (ed.), *World soil erosion and conservation*. Cambridge University Press, Cambridge: 277–92.

Pimental, D., Harvey, C., Resodudarmo, P., Sinclair, K., Kurtz, D., McNair, M., Crist, S., Spritz, L., Fitton, L., Saffouri, R. and Blair, R. 1995. Environmental and economic costs of soil erosion and conservation benefits. *Science* 267: 117–23.

Pla, I. 1977. Aggregate size and erosion control on sloping land treated with hydrophobic bitumen emulsion. In Greenland, D.J. and Lal, R. (eds), *Soil conservation and management in the humid tropics*. Wiley, London: 109–15.

Poesen, J. 1981. Rainwash experiments on the erodibility of loose sediments. *Earth Surface Processes and Landforms* 6: 285–307.

Poesen, J. 1984. The influence of slope angle on infiltration rate and Hortonian overland flow volume. *Zeitschrift für Geomorphologie Supplementband* 49: 117–31.

Poesen, J. 1985. An improved splash transport model. *Zeitschrift für Geomorphologie* 29: 193–211.

Poesen, J. 1987. Transport of rock fragments by rill flow: a field study. *Catena Supplement* 8: 35–54.

Poesen, J. 1989. Conditions for gully formation in the Belgian loam belt and some ways to control them. In Schwertmann, U., Rickson, R.J. and Auerswald, K. (eds), *Soil erosion protection measures in Europe*. Soil Technology Series 1: 39–52.

Poesen, J.W.A. 1992. Mechanisms of overland-flow generation and sediment production on loamy and sandy soils with and without rock fragments. In Parsons, A.J. and Abrahams, A.D. (eds), *Overland flow: hydraulics and erosion mechanics*. UCL Press, London: 275–305.

Poesen, J. 1993. Gully typology and gully control measures in the European loess belt. In Wicherek, S. (ed.), *Farmland erosion in temperate plains environment and hills*. Elsevier, Amsterdam: 221–39.

Poesen, J. and Ingelmo-Sanchez, F. 1992. Runoff and sediment yield from topsoils with different porosity as affected by rock fragment cover and position. *Catena* 19: 451–74.

Poesen, J., Nachtergaele, J., Verstraeten, G. and Valentin, C. 2003. Gully erosion and environmental change: importance and research needs. *Catena* 50: 91–133.

Poesen, J. and Torri, D. 1988. The effect of cup size on splash detachment and transport measurements. I. Field measurements. *Catena Supplement* 12: 133–6.

Poesen, J.W., Torri, D. and Bunte, K. 1994. Effects of rock fragments on soil erosion by water at different spatial scales: a review. *Catena* 23: 141–66.

Pope, A. and Harper, J.J. 1966. *Low-speed wind tunnel testing*. Wiley, London.

Porta, J., Poch, R.M. and Boixadera, J. 1989. Land evaluation and erosion control practices on mined soils in NE Spain. In Schwertmann, U., Rickson, R.J. and Auerswald, K. (eds), *Soil erosion protection measures in Europe*. Soil Technology Series 1: 189–206.

Pratt, D.T. and Gwynne, M.D. 1977. *Rangeland management and ecology in East Africa*. Hodder and Stoughton, London.

Presbitero, A.L., Escalante, M.C., Rose, C.W., Coughlan, K.J. and Ciesiolka, C.A. 1995. Erodibility evaluation and the effect of land management practices on soil erosion from steep slopes in Leyte, The Philippines. *Soil Technology* 8: 205–13.

Preston, N.J. and Crozier, M.J. 1999. Resistance to shallow landslide failure through root-derived cohesion in East Coast Hill Country soils, North Island, New Zealand. *Earth Surface Processes and Landforms* 24: 665–75.

Prinz, D. 1992. Towards an optimal utilisation of the soil and water resources in a semi-humid environment: an example from Thailand. In Tato, K. and Hurni, H. (eds), *Soil conservation for survival*. Soil and Water Conservation Society, Ankeny, IA: 321–31.

Prior, J.C. 1996. From technology transfer to community development: the policy implications of the Australian Landcare movement. In Sombatpanit, S., Zöbisch, M.A., Sanders, D.W. and Cook, M.G. (eds), *Soil conservation extension: from concepts to adoption*. Soil and Water Conservation Society of Thailand, Bangkok: 77–86.

Proffitt, A. and Rose, C.W. 1992. Relative contributions to soil loss by rainfall detachment and runoff entrainment. In Hurni, H. and Tato, K. (eds), *Erosion, conservation and small-scale farming*. Geographica Bernensia, Bern: 75–89.

Proffitt, A.P.B., Rose, C.W. and Hairsine, P.B. 1991. Detachment and deposition: experiments with low slopes and significant water depths. *Soil Science Society of America Journal* 55: 325–32.

Quansah, C. 1981. The effect of soil type, slope, rain intensity and their interactions on splash detachment and transport. *Journal of Soil Science* 32: 215–24.

Quansah, C. 1982. Laboratory experiments for the statistical derivation of equations for soil erosion modelling and soil conservation design. PhD Thesis, Cranfield Institute of Technology.

Quansah, C. 1985. Rate of soil detachment by overland flow, with and without rain, and its relationship with discharge, slope steepness and soil type. In El-Swaify, S.A., Moldenhauer, W.C. and Lo, A. (eds), *Soil erosion and conservation*. Soil Conservation Society of America, Ankeny, IA: 406–23.

Quansah, C. and Baffoe-Bonnie, E. 1981. The effect of soil management systems on soil loss, runoff and fertility erosion in Ghana. In Tingsanchali, T. and Eggers, H. (eds), *Southeast Asian regional symposium on problems of soil erosion and sedimentation*. Asian Institute of Technology, Bangkok: 207–17.

Quinn, N.W., Morgan, R.P.C. and Smith, A.J. 1980. Simulation of soil erosion induced by human trampling. *Journal of Environmental Management* 10: 155–65.

References

Quinton, J.N. 1994. The validation of physics-based erosion models with particular reference to EUROSEM. PhD Thesis, Cranfield University.

Quinton, J.N. 1997. Reducing predictive uncertainty in model simulations: a comparison of two methods using the European Soil Erosion Model (EUROSEM). *Catena* 30: 101–17.

Quinton, J.N. and Morgan, R.P.C. 1998. EUROSEM: an evaluation with single event data from the C5 Watershed, Oklahoma, USA. In Boardman, J. and Favis-Mortlock, D. (eds), *Modelling soil erosion by water*. NATO ASI Series 1, 55: 65–74.

Quinton, J.N. and Rodriguez, F. 1999. Impact of live barriers on soil erosion in the Pairuani sub-catchment, Bolivia. *Mountain Research and Development* 19: 292–9.

Quinton, J.N. and Veihe, A. 2000. Development and application of soil productivity index for Central America: soil erosion modelling. Individual Partner Final Report to Commission of European Communities, Research Contract ERBI 18 CT 960096.

Ramaswamy, S.D., Aziz, M.A. and Narayanan, N. 1981. Some methods of control of erosion on natural slopes in urbanizd areas. In Tingsanchali, T. and Eggers, H. (eds), *Southeast Asian regional symposium on problems of soil erosion and sedimentation*. Asian Institute of Technology, Bangkok: 327–39.

Ramser, C.E. 1945. Developments in terrace spacing. *Agricultural Engineering* 26: 285–9.

Randall, J.M. 1969. Wind profiles in an orchard plantation. *Agricultural Meteorology* 6: 439–52.

Rankilor, P.R. 1989. The reduction of soil erosion by preformed systems. In *Soil erosion and its control*. Jason Consultants, Geneva: 56–70.

Rapp, A., Axelsson, V., Berry, L. and Murray-Rust, D.H. 1972a. Soil erosion and sediment transport in the Morogoro River catchment. *Geografiska Annaler* 54-A: 125–55.

Rapp, A., Murray-Rust, D.H., Christiansson, C. and Berry, L. 1972b. Soil erosion and sedimentation in four catchments near Dodoma, Tanzania. *Geografiska Annaler* 54-A: 255–318.

Rasmussen, K.R., Sørensen, M. and Willetts, B.B. 1985. Measurement of saltation and wind strength on beaches. In Barndorff-Nielsen, O.E., Møller, J.T., Rasmussen, K.R. and Willetts, B.B. (eds), *Proceedings of international workshop on the physics of blown sand*. Memoirs No. 8, Department of Theoretical Statistics, University of Aarhus: 301–25.

Rasmussen, P.E., Albrecht, S.L. and Smiley, R.W. 1998. Soil C and N changes under tillage and cropping systems in semi-arid Pacific Northwest agriculture. *Soil and Tillage Research* 47: 197–205.

Rasmussen, W.D. 1982. History of soil conservation, institutions and incentives. In Halcrow, H.G., Heady, E.O. and Cotner, M.L. (eds), *Soil conservation policies, institutions and incentives*. Soil Conservation Society of America, Ankeny, IA: 3–18.

Rauws, G. 1987. The initiation of rills on plane beds of non-cohesive sediments. *Catena Supplement* 8: 107–18.

Rauws, G. and Govers, G. 1988. Hydraulic and soil mechanical aspects of rill generation on agricultural soils. *Journal of Soil Science* 39: 111–24.

Ree, W.O. 1949. Hydraulic characteristics of vegetation for vegetated waterways. *Agricultural Engineering* 30: 184–7, 189.

Reeder, J.D., Schuman, G.E. and Bowman, R.A. 1998. Soil C and N changes on conservation reserve program lands in the Central Great Plains. *Soil and Tillage Research* 47: 339–49.

Reid, M., Dunne, T. and Cederholm, C.J. 1981. Application of sediment budget studies to the evaluation of logging road impact. *New Zealand Journal of Hydrology* 20: 49–62.

Reij, C., Scoones, I. and Toulmin, C. 1996. *Sustaining the soil. Indigenous soil and water conservation in Africa*. Earthscan, London.

Renard, K.G., Foster, G.R., Weesies, G.A. and Porter, J.P. 1991. RUSLE: Revised Universal Soil Loss Equation. *Journal of Soil and Water Conservation* 46: 30–3.

Rice, R.M. and Krammes, J.S. 1970. Mass-wasting processes in watershed management. In *Proceedings, Symposium on interdisciplinary aspects of watershed management*. American Society of Civil Engineers, St Giles, MI: 231–60.

Richter, G. and Negendank, J.F.W. 1977. Soil erosion processes and their measurement in the German area of the Moselle river. *Earth Surface Processes* 2: 261–78.

Rickson, R.J. 1987. Small plot field studies of soil erodibility using a rainfall simulator. In Pla Sentis, I. (ed.), *Soil conservation and productivity*. Sociedad Venezolana de la Ciencia del Suolo, Maracay: 339–48.

Rickson, R.J. 1995. Simulated vegetation and geotextiles. In Morgan, R.P.C. and Rickson, R.J. (eds), *Slope stabilization and erosion control: a bioengineering approach*. E. and F.N. Spon, London: 95–131.

Riley, S.J. 1988. Soil loss from road batters in the Karuah State Forest, eastern Australia. *Soil Technology* 1: 313–32.

Ripley, E.A. and Redman, R.E. 1976. Grassland. In Monteith, J.L. (ed.), *Vegetation and the atmosphere. Volume 2. Case studies*. Academic Press, London: 349–98.

Ritchie, J.C. and Ritchie, C.A. 2001. Bibliography of publications of ^{137}caesium studies related to erosion and sediment deposition (http://hydrolab.arsusda.gov/cesium137bib.htm).

Roberts, B. 1992. *Landcare Manual*. New South Wales University Press, Kensington, NSW.

Robinson, D.A. and Blackman, J.D. 1990. Some costs and consequences of soil erosion and flooding around Brighton and Hove, autumn 1987. In Boardman, J., Foster, I.D.L. and Dearing, J.A. (eds), *Soil erosion on agricultural land*. Wiley, Chichester: 369–82.

Robinson, M. and Blyth, K. 1982. The effect of forestry drainage operations on upland sediment yields: a case study. *Earth Surface Processes and Landforms* 7: 85–90.

Roels, J.M. and Jonker, P.J. 1985. Representativity and accuracy of measurements of soil loss from runoff plots. *Transactions of the American Society of Agricultural Engineers* 28: 1458–66.

Rogers, N.W. and Selby, M.J. 1980. Mechanisms of shallow translational landsliding during summer rainstorms: North Island, New Zealand. *Geografiska Annaler* 62-A: 11–21.

Rogler, H. and Schwertmann, U. 1981. Erosivität der Niederschläge und Isoerodentkarte von Bayern. *Zeitschrift für Kulturtechnik und Flurbereiniging* 22: 99–112.

Roose, E.J. 1967. Dix années de mesure de l'érosion et du ruissellement au Sénégal. *L'Agronomie Tropicale* 22: 123–52.

Roose, E.J. 1971. *Influence des modifications du milieu naturel sur l'érosion: le bilan hydrique et chimique suite à la mise en culture sous climat tropical*. ORSTOM, Adiopodoumé, Ivory Coast.

Roose, E.J. 1975. *Erosion et ruissellement en Afrique de l'ouest: vingt années de mesures en petites parcelles expérimentales*. ORSTOM, Adiopodoumé, Ivory Coast.

Roose, E.J. 1976. *Contribution à l'étude de l'influence de la mésofaune sur la pédogenèse actuelle*. Rapports ORSTOM, Abidjan: 2–23.

Roose, E.J. 1977. Application of the Universal Soil Loss Equation of Wischmeier and Smith in West Africa. In Greenland, D.J. and Lal, R. (eds), *Soil conservation and management in the humid tropics*. Wiley, London: 177–87.

Roose, E.J. 1988. Soil and water conservation lessons from steep-slope farming in French-speaking countries of Africa. In Moldenhauer, W.C. and Hudson, N.W. (eds), *Conservation farming on steep lands*. Soil and Water Conservation Society of America, Ankeny, IA: 207–14.

Roose, E.J. 1992. Traditional and modern strategies for soil and water conservation in the Sudano-Sahelian areas of western Africa. In Hurni, H. and Tato, K. (eds), *Erosion, conservation and small-scale farming*. Geographica Bernensia, Bern: 349–65.

Rose, C.W. and Hairsine, P.B. 1988. Processes of water erosion. In Stefan, W.L., Denmead, Q.T. and White, I. (eds), *Flow and transport in the natural environment*. Springer-Verlag, Berlin: 312–26.

Rose, C.W., Williams, J.R., Sander, G.C. and Barry, D.A. 1983. A mathematical model of soil erosion and deposition process. I. Theory for a plane element. *Soil Science Society of America Journal* 47: 991–5.

Rose, S.J.C. 1989. The Three Peaks Project: tackling footpath erosion. In *Erosion knows no boundaries*. International Erosion Control Association, Steamboat Springs, CO: 369–78.

Rosewell, C.J. 1970. Investigations into the control of earth-work tunnelling. *Journal of the Soil Conservation Service NSW* 26: 188–203.

Rosewell, C.J. 1986. Rainfall kinetic energy in eastern Australia. *Journal of Climate and Applied Meteorology* 25: 1695–701.

Rowntree, K.M. 1983. Rainfall erosivity in Kenya: some preliminary considerations. In Thomas, D.B. and Senga, W.M. (eds), *Soil and water conservation in Kenya*. Institute of Development Studies and Faculty of Agriculture, University of Nairobi Occasional Paper 42: 1–19.

Runólfsson, S. 1978. Soil conservation in Iceland. In Holdgate, M.W. and Woodman, M.J. (eds), *The breakdown and restoration of ecosystems*. Plenum, New York: 231–40.

Runólfsson, S. 1987. Land reclamation in Iceland. *Arctic and Alpine Research* 19: 514–17.

Russell, M.A., Walling, D.E. and Hodgkinson, R.A. 2001. Suspended sediment sources in two small lowland agricultural catchments in the UK. *Journal of Hydrology* 252: 1–24.

Rutin, J. 1992. Geomorphic activity of rabbits on a coastal sand dune, De Blink Dunes, The Netherlands. *Earth Surface Processes and Landforms* 17: 85–94.

Salles, C., Poesen, J. and Govers, G. 2000. Statistical and physical analysis of soil detachment by raindrop impact: rain erosivity indices and threshold energy. *Water Resources Research* 36: 2721–9.

Sanders, D.W. 1988. Food and Agricultural Organization activities in soil conservation. In Moldenhauer, W.C. and Hudson, N.W. (eds), *Conservation farming on steep lands*. Soil and Water Conservation Society, Ankeny, IA: 54–62.

Savage, R.P. and Woodhouse, W.W. 1968. Creation and stabilization of coastal barrier dunes. In *Proceedings, Coastal engineering conference (London)* 1: 671–700.

Savat, J. 1975. Discharge velocities and total erosion of a calcareous loess: a comparison between pluvial and terminal runoff. *Revue de Géomorphologie Dynamique* 24: 113–22.

Savat, J. 1979. Laboratory experiments on erosion and deposition of loess by laminar sheet flow and turbulent rill flow. In Vogt, H. and Vogt, Th. (ed.), *Colloque sur l'érosion agricole des sols en milieu tempéré non Mediterranéen.* L'Université Louis Pasteur, Strasbourg: 139–43.

Savat, J. 1981. Work done by splash: laboratory experiments. *Earth Surface Processes and Landforms* 6: 275–83.

Savat, J. 1982. Common and uncommon selectivity in the process of fluid transportation: field observations and laboratory experiments on bare surfaces. *Catena Supplement* 1: 139–60.

Savat, J. and De Ploey, J. 1982. Sheetwash and rill development by surface flow. In Bryan, R. and Yair, A. (eds), *Badland geomorphology and piping.* Geo Books, Norwich: 113–26.

Schertz, D.L. 1983. The basis for soil loss tolerance. *Journal of Soil and Water Conservation* 38: 10–14.

Schiechtl, H.M. and Stern, R. 1996. *Ground bioengineering techniques for slope protection and erosion control.* Blackwell Science, Oxford.

Schjønning, P. and Rasmussen, K.J. 1989. Long term reduced cultivation. I. Soil strength and stability. *Soil and Tillage Research* 15: 79–90.

Scholten, T. 1997. Hydrology and erodibility of the soils and saprolite cover of the Swaziland Middleveld. *Soil Technology* 11: 247–62.

Schumm, S.A. 1979. Geomorphic thresholds: the concept and its applications. *Transactions of the Institute of British Geographers New Series* 4: 485–515.

Schwab, G.O., Frevert, R.K., Edminster, T.W. and Barnes, K.K. 1966. *Soil and water conservation engineering.* Wiley, New York.

Schwertmann, U., Vogl, W. and Kainz, M. 1987. *Bodenerosion durch Wasser.* Ulmer Verlag, Stuttgart.

Scoging, H., Parsons, A.J. and Abrahams, A.D. 1992. Application of a dynamic overland-flow hydraulic model to a semi-arid hillslope, Walnut Gulch, Arizona. In Parsons, A.J. and Abrahams, A.D. (eds), *Overland flow: hydraulics and erosion mechanics.* UCL Press, London: 105–45.

Scoging, H.M. and Thornes, J.B. 1979. Infiltration characteristics in a semi-arid environment. *International Association of Scientific Hydrology Publication* 128: 159–68.

Sempere-Torres, D., Salles, C., Creutin, J.D. and Delrieu, G. 1992. Quantification of soil detachment by raindrop impact: performances of classical formulae of kinetic energy in Mediterranean storms. *International Association of Hydrological Sciences Publication* 210: 115–24.

Seubert, C.E., Sanchez, P.A. and Valverde, C. 1977. Effects of land clearing methods on soil properties and soil performance in the Amazon jungle of Peru. *Tropical Agriculture* 54: 307–21.

Shah, R.B., Eusof, Z. and Zamnis, M.M. 1996. Soil conservation extension in Malaysia. In Sombatpanit, S., Zöbisch, M.A., Sanders, D.W. and Cook, M.G. (eds), *Soil conservation extension: from concepts to adoption.* Soil and Water Conservation Society of Thailand, Bangkok: 279–91.

Shakesby, R.A., Coelho, C. de A.O., Ferreira, A.D., Terry, J.P. and Walsh, R.P.D. 1993. Wildfire impacts on soil erosion and hydrology in west Mediterranean forest, Portugal. *International Journal of Wildland Fire* 3: 95–110.

Shakesby, R.A. and Whitlow, R. 1991. Perspectives on prehistoric and recent gullying in central Zimbabwe. *GeoJournal* 23: 49–58.

Sharpe, C.F.S. 1938. *Landslides and related phenomena.* Columbia University Press, New York.

Sharpley, A.N. and Smith, S.J. 1990. Phosphorus transport in agricultural runoff: the role of soil erosion. In Boardman, J., Foster, I.D.L. and Dearing, J.A. (eds), *Soil erosion on agricultural land.* Wiley, Chichester: 351–66.

Shaxson, T.F. 1981. Reconciling social and technical needs in conservation work on village farmlands. In Morgan, R.P.C. (ed.), *Soil conservation: problems and prospects.* Wiley, Chichester: 385–97.

Shaxson, T.F. 1987. Changing approaches to soil conservation. In Pla Sentis, I. (ed.), *Soil conservation and productivity.* Sociedad Venezolana de la Ciencia del Suelo, Maracay: 11–27.

Shaxson, T.F. 1988. Conserving soil by stealth. In Moldenhauer, W.C. and Hudson, N.W. (eds), *Conserving farming on steep lands.* Soil and Water Conservation Society, Ankeny, IA: 9–17.

Sheng, T.C. 1972a. A treatment-oriented land capability classification scheme for hilly marginal lands in the humid tropics. *Journal of the Scientific Research Council Jamaica* 3: 93–112.

Sheng, T.C. 1972b. Bench terracing. *Journal of the Scientific Research Council Jamaica* 3: 113–27.

Sheng, T.C. 1981. The need for soil conservation structures for steep cultivated slopes in the humid tropics. In Lal, R. and Russell, E.W. (eds), *Tropical agricultural hydrology.* Wiley, Chichester: 357–72.

Sherchan, D.P., Chand, S.P., Thapa, Y.B., Tiwari, T.P. and Gurung, G.B. 1990. Soil and nutrient losses in runoff on selected crop husbandry practices on hill slope soil of the eastern Nepal. In *Proceedings, International symposium on water erosion, sedimentation and resource conservation*. Central Soil and Water Conservation Research and Training Institute, Dehra Dun: 188–98.

Sheridan, J.M., Lowrance, R. and Bosch, D.D. 1999. Management effects on runoff and sediment transport in riparian forest buffers. *Transactions of the American Society of Agricultural Engineers* 42: 55–64.

Shields, A. 1936. Anwendung der Ähnlichkeitsmechanik und der Turbulenzforschung auf die Geschiebebewegung. *Mitteilungen der Preussischen Anstalt Wasserbau and Schiffbau* 26.

Shrestha, D.P. 1997. Assessment of soil erosion in the Nepalese Himalaya: a case study in Likhu Khola Valley, Middle Mountain Region. *Land Husbandry* 2(1): 59–80.

Sibbesen, E., Schjønning, P., Hansen, A.C., Nielsen, J.D. and Heidmann, T. 1994. Surface runoff, erosion and loss of phosphorus relative to soil physical factors as influenced by tillage and cropping systems. In Jensen, H.E., Schjønning, P., Mikkelsen, S.A. and Madsen, K.B. (eds), *Soil tillage for crop production and protection of the environment*. ISTRO, Aalborg: 245–50.

Sibbesen, E. and Sharpley, A.N. 1997. Setting and justifying upper critical limits for phosphorus in soils. In Tunney, H., Carlton, O.T., Brookes, P.C. and Johnston, A.E. (eds), *Phosphorus loss from soil to water*. CAB International, Wallingford.

Sidorchuk, A., Märker, M., Moretti, S. and Rodolfi, G. 2003. Gully erosion modelling and landscape response in the Mbuluzi River catchment of Swaziland. *Catena* 50: 507–25.

Siemens, J.C. and Oschwald, W.R. 1978. Corn-soybean tillage systems: erosion control, effects on crop production, costs. *Transactions of the American Society of Agricultural Engineers* 21: 293–302.

Sims, B.G., Rodríguez, F., Eid, M. and Espinoza, T. 1999. Biophysical aspects of vegetative soil and water conservation practices in the inter-Andean valleys of Bolivia. *Mountain Research and Development* 19: 282–91.

Singh, G., Babu, R. and Chandra, S. 1981. *Soil loss prediction research in India*. Central Soil and Water Conservation Research and Training Institute Bulletin No. T12/D9, Dehra Dun.

Singh, G., Bhardwaj, S.P. and Singh, B.P. 1979. Effect of row cropping of maize and soybean on erosion losses. *Indian Journal of Soil Conservation* 7: 43–6.

Singh, K. 1990. Enlisting people's participation in soil and water conservation programmes: lessons of India's experience. In *Proceedings, International symposium on water erosion, sedimentation and resource conservation*. Central Soil and Water Conservation Research and Training Institute, Dehra Dun: 471–81.

Skidmore, E.L. and Hagen, L.J. 1977. Reducing wind erosion with barriers. *Transactions of the American Society of Agricultural Engineers* 20: 911–15.

Skidmore, E.L. and Williams, J.P. 1991. Modified EPIC wind erosion model. In *Modeling plant and soil systems*. ASA-CSSA-SSSA Agronomy Monograph No. 31, Madison WI: 457–69.

Skidmore, E.L. and Woodruff, N.P. 1968. Wind erosion forces in the United States and their use in predicting soil loss. *USDA Agricultural Research Service Handbook* 346.

Slattery, M.C. and Bryan, R.B. 1992. Hydraulic conditions for rill incision under simulated rainfall: a laboratory experiment. *Earth Surface Processes and Landforms* 17: 127–46.

Smalley, I.J. 1970. Cohesion of soil particles and the intrinsic resistance of simple soil systems to wind erosion. *Journal of Soil Science* 21: 154–61.

Smith, D.D. 1958. Factors affecting rainfall erosion and their evaluation. *International Association of Scientific Hydrology Publication* 43: 97–107.

Smith, P., Milne, R., Powlson, D.S., Smith, J.U., Falloon, P. and Coleman, K. 2000. Revised estimates of the carbon mitigation potential of UK agricultural land. *Soil Use and Management* 16: 293–5.

Smithen, A.A. and Schulze, R.E. 1982. The spatial distribution in southern Africa of rainfall erosivity for use in the Universal Soil Loss Equation. *Water SA* 8: 74–8, 165–7.

Soane, B.D. and Pidgeon, J.D. 1975. Tillage requirement in relation to soil physical properties. *Soil Science* 119: 376–84.

Soil Survey of England and Wales 1979. *Land use capability: England and Wales*. 1 : 1,000,000 map sheets.

Soil Survey of England and Wales 1983. *Soil map of England and Wales*. 1 : 250,000 map sheets.

Sørensen, M. 1985. Estimation of some aeolian saltation transport parameters from transport rate profiles. In Barndorff-Nielsen, O.E., Møller, J.T., Rasmussen, K.R. and Willetts, B.B. (eds), *Proceedings on international workshop on the physics of blown sand*. Memoirs No. 8, Department of Theoretical Statistics, University of Aarhus: 141–90.

Spaan, W.P. and van den Abeele, G.D. 1991. Wind borne particle measurements with acoustic sensors. *Soil Technology* 4: 51–63.

Spoor, G. and Godwin, R.J. 1979. Soil deformation and shear strength characteristics of some clay soils at different moisture contents. *Journal of Soil Science* 30: 483–98.

Spoor, G., Tijink, F.G.J. and Weisskopf, P. 2003. Subsoil compaction: risk, avoidance, identification and alleviation. *Soil and Tillage Research* 73: 175–82.

References

Sreenivas, L., Johnston, J.R. and Hill, H.O. 1947. Some relationships of vegetation and soil detachment in the erosion process. *Soil Science Society of America Proceedings* 11: 471–4.

Srivastva, A.K. and Rama Mohan Rao, M.S. 1988. Crop responses under different agroforestry practices in semi arid black soil region. *Indian Journal of Soil Conservation* 16: 1–10.

Starkel, L. 1972. The role of catastrophic rainfall in the shaping of the relief of the Lower Himalaya (Darjeeling Hills). *Geographica Polonica* 21: 103–47.

Steindórsson, S. 1980. Flokkun gróðurs í gróðurfélög. *Íslenzkar Landbúnaðarrannsóknir* 12: 11–52.

Sterk, G., Jacobs, A.F.G. and van Boxel, J.H. 1998. The effect of turbulent flow structures on saltation sand transport in the atmospheric boundary layer. *Earth Surface Processes and Landforms* 23: 877–87.

Sterk, G. and Raats, P.A.C. 1996. Comparison of models describing the vertical distribution of wind-eroded sediment. *Soil Science Society of America Journal* 60: 1914–19.

Sterk, G. and Stein, A. 1997. Mapping wind-blown mass transport modelling in space and time. *Soil Science Society of America Journal* 61: 232–9.

Stevens, J.H. 1974. Stabilization of eolian sands in Saudi Arabia's Al Hasa oasis. *Journal of Soil and Water Conservation* 29: 129–33.

Stockfisch, N., Forstreuter, T. and Ehlers, W. 1999. Ploughing effects on soil organic matter after twenty years of conservation tillage in Lower Saxony, Germany. *Soil and Tillage Research* 52: 91–101.

Stocking, M.A. 1986. *The cost of soil erosion in Zimbabwe in terms of loss of three major nutrients*. FAO Consultants Working Paper 3, AGLS, Rome.

Stocking, M. and Abel, N. 1992. Labour costs: a critical element in soil conservation. In Tato, K. and Hurni, H. (eds), *Soil conservation for survival*. Soil and Water Conservation Society, Ankeny, IA: 206–18.

Stocking, M. and Clark, R. 1999. Soil productivity and erosion: biophysical and farmer-perspective assessment for hillslopes. *Mountain Research and Development* 19: 191–202.

Stocking, M.A. and Elwell, H.A. 1973a. Prediction of subtropical storm soil losses from field plot studies. *Agricultural Meteorology* 12: 193–201.

Stocking, M.A. and Elwell, H.A. 1973b. Soil erosion hazard in Rhodesia. *Rhodesian Agricultural Journal* 70: 93–101.

Stocking, M.A. and Elwell, H.A. 1976. Rainfall erosivity over Rhodesia. *Transactions of the Institute of British Geographers New Series* 1: 231–45.

Stocking, M. and Peake, L. 1987. Erosion-induced loss in soil productivity: trends in research and international cooperation. In Pla Sentis, I. (ed.), *Soil conservation and productivity*. Sociedad Venezolana de la Ciencia del Suolo, Maracay: 399–438.

Stocking, M. and Tengberg, A. 1999. Soil conservation as incentive enough – experiencies from southern Brazil and Argentina on identifying sustainable practices. In Sanders, D.W., Huszar, P.C., Sombatpanit, S. and Enters, T. (eds), *Incentives in soil conservation: from theory to practice*. Oxford and IBH Publishing, New Delhi: 69–84.

Stredňanský, J. 1977. Kritickérýchlosti vetra z hladiska erodavtelnosti pôd na južnom Slovensku. *Vysoká škola polnohospodárska*. Nitra.

Streeter, D.T. 1977. Gully restoration on Box Hill. *Countryside Recreation Review* 2: 28–40.

Styczen, M. and Høgh-Schmidt, K. 1988. A new description of splash erosion in relation to raindrop sizes and vegetation. In Morgan, R.P.C. and Rickson, R.J. (eds), *Erosion assessment and modelling*. Commission of the European Communities Report No.EUR 10860 EN: 147–84.

Styczen, M. and Nielsen, S.A. 1989. A view of soil erosion theory, process research and model building: possible interactions and future developments. *Quaderni di Scienza del Suolo* 2: 27–45.

Sulaiman, W., Maene, L.M. and Mokhtaruddin, A.M. 1981. Runoff, soil and nutrient losses from an ultisol under different legumes. In Tingsanchali, T. and Eggers, H. (eds), *Southeast Asian regional symposium on problems of soil erosion and sedimentation*. Asian Institute of Technology, Bangkok: 275–86.

Sutherland, R.A. and Ziegler, A.D. 1996. Geotextile effectiveness in reducing interrill runoff and sediment flux. In *Erosion control technology: bringing it home*. International Erosion Control Association, Steamboat Springs, CO: 393–405.

Sutherland, R.A. and Ziegler, A.D. 1998. The influence of the soil conditioner 'Agri-SC' on splash detachment and aggregate stability. *Soil and Tillage Research* 45: 373–87.

Sutherland, R.A., Ziegler, A.D. and Tran, L.T. 1997. Rolled erosion control systems and their effect on sediment redistribution by rainsplash – a laboratory investigation. In *Proceedings of Conference 28*. International Erosion Control Association, Steamboat Springs, CO: 427–43.

Suwardji, P. and Eberbach, P.L. 1998. Seasonal changes of physical properties of an Oxic Paleustalf (Red Kandosol) after 16 years of direct drilling or conventional cultivation. *Soil and Tillage Research* 49: 65–77.

Swanson, F.J. and Dyrness, C.T. 1975. Impact of clearcutting and road construction on soil erosion by landslides in the West Cascade Range, Oregon. *Geology* 3: 393–6.

Swanson, L.E., Camboni, S.M. and Napier, T.L. 1986. Barriers to the adoption of soil conservation practices on farms. In Lovejoy, S.B. and Napier, T.L. (eds), *Conserving soil: insights from socioeconomic research.* Soil Conservation Society of America, Ankeny, IA: 108–20.

Swanson, N.P. 1965. Rotating-boom rainfall simulator. *Transactions of the American Society of Agricultural Engineering* 8: 71–2.

Święchowicz, J. 2002. Linkage of slope wash and sediment and solute export from a foothill catchment in the Carpathian foothills of south Poland. *Earth Surface Processes and Landforms* 27: 1389–1413.

Tackett, J.L. and Pearson, R.W. 1965. Some characteristics of soil crust formed by simulated rainfall. *Soil Science* 99: 407–13.

Talsma, T. 1969. *In situ* measurement of sorptivity. *Australian Journal of Soil Research* 17: 269–76.

Tang, K., Zhou, H., Hou, X. and Liu, Y. 1987. The influence of destruction and reconstruction of vegetation on soil erosion and its control in the loess plateau of China. In Pla Sentis, I. (eds), *Soil conservation and productivity.* Sociedad Venezolana de la Ciencia del Suelo, Maracay: 963–72.

Tebrügge, F. and Düring, R.A. 1999. Reducing tillage intensity – a review of results from a long-term study in Germany. *Soil and Tillage Research* 53: 15–28.

Tejwani, K.G. 1981. Watershed management as a basis for land development and management in India. In Lal, R. and Russell, E.W. (eds), *Tropical agricultural hydrology.* Wiley, Chichester: 239–55.

Tejwani, K.G. 1992. Training, education and demonstration in soil and water conservation. In Tato, K. and Hurni, H. (eds), *Soil conservation for survival.* Soil and Water Conservation Society, Ankeny, IA: 269–76.

Temple, D.M. 1982. Flow retardance of submerged grass channel linings. *Transactions of the American Society of Agricultural Engineers* 25: 1300–3.

Temple, D.M. 1991. Changes in vegetal flow retardance during long-duration flows. *Transactions of the American Society of Agricultural Engineers* 34: 1769–74.

Temple, D.M. and Alspach, D. 1992. Failure and recovery of a grass-lined channel. *Transactions of the American Society of Agricultural Engineers* 35: 171–3.

Temple, P.H. 1972. Soil and water conservation policies in the Uluguru Mountains, Tanzania. *Geografiska Annaler* 54-A: 110–23.

Temple, P.H. and Rapp, A. 1972. Landslides in the Mgeta area, western Uluguru Mountains, Tanzania. *Geografiska Annaler* 54-A: 157–93.

Tengberg, A., Ellis-Jones, J., Kiome, R.M. and Stocking, M. 1998. Applying the concept of agrodiversity to indigenous soil and water conservation practices in eastern Kenya. *Agriculture, Ecosystems and Environment* 70: 259–72.

Tengberg, A., Muriithi, L. and Okoba, B. 1999. Land management on semi-arid hillsides in eastern Kenya: learning from farmers' diverse practices. *Mountain Research and Development* 19: 354–63.

Terry, J.P. and Shakesby, R.A. 1993. Soil hydrophobicity effects on rainsplash: simulated rainfall and photographic evidence. *Earth Surface Processes and Landforms* 18: 519–25.

Terwilliger, V.J. 1990. Effects of vegetation on soil slippage by pore pressure modification. *Earth Surface Processes and Landforms* 15: 553–70.

Thapa, B.B., Cassel, D.K. and Garrity, D.P. 1999. Ridge tillage and contour natural grass barrier strips reduce tillage erosion. *Soil and Tillage Research* 51: 341–56.

Theocharopoulos, S.P., Florou, H., Walling, D.E., Kalantzakos, H., Christou, M., Tountas, P. and Nikolaou, T. 2003. Soil erosion and deposition rates in a cultivated catchment area in central Greece, estimated using the ^{137}Cs technique. *Soil and Tillage Research* 69: 153–62.

Thomas, D.B. 1996. Soil conservation extension: constraints to progress and lessons learned in East Africa. In Sombatpanit, S., Zöbisch, M.A., Sanders, D.W. and Cook, M.G. (eds), *Soil conservation extension: from concepts to adoption.* Soil and Water Conservation Society of Thailand, Bangkok: 245–56.

Thomas, D.B. and Biamah, E.K. 1989. Origin, application and design of the fanya juu terrace. In Moldenhauer, W.C., Hudson, N.W., Sheng, T.C. and Lee, S.W. (eds), *Development of conservation farming on hillslopes.* Soil and Water Conservation Society, Ankeny, IA: 185–94.

Thornes, J.B. 1976. *Semi-arid erosion systems: case studies from Spain.* London School of Economics Geographical Papers 7.

Thorp, J. 1949. Effect of certain animals that live in soils. *Science Monthly* 68: 180–91.

Thorsteinsson, I. 1980a. Nýting útaga: beitarþungi. *Íslenzkar Landbúnaðarrannsóknir* 12(2): 113–22.

Thorsteinsson, I. 1980b. Beitagildi gróðurlenda. *Íslenzkar Landbúnaðarrannsóknir* 12(2): 123–5.

Thorsteinsson, I., Ólafsson, G. and van Dyne, G.M. 1971. Range resources of Iceland. *Journal of Range Management* 24: 86–93.

Tiffen, M., Mortimore, M. and Gichuki, F. 1994. *More people, less erosion. Environmental recovery in Kenya.* Wiley, Chichester.

References

Till, R. 1973. The use of linear regression in geomorphology. *Area* 5: 303–8.

Tiwari, A.K., Risse, L.M. and Nearing, M.A. 2000. Evaluation of WEPP and its comparison with USLE and RUSLE. *Transactions of the American Society of Agricultural Engineers* 43: 1129–35.

Tjernström, R. 1992. Yields from terraced and non-terraced fields in the Machakos District of Kenya. In Tato, K. and Hurni, H. (eds), *Soil conservation for survival.* Soil and Water Conservation Society, Ankeny, IA: 251–65.

Tobiason, S., Jenkins, D., Molash, E. and Rush, S. 2001. Polymer use and testing for erosion and sediment control on construction sites. *Erosion Control* 8(1): 90–101.

Torri, D. and Borselli, L. 1991. Overland flow and soil erosion: some processes and their interactions. *Catena Supplement* 19: 129–37.

Torri, D. and Poesen, J. 1988. Incipient motion conditions for single rock fragments in simulated rill flow. *Earth Surface Processes and Landforms* 13: 225–37.

Torri, D. and Poesen, J. 1992. The effect of soil surface slope on raindrop detachment. *Catena* 19: 561–78.

Torri, D., Regüés, D., Pelligrini, S. and Bazzoffi, P. 1999. Within-storm soil surface dynamics and erosive effects of rainstorms. *Catena* 38: 131–50.

Torri, D. and Sfalanga, M. 1986. Some problems on soil erosion modelling. In Giorgini, A. and Zingales, F. (eds), *Agricultural nonpoint source pollution: model selection and applications.* Elsevier, Amsterdam: 161–71.

Torri, D., Sfalanga, M. and Chisci, G. 1987a. Threshold conditions for incipient rilling. *Catena Supplement* 8: 97–105.

Torri, D., Sfalanga, M. and Del Sette, M. 1987b. Splash detachment: runoff depth and soil cohesion. *Catena* 14: 149–55.

Trabaud, L. and Lepart, J. 1980. Diversity and stability in garrigue ecosystems after fire. *Vegetation* 43: 49–57.

Tracy, F.C., Renard, K.G. and Fogel, M.M. 1984. Rainfall energy characteristics for southeastern Arizona. In *Water today and tomorrow.* Irrigation and Drainage Division ASCE, New York: 559–66.

Trimble, S.W. 1983. A sediment budget for Coon Creek basin in the Driftless Area, Wisconsin, 1853–1977. *American Journal of Science* 283: 454–74.

Troeh, F.R., Hobbs, J.A. and Donohue, R.L. 1980. *Soil and water conservation for productivity and environmental protection.* Prentice Hall, Englewood Cliffs, NJ.

Truman, C.C., Bradford, J.M. and Ferris, J.E. 1990. Antecedent water content and rainfall energy influence on soil aggregate breakdown. *Soil Science Society of America Journal* 54: 1385–92.

Trustrum, N.A., Thomas, V.J. and Lambert, M.G. 1984. Soil slip erosion as a constraint to hill country pasture production. *Proceedings of the New Zealand Grassland Association* 45: 66–76.

Turkelboom, F., Poesen, J., Ohler, I. and Ongprasert, S. 1999. Reassessment of tillage erosion rates by manual tillage on steep slopes in northern Thailand. *Soil and Tillage Research* 51: 245–59.

Uchijima, Z. 1976. Maize and rice. In Monteith, J.L. (ed.), *Vegetation and the atmosphere. Volume 2. Case studies.* Academic Press, London: 33–64.

United States Department of Agriculture 1979. Field manual for research in agricultural hydrology. *USDA Agricultural Handbook* 224.

Uri, N.D. and Lewis, J.A. 1998. The dynamics of soil erosion in US agriculture. *Science of the Total Environment* 218: 45–58.

Usón, A. and Ramos, M.C. 2001. An improved rainfall erosivity index obtained from experimental interrill soil losses in soils with a Mediterranean climate. *Catena* 43: 293–305.

Van Asch, Th.W.J. 1983. Water erosion on slopes in some land units in a Mediterranean area. *Catena Supplement* 4: 129–40.

Vandaele, K. and Poesen, J. 1995. Spatial and temporal patterns of soil erosion rates in an agricultural catchment, central Belgium. *Catena* 25: 213–26.

Vandekerckhove, L., Poesen, J., Oostwoud Wijdenes, D., Nachtergaele, J., Kosmas, C., Roxo, M.J. and de Figueiredo, T. 2000. Thresholds for gully initiation and sedimentation in Mediterranean Europe. *Earth Surface Processes and Landforms* 25: 1201–20.

Van Dijk, A.I.J.M., Bruijnzeel, L.A. and Rosewell, C.J. 2002. Rainfall intensity–kinetic energy relationships: a critical literature review. *Journal of Hydrology* 261: 1–23.

Van Dijk, P.M., Kwaad, F.J.P.M. and Klapwijk, M. 1996. Retention of water and sediment by grass strips. *Hydrological Processes* 10: 1069–80.

Van Muysen, W. and Govers, G. 2002. Soil displacement and tillage erosion during secondary tillage operations: the case of rotary harrow and seeding equipment. *Soil and Tillage Research* 65: 185–91.

Van Muysen, W., Govers, G., Bergkamp, G., Roxo, M. and Poesen, J. 1999. Measurement and modelling of the effects of initial soil condition and slope gradient on soil translocation during tillage. *Soil and Tillage Research* 51: 303–16.

Van Muysen, W., Govers, G. and Van Oost, K. 2002. Identification of important factors in the process of tillage erosion: the case of mouldboard tillage. *Soil and Tillage Research* 65: 77–93.

Van Oost, K., Govers, G. and Van Muysen, W. 2003. A process-based conversion model for caesium-137 derived erosion rates on agricultural land: an integrated spatial approach. *Earth Surface Processes and Landforms* 28: 187–207.

Vatn, A., Bakken, L.R., Bleken, M.A., Botterweg, P., Lundeby, H., Romstad, E., Rørstad, P.K. and Vold, A. 1996. Policies for reduced nutrient losses and erosion from Norwegian agriculture. *Norwegian Journal of Agricultural Science Supplement* 23.

Veihe, A. and Quinton, J. 2000. Sensitivity analysis of EUROSEM using Monte Carlo simulation. I. Hydrological, soil and vegetation parameters. *Hydrological Processes* 14: 915–26.

Veihe, A., Rey, J., Quinton, J.N., Strauss, P., Sancho, F.M. and Somarriba, M. 2001. Modelling of event-based soil erosion in Costa Rica, Nicaragua and Mexico: evaluation of the EUROSEM model. *Catena* 44: 187–203.

Verbyla, D.L. and Richardson, C.A. 1996. Remote sensing clearcut areas within a forested watershed: comparing SPOT HRV Panchromatic, SPOT HRV multispectral, and Landsat Thematic Mapper data. *Journal of Soil and Water Conservation* 51: 423–7.

Verstraeten, G. and Poesen, J. 1999. The nature of small-scale flooding, muddy floods and retention pond sedimentation in central Belgium. *Geomorphology* 29: 275–92.

Verstraeten, G. and Poesen, J. 2001. Modelling the long-term sediment trap efficiency of small ponds. *Hydrological Processes* 15: 2797–819.

Vis, M. 1986. Interception, drop size distributions and rainfall kinetic energy in four Colombian forest ecosystems. *Earth Surface Processes and Landforms* 11: 591–603.

Vittorini, S. 1972. The effect of soil erosion in an experimental station in the Pliocene clay of the Val d'Era (Tuscany) and its influence on the evolution of the slopes. *Acta Geographica Debrecina* 10: 71–81.

Vogel, H. 1990. Deterioration of a mountainous agro-ecosystem in the Third World due to emigration of rural labour. In Messerli, B. and Hurni, H. (eds), *African mountains and highlands: problems and perspectives*. African Mountains Association, Bern: 389–406.

Vogel, H. 1994. *Conservation tillage in Zimbabwe. Evaluation of several techniques for the development of sustainable crop production systems in smallholder farming.* Geographica Bernensia, African Studies Series A11.

Voorhees, W.B. and Lindstrom, M.J. 1984. Long-term effects of tillage method on soil tilth independent of wheel traffic compaction. *Soil Science Society of America Journal* 48: 152–5.

Voroney, R.P., van Veen, J.A. and Paul, E.A. 1981. Organic carbon dynamics in grassland soils. II. Model validation and simulation of the long-term effects of cultivation and rainfall erosion. *Canadian Journal of Soil Science* 61: 211–24.

Voznesensky, A.S. and Artsruui, A.B. 1940. Laboratoriya metod opretseleniy protivoerozionnoy ustoychivosti pochv. In *Voprosi protivoerozionnoy ustoychivosti pochv*. Izd. Zakavk. NIIVKH, Tbilisi.

Wallace, G.A. and Wallace, A. 1986. Control of soil erosion by polymeric soil conditioners. *Soil Science* 141: 363–7.

Walling, D.E. 1983. The sediment delivery problem. *Journal of Hydrology* 69: 209–37.

Walling, D.E. and He, Q. 1999a. Improved models for estimating soil erosion rates from cesium-137 measurements. *Journal of Environmental Quality* 28: 611–22.

Walling, D.E. and He, Q. 1999b. Use of fallout lead-210 measurements to estimate soil erosion on cultivated land. *Soil Science Society of America Journal* 63: 1404–12.

Walling, D.E., He, Q. and Appleby, P.G. 2002a. Conversion models for use in soil-erosion, soil-redistribution and sedimentation investigations. In Zapata, F. (ed.), *Handbook for the assessment of soil erosion and sedimentation using environmental radionuclides*. Kluwer, Dordrecht: 111–64.

Walling, D.E., He, Q. and Blake, W. 2000. Use of ^{7}Be and ^{137}Cs measurements to document short- and medium-term rates of water-induced soil erosion on agricultural land. *Water Resources Research* 35: 3865–74.

Walling, D.E. and Kleo, A.H.A. 1979. Sediment yield of rivers in areas of low precipitation: a global view. *International Association of Scientific Hydrology Publication* 128: 479–93.

Walling, D.E. and Quine, T.A. 1990. Use of caesium-137 to investigate patterns and rates of soil erosion on arable fields. In Boardman, J., Foster, I.D.L. and Dearing, J.A. (eds), *Soil erosion in agricultural land*. Wiley, Chichester: 33–53.

Walling, D.E., Russell, M.A., Hodgkinson, R.A. and Zhang, Y. 2002b. Establishing sediment budgets for two small lowland agricultural catchments in the UK. *Catena* 47: 323–53.

Walling, D.E. and Webb, B.W. 1983. Patterns of sediment yield. In Gregory, K.J. (ed.), *Background to palaeohydrology*. Wiley, Chichester: 69–100.

Walling, D.E., Webb, B.W. and Woodward, J.C. 1992. Some sampling considerations in the design of effective strategies for monitoring sediment-associated transport. *International Association of Hydrological Sciences Publication* 210: 279–88.

Wan Yusoff, W.A. 1988. Forest management and conservation practices for controlling soil erosion in the natural forest of Peninsular Malaysia. In Rimwanich, S. (ed.), *Land conservation for future generations*. Department of Land Development, Bangkok: 869–75.

Wells, N.A. and Andriamihaja, B. 1991. Growth of gullies on laterite in Madagascar: non-intuitive implications for control. In *Erosion control: a global perspective*. International Erosion Control Association, Steamboat Springs, CO: 379–401.

Wells, N.A., Andriamijaha, B. and Rakotovololona, H.F.S. 1991. Patterns of development of lavaka, Madagascar's unusual gullies. *Earth Surface Processes and Landforms* 16: 189–206.

Wen, D. 1993. Soil erosion and conservation in China. In Pimental, D. (ed.), *World soil erosion and conservation*. Cambridge University Press, Cambridge: 63–85.

Wenner, C.C. 1981. *Soil conservation in Kenya*. Ministry of Agriculture, Nairobi.

Wenner, C.C. 1988. The Kenyan model of soil conservation. In Moldenhauer, W.C. and Hudson, N.W. (eds), *Conservation farming on steep lands*. Soil and Water Conservation Society, Ankeny, IA: 197–206.

Whitlow, R. and Bullock, A. 1986. Rapid gully development in Mangwende Communal Land. *Zimbabwe Agricultural Journal* 83: 149–60.

Whitmore, T.C. and Burnham, C.P. 1969. The altitudinal sequence of forests and soils on granite near Kuala Lumpur. *Malayan Nature Journal* 22: 99–118.

Wiersum, K.F., Budirijanto, P. and Rhomdoni, D. 1979. Influence of forests on erosion. Seminar on the erosion problem in the Jatiluhur area. Institute of Ecology, Padjadjaran University, Bandung, Report 3.

Wiggins, S.L. 1981. The economics of soil conservation in the Acelhuate River Basin, El Salvador. In Morgan, R.P.C. (ed.), *Soil conservation: problems and prospects*. Wiley, Chichester: 399–417.

Willetts, B.B. and Rice, M.A. 1985. Inter-saltation collision. In Barndorff-Nielsen, O.E., Møller, J.T., Rasmussen, K.R. and Willetts, B.B. (eds), *Proceedings of international workshop on the physics of blown sand*. Memoir No. 8, Department of Theoretical Statistics, University of Aarhus: 83–100.

Willetts, B.B. and Rice, M.A. 1988. Particle dislodgement from a flat sand bed by wind. *Earth Surface Processes and Landforms* 13: 717–28.

Williams, A.R. and Morgan, R.P.C. 1976. Geomorphological mapping applied to soil erosion evaluation. *Journal of Soil and Water Conservation* 31: 164–8.

Williams, C.N. and Joseph, K.T. 1970. *Climate, soil and crop production in the humid tropics*. Oxford University Press, Kuala Lumpur.

Williams, J.R., Jones, C.A. and Dyke, P.T. 1984. A modeling approach to determining the relationship between erosion and soil productivity. *Transactions of the American Society of Agricultural Engineers* 27: 129–44.

Wischmeier, W.H. 1973. Conservation tillage to control water erosion. In *Proceedings, National conservation tillage conference*. Soil Conservation Society of America, Ankeny, IA: 133–41.

Wischmeier, W.H. 1975. Estimating the soil loss equation's cover and management factor for undisturbed area. In *Present and prospective technology for predicting sediment yields and sources*. USDA ARS Publication ARS-S-40: 118–24.

Wischmeier, W.H. 1978. Use and misuse of the Universal Soil Loss Equation. *Journal of Soil and Water Conservation* 31: 5–9.

Wischmeier, W.H., Johnson, C.B. and Cross, B.V. 1971. A soil erodibility nomograph for farmland and construction sites. *Journal of Soil and Water Conservation* 26: 189–93.

Wischmeier, W.H. and Mannering, J.V. 1969. Relation of soil properties to its erodibility. *Soil Science Society of America Proceedings* 23: 131–7.

Wischmeier, W.H. and Smith, D.D. 1958. Rainfall energy and its relationship to soil loss. *Transactions of the American Geophysical Union* 39: 285–91.

Wischmeier, W.H. and Smith, D.D. 1978. Predicting rainfall erosion losses. *USDA Agricultural Research Service Handbook* 537.

Wise, S.M., Thornes, J.B. and Gilman, A. 1982. How old are the badlands? A case study from southeast Spain. In Bryan, R. and Yair, A. (eds), *Badland geomorphology and piping*. Geo Books, Norwich: 259–77.

Withers, B. and Vipond, S. 1974. *Irrigation: design and practice*. Batsford, London.

WMS Associates 1988. *Gully erosion in Swaziland*. Final Report. WMS, Fredericton, NB.

Wolde, F.T. and Thomas, D.B. 1989. The effect of narrow grass strips in reducing soil loss and runoff in a Kabete nitosol, Kenya. In Thomas, D.B., Biamah, E.K., Kilewe, A.M., Lundgren, L. and Mochoge, B.O. (eds), *Soil and water conservation in Kenya*. Department of Agricultural Engineering, University of Nairobi/SIDA: 176–94.

Wolfe, S.A. and Nickling, W.G. 1996. Shear stress partitioning in sparsely vegetated desert canopies. *Earth Surface Processes and Landforms* 21: 607–19.

Wolman, M.G. 1967. A cycle of sedimentation and erosion in urban river channels. *Geografiska Annaler* 49-A: 385–95.

Wood, A.P. 1992. Zambia's soil conservation heritage: a review of policies and attitudes towards soil conservation from colonial times to the present. In Tato, K. and Hurni, H. (eds), *Soil conservation for survival*. Soil and Water Conservation Society, Ankeny, IA: 156–71.

Woodburn, R. and Kozachyn, J. 1956. A study of relative erodibility of a group of Mississippi gully soils. *Transactions of the American Geophysical Union* 37: 749–53.

Woodruff, N.P., Chepil, W.S. and Lynch, R.D. 1957. Emergency chiseling to control wind erosion. *Kansas Agricultural Experimental Station Technical Bulletin* 90.

Woodruff, N.P. and Siddoway, F.H. 1965. A wind erosion equation. *Soil Science Society of America Proceedings* 29: 602–8.

Woodruff, N.P. and Zingg, A.W. 1952. Wind-tunnel studies of fundamental problems related to windbreaks. *US Soil Conservation Service Publication* SCS-TP-112.

Wright, J.L. and Brown, K.W. 1967. Comparison of momentum and energy balance methods of computing vertical transfer within in a crop. *Agronomy Journal* 59: 427–32.

Wu, C.C. and Wang, A.B. 1998. Soil loss and soil conservation measures on steep sloping orchards. *Advances in GeoEcology* 31: 383–7.

Wu, T.H. 1995. Slope stabilization. In Morgan, R.P.C. and Rickson, R.J. (eds), *Slope stabilization and erosion control: a bioengineering approach*. E. and F.N. Spon, London: 221–64.

Yair, A. and Rutin, J. 1981. Some aspects of the regional variation in the amount of the available sediment produced by isopods and porcupines, northern Negev, Israel. *Earth Surface Processes and Landforms* 6: 221–34.

Young, A. 1969. Present rate of land erosion. *Nature* 224: 851–2.

Young, A. 1976. *Tropical soils and soil survey*. Cambridge University Press, Cambridge.

Young, A. 1998. *Land resources: now and for the future*. Cambridge University Press, Cambridge.

Young, R.A., Onstad, C.A., Bosch, D.D. and Anderson, W.P. 1989. AGNPS: a nonpoint-source pollution model for evaluating agricultural watersheds. *Journal of Soil and Water Conservation* 44: 168–73.

Zachar, D. 1982. *Soil erosion*. Elsevier, Amsterdam.

Zanchi, C. 1989. Drainage as a soil conservation and soil stabilizing practice on hilly slopes. In Schwertmann, U., Rickson, R.J. and Auerswald, K. (eds), *Soil protection measures in Europe*. Soil Technology Series 1: 73–82.

Zanchi, C., Bazzoffi, P., D'Egidio, G. and Nistri, L. 1981. Field rainfall simulator. In *Field excursion guide, Proceedings of the Florence Symposium*. International Association of Scientific Hydrology: 46–9.

Zanchi, C. and Torri, D. 1980. Evaluation of rainfall energy in central Italy. In De Boodt, M. and Gabriels, D. (eds), *Assessment of erosion*. Wiley, London: 133–42.

Zaruba, Q. and Mencl, V. 1969. *Landslides and their control*. Elsevier, Amsterdam.

Zhang, X.C., Nearing, M.A., Risse, L.M. and McGregor, K.C. 1996. Evaluation of runoff and soil loss predictions using natural runoff plot data. *Transactions of the American Society of Agricultural Engineers* 39: 855–63.

Zhang, X., Zhang, Y., Wen, A. and Feng, M. 2003. Assessment of soil losses on cultivated land by using the ^{137}Cs technique in the upper Yangtze River basin of China. *Soil and Tillage Research* 69: 99–106.

Zhu, T.X. 2003. Tunnel development over a 12 year period in a semi-arid catchment of the loess plateau, China. *Earth Surface Processes and Landforms* 28: 507–25.

Zhu, T.X., Luk, S.H. and Cai, Q.G. 2002. Tunnel erosion and sediment production in the hilly loess region, North China. *Journal of Hydrology* 257: 78–90.

Zingg, A.W. 1940. Degree and length of land slope as it affects soil loss in runoff. *Agricultural Engineering* 21: 59–64.

Zingg, A.W. 1951. A portable wind tunnel and dust collector developed to evaluate the erodibility of field surfaces. *Agronomy Journal* 43: 189–91.

Acknowledgements

The authors and publishers gratefully acknowledge the following for permission to reproduce copyright material:

Fig. 1.1 Walling, D.E. and Kleo, A.H.A. 1979. Sediment yield of rivers in areas of low precipitation: a global view. In: The Hydrology of Areas of Low Precipitation. *International Association of Scientific Hydrology Publication* 128: 479–93. Reproduced with permission from IAHS Press, Wallingford, UK.

Fig. 1.2 Walling, D.E. and Webb, B.W. 1983. Patterns of sediment yield. In Gregory, K.J. (ed.), *Background to Palaeohydrology.* Wiley, Chichester: 69–100. Reproduced with permission from John Wiley & Sons Ltd.

Fig. 1.4 Kirkby, M.J. 1980. The problem. In Kirkby, M.J. and Morgan, R.P.C. (eds), *Soil Erosion.* Wiley, Chichester: 1–16. Reproduced with permission from John Wiley & Sons Ltd.

Fig. 1.5 Wolman, M.G. 1967. A cycle of sedimentation and erosion in urban river channels. *Geografiska Annaler* 49A: 385–95. Figure 1, p. 386. Reproduced with permission from Blackwell Publishing Ltd.

Fig. 2.2 Poesen, J. 1985. An improved splash transport model. *Zeitschrift für Geomorphologie* 29: 193–211. Figure 1, p. 196. Reproduced with permission from Gebrüder Borntraeger Verlagsbuchhandlung.

Fig. 2.5 Poesen, J., Nachtergaele, J., Verstraeten, G. and Valentin, C. 2003. Gully erosion and environmental change: importance and research needs. *Catena* 50: 91–133. Figure 8. Reproduced with permission from Elsevier.

Fig. 4.1 Stocking, M.A. and Elwell, H.A. 1976. Rainfall erosivity over Rhodesia. *Transactions of the Institute of British Geographers New Series* 1: 231–45. Figure 6, p. 244. Reproduced with permission from Blackwell Publishing Ltd.

Fig. 4.2 Hudson, N.W. 1981. *Soil Conservation.* Batsford, London (2nd edn). Reproduced with permission from Chrysalis Books Group.

Fig. 4.7 Morgan, R.P.C. 1980. Soil erosion and conservation in Britain. *Progress in Physical Geography* 4: 24–47. Figure 1, p. 36. Reproduced with permission from Arnold Publishers, a member of the Hodder Headline Group.

Fig. 4.12 Hudson, N.W. 1981. *Soil Conservation.* Batsford, London (2nd edn). Reproduced with permission from Chrysalis Books Group.

Fig. 4.14 Morgan, R.P.C. 1985. Assessment of soil erosion risk in England and Wales. *Soil Use and Management* 1: 127–31. Figure 1, p. 129. Reproduced with permission from the British Soil Science Society.

B4.1 Arnalds, O., Thorarinsdottir, E.F., Metusalemsson, S., Jonsson, A., Gretarsson, E. and Arnason, A. 2001. *Soil Erosion in Iceland.* The Soil Conservation Research Service and the Agricultural Research Institute, Iceland. Reproduced with permission from the Agricultural Research Institute of Iceland.

Table 4.1 Arnalds, O., Thorarinsdottir, E.F., Metusalemsson, S., Jonsson, A., Gretarsson, E. and Arnason, A. 2001. *Soil Erosion in Iceland.* The Soil Conservation Research Service and the Agricultural Research Institute, Iceland. Reproduced with permission from the Agricultural Research Institute of Iceland.

Fig. 5.4 Morgan, R.P.C. 1981. Field measurement of splash erosion. In: Erosion and Sediment Transport Measurement. *International Association of Scientific Hydrology Publication* 133: 373–82. Figure 1, p. 375. Reproduced with permission from IAHS Press, Wallingford, UK.

Fig. 5.5 Walling, D.E. and Quine, T.A. 1990. Use of caesium-137 to investigate patterns and rates of soil erosion on arable fields. In Boardman, J., Foster, I.D.L. and Dearing, J.A. (eds), *Soil Erosion in Agricultural Land.* Wiley, Chichester: 33–53. Figure 4.2, p. 36. Reproduced with permission from John Wiley & Sons Ltd.

Fig. 6.1 Meyer, L.D. and Wischmeier, W.H. 1969. Mathematical simulation of the process of soil erosion by water. *Transactions of the American Society of Agricultural Engineers* 12: 754–58, 762. Figure 1, p. 755. Reproduced with permission from the American Society of Agricultural Engineers.

Table B6.1 Brazier, R.E., Beven, K.J., Anthony, S.G. and Rowan J.S. 2000. Implications of model uncertainty for the mapping of hillslope-scale soil erosion prediction. *Earth Surface Processes and Landforms* 26. Table II, p. 1343. Reproduced with permission from John Wiley & Sons Ltd.

Fig. 8.3 Brierley, J.S. 1976. Kitchen gardens in the West Indies with a contemporary study from Grenada. *Journal of Tropical Geography* 43: 30–40. Figure 2, p. 37. Reproduced with permission from Blackwell Publishing Ltd.

Fig. 8.4 Jiang, D., Qi, L. and Tan, J. 1981. Soil erosion and conservation in the Wuding River Valley, China. In Morgan, R.P.C. (ed), *Soil Conservation: Problems and Prospects.* Wiley, Chichester: 461–79. Figure 11, p. 472. Reproduced with permission from John Wiley & Sons Ltd.

Table B8.1 Coppin, N.J. and Richards, I.G. 1990. *Use of Vegetation in Civil Engineering.* CIRIA/Butterworths, London. Reproduced with permission from CIRIA.

Fig. 10.8 Agate, E. 1983. *Footpaths.* British Trust for Conservation Volunteers, Wallingford. Reproduced with permission from the BTCV.

Table 10.1 Hudson, N.W. 1981. *Soil Conservation.* Batsford, London (2nd edn). Reproduced with permission from Chrysalis Books Group.

Table 10.6 Temple, D.M. 1982. Flow retardance of submerged grass channel linings. *Transactions of the American Society of Agricultural Engineers* 25: 1300–3. Table 1, p. 755. Reproduced with permission from the American Society of Agricultural Engineers.

The publishers apologize for any errors or omissions in the above list and would be grateful to be notified of any corrections that should be incorporated in the next edition or reprint of this book.

Acknowledgements

Index

training   247, 250, 252, 255, 260
Training and Visit System   246
trampling   11, 58, 104, 162, 168, 196
transport   11, 12, 20, 24–6, 38, 41, 50, 58, 63, 129,
    153, 154
transport capacity   24, 25, 26, 29, 41, 59, 114, 129,
    131, 132, 133, 140, 142, 145, 146, 149, 241
transport-limited erosion   11, 131
treatment-oriented land classification   78
tree crops   187
tunnel erosion   27, 30, 31, 32, 36, 191, 208, 209
turbidity recorder   103

uncertainty, in model predictions   149–51
undergrazing   177
Universal Soil Loss Equation (USLE)   85, 120–5,
    129, 131, 133, 139, 141, 149, 188
urban land use   7, 84, 169, 261

vegetation   3, 26, 27, 41, 59, 60, 63, 64, 65, 72, 84,
    89, 91, 133, 141, 143, 147, 149, 166, 171, 177,
    197

vetiver grass   184
volunteers   249

Water Erosion Prediction Project (WEPP)
    139–41, 142, 146, 147, 148, 149, 150, 151
water quality   10, 253, 255
waterways   170, 218–27
wattling   170, 171, 228
weathering   11, 16, 17, 30
wildlife   101, 178
wind, effect on rainfall energy   20
wind breaks   2, 153, 236–41
wind erosion   38–41, 53, 58, 93, 107, 111, 113,
    181, 184, 188, 189, 245, 253
  control   190, 204, 207, 208, 210
  measurement of   108
Wind Erosion Prediction Equation   133
Wind Erosion Prediction System (WEPS)   146
wind erosivity   50
wind tunnels   53, 63, 111–12, 113, 238